T0260881

MATHEMATICAL
THEORY OF OPTICS

R. K. Luneburg

MATHEMATICAL THEORY OF OPTICS

by
R. K. LUNEBURG

Foreword by
EMIL WOLF

Supplementary Notes by
M. HERZBERGER

UNIVERSITY OF CALIFORNIA PRESS
BERKELEY AND LOS ANGELES
1966

University of California Press
Berkeley and Los Angeles, California

Second Printing, 1966

Cambridge University Press
London, England

Library of Congress Catalog Card Number: 64-19010.

FOREWORD

During the Summer of 1944, Dr. Rudolf K. Luneburg presented a course of lectures on the Mathematical Theory of Optics at Brown University. The lecture material was later collected in a volume which was issued by Brown University in the form of mimeographed notes. These notes were by no means a compilation of generally available knowledge. They contained a highly original, thorough, and systematic account of the foundations of several branches of optics and numerous new and important results.

The supply of copies of Luneburg's notes was soon exhausted, but demand for them has continued. The University of California Press is providing a real service to the scientific community by issuing a printed version of these notes. Fate has prevented Dr. Luneburg from seeing this volume. He died in 1949, at a time when the importance of his work was just beginning to be generally recognized.

The chief contribution which Luneburg has made through these notes lies in having shown how the two main mathematical disciplines of instrumental optics, namely geometrical optics and the scalar diffraction optics, may be developed in a systematic manner from the basic equations of Maxwell's electromagnetic theory. Prior to Luneburg's work these two disciplines were, by and large, treated as self-contained fields, with little or no contact with electromagnetic theory.

The starting point of Luneburg's investigation was the observation of the formal equivalence of the basic equation of geometrical optics (the eikonal equation) and the equation that governs the propagation of discontinuous solutions of Maxwell's equations (the equation of characteristics). By boldly identifying the geometrical optics field with the electromagnetic field on a moving discontinuity surface, Luneburg was led to a complete formulation of geometrical optics as a particular class of exact solutions of Maxwell's equations. This formulation is by no means based on traditional ideas; for traditionally geometrical optics is regarded as the short wavelength limit (or, more precisely, as the asymptotic approximation for large wave numbers) of the monochromatic solution of the wave equation. Luneburg was, of course, aware of this more traditional viewpoint and he touches briefly on it in §16. In fact, in a course of lectures which he later presented at New York University (during the academic year 1947-1948) Luneburg devoted considerable time to the interrelation between the two approaches. Some of the ideas outlined in the two courses have become the nucleus from which a systematic theory of asymptotic series solutions of Maxwell's equations is gradually being developed. An account of the material presented by Luneburg in his New York

v

lectures and of related more recent developments will soon be published by
Drs. M. Kline and I.W. Kay in a book entitled <u>Electromagnetic Theory and
Geometrical Optics</u> (J. Wiley and Sons, New York).

Chapter I of the present work contains the derivation of the basic laws
of geometrical optics from Maxwell's equations. Amongst the many new results
which this chapter contains, the transport equations, eq. (11.38), relating to
propagation of the electric and the magnetic field vectors along geometrical
rays, are of particular significance. In Chapter II Hamilton's theory of
geometrical optics is formulated and in the chapter which follows it is applied
to special problems. Amongst results which seem to make their first appearance in the scientific literature are some of the formulae of §24 relating to
final corrections of optical instruments by aspheric surfaces; some new
theorems relating to perfect optical instruments (§28.4); and the introduction
in §29 of a new "perfect lens," which images stigmatically onto each other two
spherical surfaces which are situated in a homogeneous medium. This is the
now well known "Luneburg lens" which has found valuable applications as a
microwave antenna. First and third order theories of optical systems are
discussed in Chapters IV and V and, like all the other chapters, they contain
a wealth of information.

Chapter VI deals with the diffraction theory of optical instruments. The
first sections of this chapter are devoted to the derivation, in a mathematically
consistent way, of expressions for the electromagnetic field in the image
region of an optical system suffering from any prescribed aberrations. A
solution of this difficult problem (naturally somewhat idealized) is embodied
in formulae (47.33), now known as the Luneburg diffraction integrals. These
formulae are an important and elegant generalization of certain classical
results of P. Debye and J. Picht. Section 48 deals with another important
problem, often ignored in other treatises, namely with a systematic derivation
of the scalar theory for the description of certain diffraction phenomena with
unpolarized light. The concluding sections deal with problems of resolution
and contain a discussion of the possibility of improvements in resolution by a
suitable choice of the pupil function. These investigations are amongst the
first in a field that has attracted a good deal of attention in recent years.

In two appendices formulae are summarized relating to vector analysis
and to ray tracing in a system of plane surfaces. They are followed by
supplementary notes on electron optics, prepared by Dr. A. Blank and based
on lectures of Dr. N. Chako. The volume concludes with supplementary notes
by Dr. M. Herzberger, based on his lectures dealing with optical qualities of
glass, with mathematics and geometrical optics and with symmetry and
asymmetry in optical images.

It is evident that Luneburg's <u>Mathematical Theory of Optics</u> is a highly original contribution to the optical literature. I consider it to be one of the most important publications on optical theory that has appeared within the last few decades.

Emil Wolf

Department of Physics and Astronomy
University of Rochester,
Rochester 27, New York
May, 1964

PUBLISHER'S NOTE

The present edition has been reproduced from mimeographed notes issued by Brown University in 1944. It is reprinted by permission of the Brown University Press.

The University of California Press extends gratitude for help in making this edition possible to Dr. A. A. Blank, Dr. Max Herzberger, Mrs. R. K. Luneburg, Dr. Gordon L. Walker, and Dr. Emil Wolf.

The author's name was misspelled in the original edition. This has, of course, been corrected, and a number of typographical errors, almost all of which were listed originally in the Errata of the mimeographed version, have also been corrected. Dr. Blank has clarified the last section of Chapter V on the basis of the Errata. No other changes have been made in this edition, which presents Luneburg's work as he left it.

BIOGRAPHICAL NOTE

Dr. Rudolf Karl Luneburg

Born in Volkersheim, Germany, June 30, 1903; resident in United States of America since 1935; naturalized U. S. citizen 1944. Ph.D. University of Göttingen, 1930.

Research Associate in Mathematics, University of Göttingen, 1930-1933; Research Fellow in Physics, University of Leiden, 1934-1935. Research Associate in Mathematics, New York University, 1935-1938. Mathematician, Research Department of Spencer Lens Company (subsidiary of American Optical Company), Buffalo, New York, 1938-1945. Visiting Lecturer, Brown University, Summer, 1944. Mathematical Consultant, Dartmouth Eye Institute 1946. Research Mathematician, Institute of Mathematics and Mechanics (now Courant Institute of Mathematical Sciences), New York University, 1946-1948. Visiting Lecturer, University of Marburg and Darmstadt Institute of Technology, 1948-1949. Associate Professor of Mathematics, University of Southern California, 1949.

Died at Great Falls, Montana, August 19, 1949.

PUBLICATIONS †

1. Das Problem der Irrfahrt ohne Richtungbeschränkung und die Randwertaufgabe der Potentialtheorie. Math. Annalen 104, 45 (1931).

2. Eine Bemerkung zum Beweise eines Satzes über fastperiodische Funktionen. Copenhagen, Hovedkommissionaer: Levin & Munksgaard, B. Lunos boktrykkeri a/s, 1932.

3. On multiple scattering of neutrons. I. Theory of albedo and of a plane boundary (with O. Halpern and O. Clark). Phys. Rev. 53, 173 (1938).

4. Mathematical theory of optics (mimeographed lecture notes). Brown University, Providence, Rhode Island, 1944.

5. Mathematical analysis of binocular vision. Princeton, New Jersey: Princeton University Press, for the Hanover Institute, 1947.

---

† Research reports are not included in this bibliography.

6.   Metric studies in binocular vision perception (Studies and Essays
      presented to R. Courant on his 60th birthday, January 8, 1948).
      New York: Interscience Publishers, Inc., 1948.

7.   Propagation of electromagnetic waves (mimeographed lecture notes).
      New York University, New York, New York, 1948.

8.   Multiple scattering of neutrons. II. Diffusion in a plane of finite
      thickness (with O. Halpern). Phys. Rev. 76, 1811 (1949).

9.   The metric of binocular visual space. J. Opt. Soc. Amer. 40, 627
      (1950).

ACKNOWLEDGMENT

These notes cover a course in Optics given at Brown University in the summer of 1944. The preparation for mimeographing [the original copies] was possible only through the assistance of Dr. Nicholas Chako and of my students, Miss Helen Clarkson, Mr. Albert Blank and Mr. Herschel Weil. I wish to express my thanks for their excellent cooperation.

The supplementary note on Electron Optics has been written by Mr. Blank with the aim of giving a short derivation of the main results by methods similar to those applied in the other parts of the course. This note has its source in lectures given by Dr. Chako on the physical side of this topic; in the mathematical approach it differs, however, from his presentation.

To Dr. Max Herzberger of the Eastman Kodak Company I am greatly indebted for contributing the supplementary notes II, III, IV from his recent research. These notes are a record of his three lectures.

I wish also to express my gratitude to Dean R.G.D. Richardson for his interest in the course and to his staff for unfailing help and cooperation in the task of preparing these notes.

R. K. L.

CONTENTS

CHAPTER I

WAVE OPTICS AND GEOMETRICAL OPTICS

CHAPTER II

HAMILTON'S THEORY OF GEOMETRICAL OPTICS

CHAPTER IV

FIRST ORDER OPTICS

CHAPTER V

THE THIRD ORDER ABERRATIONS IN SYSTEMS OF
ROTATIONAL SYMMETRY

CHAPTER VI

DIFFRACTION THEORY OF OPTICAL INSTRUMENTS

SUPPLEMENTARY NOTES

CHAPTER 1

WAVE OPTICS AND GEOMETRICAL OPTICS

In this course we shall be concerned with the propagation of light in a transparent medium. We shall not consider absorbing media or non-isotropic media, such as metals or crystals; but we will allow the medium to be non-homogeneous. The optical properties of such a medium can be characterized by a scalar function

$$n = n(x, y, z) \, ,$$

the refractive index of the medium. In ordinary optical instruments this function is sectionally constant and discontinuous on certain surfaces.

The mathematical treatment of the propagation of light can be based on two theories: The wave theory of light (Physical Optics) and the theory of light rays (Geometrical Optics). Both theories seem to be fundamentally different and can be developed independent of each other. Actually, however, they are intimately connected. Both points of view are needed, even in problems of practical optical design. The design of an optical objective is carried out in general on the basis of Geometrical Optics, but for the interpretation or prediction of the performance of the objective it becomes necessary to investigate the propagation of waves through the lens system.

In view of this fact, these theories will be developed simultaneously. The wave theory is considered as the general theory, and Geometrical Optics will be shown to be that special part of the wave theory which describes the propagation of light signals, i.e., of sudden discontinuities. On the other hand, in the important case of periodic waves, it represents an approximate solution of the differential equations of wave optics. This approximate solution can be used in a method of successive approximation to develop the diffraction theory of optical instruments, as will be shown in the later parts of this course.

§1. THE ELECTROMAGNETIC EQUATIONS.

1.1 The wave optical part of this course is based upon Maxwell's electromagnetic theory of light. The phenomenon of light is identified with an electromagnetic field.

The location in space is determined by three coordinates x, y, z, the unit of length being 1 cm. The time is determined by the coordinate t; the unit of this variable being 1 sec. The electromagnetic field is represented by two vectors:

the <u>electric vector</u>: $E(x,y,z,t) = (E_1, E_2, E_3)$,

the <u>magnetic vector</u>: $H(x,y,z,t) = (H_1, H_2, H_3)$.   (1.10)

The components, (E_1, E_2, E_3), of the electric vector are functions of x,y,z,t; the unit of these components is 1 electrostatic unit of E. The unit of the components (H_1, H_2, H_3) of the magnetic vector is 1 electromagnetic unit of H.

The properties of the medium can be characterized by two scalar functions of x,y,z (the medium thus is assumed not to change with the time):

the <u>dielectric constant</u>: $\epsilon = \epsilon(x,y,z)$,

the <u>magnetic permeability</u> $\mu = \mu(x,y,z)$.   (1.11)

1.2 The electromagnetic vectors satisfy a system of partial differential equations which, with the above choice of units, assumes the form:

$$\operatorname{curl} H - \frac{\epsilon}{c} E_t = 0 ,$$

$$\operatorname{curl} E + \frac{\mu}{c} H_t = 0 .$$ (1.20)

The constant c is the velocity of light, in our units numerically equal to

$$c = 3 \cdot 10^{10} .$$

If the components of E and H are introduced, the above vector equations yield a system of six linear differential equations of first order:

$$\frac{\partial H_3}{\partial y} - \frac{\partial H_2}{\partial z} - \frac{\epsilon}{c} \frac{\partial E_1}{\partial t} = 0 \qquad \frac{\partial E_3}{\partial y} - \frac{\partial E_2}{\partial z} + \frac{\mu}{c} \frac{\partial H_1}{\partial t} = 0$$

$$\frac{\partial H_1}{\partial z} - \frac{\partial H_3}{\partial x} - \frac{\epsilon}{c} \frac{\partial E_2}{\partial t} = 0 \qquad \frac{\partial E_1}{\partial z} - \frac{\partial E_3}{\partial x} + \frac{\mu}{c} \frac{\partial H_2}{\partial t} = 0 \qquad (1.21)$$

$$\frac{\partial H_2}{\partial x} - \frac{\partial H_1}{\partial y} - \frac{\epsilon}{c} \frac{\partial E_3}{\partial t} = 0 \qquad \frac{\partial E_2}{\partial x} - \frac{\partial E_1}{\partial y} + \frac{\mu}{c} \frac{\partial H_3}{\partial t} = 0$$

In the group of optical problems to be considered in this course, we can assume $\mu = 1$, since our medium (glass) is not magnetic. The dielectric constant, $\epsilon = \epsilon(x,y,z)$, will be replaced by the <u>index of refraction</u> of the substance, according to the equation

$$n = \sqrt{\epsilon} \ . \tag{1.22}$$

This relation between two different properties of a medium is actually far from being satisfied by the substances we are mainly interested in. However, experience shows that the predictions of the electromagnetic theory are in excellent agreement with observation if in theoretical results the quantity $\sqrt{\epsilon}$ is replaced by the index of refraction, measured by optical methods. Furthermore it is possible to give a satisfactory explanation of the above discrepancy by molecular considerations.

We prefer in the following sections to leave Maxwell's equations in the above forms, (1.20) and (1.21). The symmetrical structure of these equations will often allow us to find from one relation another one simply by interchanging the letters ϵ and μ, and replacing E by -H and H by E.

It is customary to add two more equations to the equations (1.20), namely:

$$\text{div}(\epsilon E) = 0, \ \text{or} \ \frac{\partial}{\partial x}(\epsilon E_1) + \frac{\partial}{\partial y}(\epsilon E_2) + \frac{\partial}{\partial z}(\epsilon E_3) = 0 \ ,$$

$$\tag{1.23}$$

$$\text{div}(\mu H) = 0, \ \text{or} \ \frac{\partial}{\partial x}(\mu H_1) + \frac{\partial}{\partial y}(\mu H_2) + \frac{\partial}{\partial z}(\mu H_3) = 0 \ .$$

These state that the electromagnetic field does not contain a source of electricity or magnetism. However, these equations are not independent of (1.20). Indeed, since div curl A = 0 for an arbitrary vector field A(x,y,z,t), it follows

$$\frac{\partial}{\partial t}(\text{div} \ \epsilon E) = \frac{\partial}{\partial t}(\text{div} \ \mu H) = 0 \ , \tag{1.24}$$

i.e., both $\text{div}(\epsilon E)$ and $\text{div}(\mu H)$ are identically zero if they are zero at any particular time.

1.3 <u>Energy</u>. If we form the scalar product of E with the first of the equations (1.20) and of H with the second one and subtract both results we obtain

$$E \cdot \text{curl} \ H - H \cdot \text{curl} \ E - \frac{1}{c}(\epsilon E \cdot E_t + \mu H \cdot H_t) = 0 \ . \tag{1.30}$$

On account of the identity

$$H \cdot \text{curl} \ E - E \cdot \text{curl} \ H = \text{div}(E \times H) \tag{1.31}$$

this gives

$$c \, \text{div}(E \times H) + \frac{1}{2} \frac{\partial}{\partial t} (\epsilon E^2 + \mu H^2) = 0 \qquad (1.32)$$

or

$$\text{div} \frac{c}{4\pi} (E \times H) + \frac{\partial}{\partial t} \frac{1}{8\pi} (\epsilon E^2 + \mu H^2) = 0 \, . \qquad (1.33)$$

The function

$$W(x,y,z,t) = \frac{1}{8\pi} (\epsilon E^2 + \mu H^2) \qquad (1.34)$$

measures the distribution of electromagnetic energy in the field. It determines the light density in Optics. The vector

$$S(x,y,z,t) = \frac{c}{4\pi} (E \times H) \qquad (1.35)$$

is called Poynting's radiation vector, and the relation between W and S is given by the equation

$$\frac{\partial W}{\partial t} + \text{div } S = 0. \qquad (1.36)$$

Let us integrate this equation over a domain D of the x,y,z space enclosed by a closed surface Γ. From Gauss' integral theorem:

$$\frac{\partial}{\partial t} \iiint\limits_D W \, dx \, dy \, dz + \iint\limits_\Gamma S_\nu \, do = 0, \qquad (1.37)$$

S_ν being the normal component of S on Γ. The first integral represents the change of the total energy of the domain D per unit time. The surface integral thus gives the amount of energy which has left the domain D through the surface. Hence we interpret the vector field

$$S = \frac{c}{4\pi} (E \times H)$$

as the vector field (or better, tensor field) of energy flux.

Let do be the area of a surface element at a point x,y,z, and N a unit vector normal to it. Then the energy flux through this surface element is given by

$$dF = S_N do$$

where $S_N = S \cdot N$ is the normal component of the Poynting vector, S.

In Optics, the energy flux per unit area is called the <u>illumination</u> of the surface element. We have

$$I = S_N \qquad\qquad (1.38)$$

1.4 <u>Boundary Conditions</u>. The vector functions E and H are of course not uniquely determined by the differential equations (1.20), unless certain boundary conditions are added. For optical problems, the following problem types are significant:

1. To find a solution of the equations (1.20), i.e., two vector fields, E(x,y,z,t) and H(x,y,z,t), if the electromagnetic field E(x,y,z,0) and H(x,y,z,0), at the time t = 0 is given and satisfies at this time the conditions div(ϵ E) = div(μH) = 0.

2. Let us assume that on the plane z = 0, the electromagnetic field is a known function of x,y and t when t > 0, and that certain homogeneous boundary conditions are satisfied on another plane, z = L; i.e.,

E = E(x,y,0,t) given for t > 0 and, for example,

E(x,y,L,t) = 0 on z = L (Figure 1).

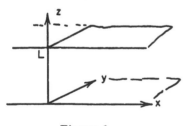

Figure 1

Let furthermore E = H = 0 for t = 0. To find a solution, E and H, in the half-space z > 0 which satisfies these boundary conditions.

3. Of greater practical importance is the case for which the electromagnetic field is established under the influence of a periodic oscillator. Let us assume that an electric dipole is oscillating at a given point in space, for example, in front of an optical objective (Figure 2). Under the influence of this point source, an electric field is established which represents the light wave which travels through the objective. The problem is to determine these forced vibrations of the space as solutions of Maxwell's equations.

§2. PERIODIC FIELDS.

2.1 We can expect that the electromagnetic field which in the end is established by a point source periodic in the time, will be periodic in time itself and, that its frequency equals the frequency of the oscillator. On the

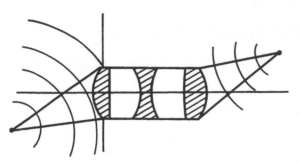

basis of this expectation,
one is led to consider
special solutions of
Maxwell's equations which
have the form

$$E = u(x,y,z)e^{-i\omega t} ,$$

$$H = v(x,y,z)e^{-i\omega t} , \quad (2.11)$$

where u and v are vectors
independent of t. The
quantity

Figure 2

$$\frac{\omega}{2\pi} = \frac{c}{\lambda} \qquad\qquad (2.12)$$

is the <u>frequency</u> of the oscillator and λ the <u>wave length</u>.

2.2 The above complex notation is chosen on account of its mathematical advantages. The vectors u and v are in general complex vectors

$$u = a + ia^* ,$$

$$v = b + ib^* ,$$

i.e., complex combinations of real vectors a, a* and b, b*. Calculations involving these complex vectors can be carried out in the same way as those involving real vectors only, when i is considered a scalar quantity, with $i^2 = -1$. For example:

<u>Scalar product</u>: $u \cdot v = (a \cdot b - a^* \cdot b^*) + i(a \cdot b^* + a^* \cdot b).$ (2.21)

<u>Vector product</u>: $u \times v = (a \times b - a^* \times b^*) + i(a \times b^* + a^* \times b).$ (2.22)

The <u>absolute value</u> of a complex vector: $u \cdot \bar{u} = a^2 + (a^*)^2 .$ (2.23)

Two complex vectors u and v are called <u>orthogonal</u> if $u \cdot \bar{v} = \bar{u} \cdot v = 0$, i.e., if

$$a \cdot b + a^* \cdot b^* = 0 ,$$

$$\qquad\qquad\qquad\qquad (2.24)$$

$$ab^* - a^* \cdot b = 0 .$$

If two complex vectors E and H satisfy Maxwell's equations then both the real and imaginary parts of E and H are solutions. The real parts of the vectors (2.11), for example, are given by

$$E = a \cos \omega t + a^* \sin \omega t ,$$

$$H = b \cos \omega t + b^* \sin \omega t ,$$

(2.25)

and will be considered in the following as representing the electromagnetic field.

2.3　We now introduce the expressions (2.11) into Maxwell's equations. This yields

$$\text{curl } v + \frac{i\omega}{c} \epsilon u = 0 ,$$

$$\text{curl } u - \frac{i\omega}{c} \mu v = 0 ,$$

(2.31)

i.e., a system of partial differential equations without the time variable. By introducing the constant

$$k = \frac{\omega}{c} = \frac{2\pi}{\lambda}$$

(2.32)

we obtain

$$\text{curl } v + ik \epsilon u = 0 ,$$

$$\text{curl } u - ik \mu v = 0 .$$

(2.33)

It follows that

$$\text{div}(\epsilon u) = \text{div}(\mu v) = 0$$

(2.34)

so that it is unnecessary to add these conditions explicitly, as in (1.23).

2.4　Energy. The period, $T = \frac{2\pi}{\omega} = \frac{\lambda}{c}$, of the functions (2.11) is so extremely short in optical problems that we are unable to observe the actual fluctuation of the electromagnetic field. Indeed, in case of sodium light, for example, we have

$$\lambda = 0.6 \times 10^{-4} \text{ cm., hence } T = 2 \times 10^{-15} \text{ sec.}$$

The same is true for the extremely rapid fluctuations of the light density

$$W(x,y,z,t) = \frac{1}{8\pi} \left[\epsilon \, (a \cos \omega t + a* \sin \omega t)^2 + \mu (b \cos \omega t + b* \sin \omega t)^2 \right]. \quad (2.41)$$

We are, however, able to observe the average value of this energy, which is given by the integral

$$\overline{W} = \frac{1}{T} \int_0^T W \, dt = \frac{1}{16\pi} \left[\epsilon \, (a^2 + a*^2) + \mu (b^2 + b*^2) \right]. \quad (2.42)$$

We can express this result in terms of the original complex vectors u and v and obtain

$$\overline{W} = \frac{1}{16\pi} \left[\epsilon \, u \cdot \overline{u} + \mu \, v \cdot \overline{v} \right] \quad (2.43)$$

as an expression for the observable light density at the point x,y,z.

2.5 <u>Flux</u>. Similar considerations may be applied to the Poynting vector, S. By introducing the expressions (2.25) into the definition of S, (1.36), it follows that

$$S = \frac{c}{4\pi} (a \cos \omega t + a* \sin \omega t) \times (b \cos \omega t + b* \sin \omega t) \quad (2.51)$$

which is also a periodic function with the small period, T. Again, only the average value,

$$\overline{S} = \frac{1}{T} \int_\phi^T S \, dt \, ,$$

can be considered as physically significant. We obtain

$$\overline{S}(x,y,z) = \frac{c}{8\pi} (a \times b + a* \times b*) \, , \quad (2.52)$$

or in terms of the complex vectors, u and v,

$$\overline{S}(x,y,z) = \frac{c}{16\pi} (u \times \overline{v} + \overline{u} \times v) \, . \quad (2.53)$$

We can show that the vector field, $\overline{S}$, of average flux is a solenoidal field, i.e., div $\overline{S} = 0$.

For the complex vectors u and v satisfy the equations

$$\text{curl } v + ik\epsilon u = 0 ,$$
$$\text{curl } u - ik\mu v = 0 . \tag{2.54}$$

The conjugate complex vectors $\bar{u}$ and $\bar{v}$, consequently, satisfy

$$\text{curl } \bar{v} - ik\epsilon \bar{u} = 0 ,$$
$$\text{curl } \bar{u} + ik\mu \bar{v} = 0 . \tag{2.55}$$

It follows that

$$\bar{u} \text{ curl } v - v \text{ curl } \bar{u} + ik (\epsilon u \cdot \bar{u} - \mu v \cdot \bar{v}) = 0 ,$$
$$\bar{v} \text{ curl } u - u \text{ curl } \bar{v} + ik (\epsilon u \cdot \bar{u} - \mu v \cdot \bar{v}) = 0 , \tag{2.56}$$

or

$$\text{div } (u \times \bar{v}) + ik (\epsilon u \cdot \bar{u} - \mu v \cdot \bar{v}) = 0 ,$$
$$\text{div } (\bar{u} \times v) - ik (\epsilon u \cdot \bar{u} - \mu v \cdot \bar{v}) = 0 . \tag{2.57}$$

Hence div $(u \times \bar{v} + \bar{u} \times v) = 0$; i.e., div $\bar{S} = 0$.

2.6 <u>Polarization</u>. The vectors E and H given by (2.25) describe certain closed curves in space. The type of these curves determines the state of polarization of the wave at the point x,y,z, and this again represents an observable characteristic of the field. In general, the electric vector is considered as the vector which gives the polarization of the light.

We have $E = a \cos \omega t + a^* \sin \omega t$, or in components

$$E_1 = a_1 \cos \omega t + a_1^* \sin \omega t ,$$
$$E_2 = a_2 \cos \omega t + a_2^* \sin \omega t ,$$
$$E_3 = a_3 \cos \omega t + a_3^* \sin \omega t . \tag{2.61}$$

The curve described by E is plane since E is a linear combination of the vectors a and a*. We can show easily that this curve is an ellipse. Let us introduce $\xi = \cos \omega t$ and $\eta = \sin \omega t$. By squaring the components of E we find

$$E_1^2 = a_1^2 \xi^2 + a_1^{*2} \eta^2 + 2a_1 a_1^* \xi\eta ,$$

$$E_2^2 = a_2^2 \xi^2 + a_2^{*2} \eta^2 + 2a_2 a_2^* \xi\eta ,$$

$$E_3^2 = a_3^2 \xi^2 + a_3^{*2} \eta^2 + 2a_3 a_3^* \xi\eta .$$

These equations, together with the relation $\xi^2 + \eta^2 = 1$ represent four linear equations for the three quantities ξ^2, η^2, and $2\xi\eta$. Their determinant thus must be zero:

$$
\begin{vmatrix}
1 & 1 & 1 & 0 \\
E_1^2 & a_1^2 & a_1^{*2} & a_1 a_1^* \\
E_2^2 & a_2^2 & a_2^{*2} & a_2 a_2^* \\
E_3^2 & a_3^2 & a_3^{*2} & a_3 a_3^*
\end{vmatrix} = 0
\tag{2.62}
$$

This is an equation of the type $AE_1^2 + BE_2^2 + CE_3^2 = D$, which means that the curve of E lies on a surface of second order. The intersection curve of a plane and a surface of second order, however, is a conic. It must be an ellipse because it is closed.

The equations (2.61) show that the ellipse is symmetrical with respect to the origin, i.e., to the point x,y,z in question. Thus we can find the length and direction of the axes by determining the extreme lengths of the vector, E, i.e., the extreme values of the quadratic form

$$|E|^2 = a \cdot a\, \xi^2 + 2a \cdot a^*\, \xi\eta + a^* \cdot a^*\, \eta^2 ,$$

under the condition $\xi^2 + \eta^2 = 1$. In other words, the axes are equal to the characteristic values of the above quadratic form and are given by the two solutions, λ_1 and λ_2, of the quadratic equation

$$
\begin{vmatrix}
a \cdot a - \lambda & a \cdot a^* \\
a \cdot a^* & a^* \cdot a^* - \lambda
\end{vmatrix} = 0 .
\tag{2.63}
$$

Hence,

$$\lambda = \frac{1}{2}\left[a^2 + a^{*2} \mp \sqrt{(a^2 - a^{*2})^2 + 4(a \cdot a^*)^2} \right] .
\tag{2.64}$$

The characteristic values, λ, are real, since the expression under the radical is not negative. The characteristic values cannot be negative; for, with the aid of the inequality $(a \cdot a^*)^2 \leq a^2 \cdot (a^*)^2$, one can see that

$$\sqrt{(a^2 - a^{*2})^2 + 4(a \cdot a^*)^2} \leqq \sqrt{(a \cdot a + a^* \cdot a^*)^2} = a^2 + a^{*2} .$$

We illustrate three types of polarization:

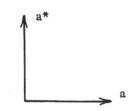

Figure 3. Elliptical polarization.

Two different characteristic values, $\lambda_1 \neq \lambda_2$, both different from zero. The electric vector describes an ellipse.

The characteristic values are equal, $\lambda_1 = \lambda_2$. This implies

$$a \cdot a = a^* \cdot a^*,$$
$$a \cdot a^* = 0. \tag{2.65}$$

The two components a and a* of u are orthogonal and equal in length. The electric vector describes a circle.

Figure 4. Circular polarization.

The smaller one of the characteristic values, λ, is zero. The electric vector describes a straight line. The two vector components of u have the same direction,

$$a \times a^* = 0. \tag{2.66}$$

Figure 5. Linear polarization.

We can express these results again by using the complex vector, $u = a + ia^*$, directly. The quadratic equation for λ may be written as follows:

$$\lambda^2 - (u \cdot \overline{u}) \lambda - \frac{1}{4}(u \times \overline{u})^2 = 0 ;$$

and this has the solution,

$$\lambda = \frac{1}{2} \left[u \cdot \overline{u} \mp \sqrt{(u)^2 (\overline{u})^2} \right]. \tag{2.67}$$

The ellipticity ϵ of the polarization, i.e., the ratio of the lengths of the axes, is thus given by the expression

$$\epsilon^2 = \frac{\lambda_1}{\lambda_2} = \frac{u \cdot \overline{u} - \sqrt{(u)^2 (\overline{u})^2}}{u \cdot \overline{u} + \sqrt{(u)^2 (\overline{u})^2}} . \tag{2.68}$$

Hence,

for linear polarization: $u \times \overline{u} = 0$,

for circular polarization: $u^2 = u \cdot u = 0$. \tag{2.69}

§3. DIFFERENTIAL EQUATIONS FOR E AND H.

3.1 If we eliminate one of the vectors, E or H, from Maxwell's equations, (1.20), we obtain second order equations for either E or H. By differentiation with respect to t:

$$\text{curl } H_t - \frac{\epsilon}{c} E_{tt} = 0 ,$$

$$\text{curl } E_t + \frac{\mu}{c} H_{tt} = 0 .$$

We introduce $H_t = -\frac{c}{\mu} \text{curl } E$ in the first of these equations, and $E_t = \frac{c}{\epsilon}$ curl H in the second. The results are

$$\mu \text{ curl } \left(\frac{1}{\mu} \text{ curl } E \right) + \frac{\epsilon\mu}{c^2} E_{tt} = 0 ,$$

$$\epsilon \text{ curl } \left(\frac{1}{\epsilon} \text{ curl } H \right) + \frac{\epsilon\mu}{c^2} H_{tt} = 0 . \tag{3.11}$$

We apply the following vector identity, which holds for an arbitrary scalar function, f(x,y,z), and an arbitrary vector field, A(x,y,z), with continuous derivatives of the second order:

$$\text{curl (f curl A)} = -f \Delta A + f \text{ grad (div A)} + (\text{grad f}) \times (\text{curl A}) , \tag{3.12}$$

where $\Delta A = A_{xx} + A_{yy} + A_{zz}$. Equations (3.11) become

$$\frac{\epsilon\mu}{c^2} E_{tt} - \Delta E = (\text{curl E}) \times \left(\mu \text{ grad } \frac{1}{\mu} \right) - \text{grad (div E)},$$

$$\tag{3.13}$$

$$\frac{\epsilon\mu}{c^2} H_{tt} - \Delta H = (\text{curl H}) \times \left(\epsilon \text{ grad } \frac{1}{\epsilon} \right) - \text{grad (div H)} .$$

From the second pair of Maxwell's equations (1.23), it follows that

$$\begin{aligned} \text{div } \epsilon E &= \epsilon \text{ div } E + E \cdot \text{grad } \epsilon = 0 , \\ \text{div } \mu H &= \mu \text{ div } H + H \cdot \text{grad } \mu = 0 ; \end{aligned} \tag{3.14}$$

i.e., div E = -E·p and div H = -H·q, where the vectors p and q are defined by

$$p = \frac{1}{\epsilon} \text{ grad } \epsilon = \text{grad (log } \epsilon) ,$$

$$\tag{3.15}$$

$$q = \frac{1}{\mu} \text{ grad } \mu = \text{grad (log } \mu) .$$

We introduce

$$n = \sqrt{\epsilon \mu} \, , \qquad\qquad (3.16)$$

and obtain from (3.11) the equations

$$\frac{n^2}{c^2} E_{tt} - \Delta E = \text{grad} \, (p \cdot E) + q \times \text{curl} \, E \, ,$$

$$\qquad\qquad (3.17)$$

$$\frac{n^2}{c^2} H_{tt} - \Delta H = \text{grad} \, (q \cdot H) + p \times \text{curl} \, H \, .$$

The vectors p and q and the function n are given by the properties of the medium; they are not independent of each other but are related by the equation

$$\frac{1}{2} (p + q) = \text{grad} \, (\log n) \, . \qquad\qquad (3.18)$$

In the special case of a homogeneous medium, both p and q are zero, and $n = \sqrt{\epsilon \mu}$ is a constant. The equations (3.17) become

$$\frac{n^2}{c^2} E_{tt} - \Delta E = 0 \, ,$$

$$\qquad\qquad (3.19)$$

$$\frac{n^2}{c^2} H_{tt} - \Delta H = 0 \, .$$

Each component of E and H satsifies the ordinary wave equation. The velocity of the waves is given by the quantity

$$v = c/n \, ,$$

which allows us to regard the quantity $n = \sqrt{\epsilon \mu}$ as the index of refraction of the medium, defined by the ratio $n = c/v$ of the velocity of light in a vacuum to the velocity in the medium.

In the case of a non-homogeneous medium, a more complicated set of equations is obtained. Since n is now a function of x,y,z, the six equations, (3.17), no longer yield one equation in each component, for the first order operators on the right sides involve all the components of the vectors in each equation. However, it is still true that the wave velocity, v, is given by the ratio c/n. Indeed, we shall see that for the propagation of a light signal, i.e., a sudden disturbance of the electric field, only the second order terms in (3.17) are significant. These terms lead to a generalized wave equation in which the coefficient, n, is not constant.

3.2 Stratified media. Let us consider as an example the case of a stratified medium, in which the functions ϵ and μ depend only on one variable, for instance on z. This case is of considerable practical interest, since the propagation of waves through thin, multilayer films, evaporated on glass, leads to a problem of this type. Let $\mu = 1$ and $\sqrt{\epsilon} = n(z)$. It follows that

$$p = 2 \text{ grad (log } n) = \left(0, 0, 2\frac{n'}{n}\right) ,$$

$$q = 0 . \tag{3.21}$$

Hence

$$p \cdot E = 2\frac{n'}{n} E_3$$

$$\text{grad } (p \cdot E) = \left(2\frac{n'}{n}\frac{\partial E_3}{\partial x}; \ 2\frac{n'}{n}\frac{\partial E_3}{\partial y}; \ 2\frac{\partial}{\partial z}\left(\frac{n'}{n} E_3\right)\right) \tag{3.22}$$

$$- p \times \text{curl } H = 2\frac{n'}{n}\left(\frac{\partial H_1}{\partial z} - \frac{\partial H_3}{\partial x}, \ -\frac{\partial H_3}{\partial y} + \frac{\partial H_2}{\partial z}, 0\right)$$

The differential equations (3.17) become

$$\frac{n^2}{c^2}\frac{\partial^2 E_1}{\partial t^2} - \Delta E_1 = 2\frac{n'}{n}\frac{\partial E_3}{\partial x}$$

$$\frac{n^2}{c^2}\frac{\partial^2 E_2}{\partial t^2} - \Delta E_2 = 2\frac{n'}{n}\frac{\partial E_3}{\partial y}$$

$$\frac{n^2}{c^2}\frac{\partial^2 E_3}{\partial t^2} - \Delta E_3 = 2\frac{\partial}{\partial z}\left(\frac{n'}{n} E_3\right)$$

$$\frac{n^2}{c^2}\frac{\partial^2 H_1}{\partial t^2} - \Delta H_1 + 2\frac{n'}{n}\frac{\partial H_1}{\partial z} = +2\frac{n'}{n}\frac{\partial H_3}{\partial x} \tag{3.23}$$

$$\frac{n^2}{c^2}\frac{\partial^2 H_2}{\partial t^2} - \Delta H_2 + 2\frac{n'}{n}\frac{\partial H_2}{\partial z} = +2\frac{n'}{n}\frac{\partial H_3}{\partial y}$$

$$\frac{n^2}{c^2}\frac{\partial^2 H_3}{\partial t^2} - \Delta H_3 = 0 .$$

We thus obtain two partial differential equations, namely, those for E_3 and H_3, in which none of the other components appear. After E_3 and H_3 have been determined from these two equations, they are substituted in the remaining equations of (3.23); and these equations then become modified wave equations for E_1, E_2, H_1, and H_2 modified in the sense that the right side is not zero, but a known function. As a result of this simplification it is possible to find explicit solutions for many problems connected with stratified media, especially with films producing low reflection.

§4. INTEGRAL FORM OF MAXWELL'S EQUATIONS.

The functions $\epsilon(x,y,z)$ and $\mu(x,y,z)$ are not necessarily continuous functions. We assume, however, that they are sectionally smooth, i.e., every finite domain of the x,y,z space can be divided into a finite number of parts in which ϵ and μ are continuous and have continuous derivatives.

The differential equations (1.20) represent conditions for the electro-magnetic field in every part of the space where ϵ, μ, and E, H are continuous and have continuous derivatives. They are, however, not sufficient to establish conditions for the boundary values of E and H on a surface of discontinuity. This is the reason why it is advantageous to replace the differential equations (1.20) by certain integral relations. These integral equations are equivalent to the differential equations if ϵ, μ, and E, H are continuous and have con-tinuous derivatives. They are more general, on the other hand, since they apply equally well to the case of discontinuous functions ϵ, μ; E, H and establish definite conditions for the electromagnetic field in this case.

4.1 Let us consider, in the four-dimensional x,y,z,t space, a domain D which is bounded by a closed three-dimensional hypersurface Γ. We assume that the hypersurface Γ consists of a finite number of sections in which the outside normal N of the hypersurface varies continuously. This normal N is a unit vector in the x,y,z,t space given by

$$N = \lambda(\varphi_x, \varphi_y, \varphi_z, \varphi_t)$$

$$\lambda = \pm \frac{1}{\sqrt{\varphi_x^2 + \varphi_y^2 + \varphi_z^2 + \varphi_t^2}}$$

if the surface Γ is represented by the equation $\varphi(x,y,z,t) = 0$. In general we denote the components of the unit vector N by

$$N = (x_N, y_N, z_N, t_N) \tag{4.11}$$

and call these components the direction cosines of N with respect to the four coordinate axes.

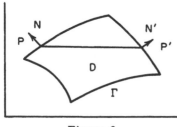

Figure 6

Let $F(x,y,z,t)$ be a function which has continuous derivatives in D. We consider the integral of F_x over D and carry out the integration with respect to x:

$$\iiiint_D F_x \, dx \, dy \, dz \, dt$$

$$= \iiint_\Gamma \Big(F(P') - F(P)\Big) \, dy \, dz \, dt.$$

The integral on the right side is a surface integral over the hypersurface Γ. We introduce

$$\text{at P': } dy\ dz\ dt\ =\ x_{N'}\ do'$$

$$\text{at P : } dy\ dz\ dt\ =\ -\ x_N\ do$$

where $x_{N'}$ and x_N are the x-components of the unit vector N at P' and P, respectively, and do' and do, differentials independent of the choice of the coordinate system. We call do the surface element of the hypersurface Γ. With this notation, we obtain

$$\iiiint_D F_x\ dx\ dy\ dz\ dt\ =\ \iiint_\Gamma F\ x_N\ do\ . \tag{4.12}$$

In the same way, we find

$$\iiiint_D F_y\ dx\ dy\ dz\ dt\ =\ \iiint_\Gamma F\ y_N\ do\ ,$$

$$\iiiint_D F_z\ dx\ dy\ dz\ dt\ =\ \iiint_\Gamma F\ z_N\ do\ , \tag{4.13}$$

$$\iiiint_D F_t\ dx\ dy\ dz\ dt\ =\ \iiint_\Gamma F\ t_N\ do\ .$$

These formulae allow us to transform equations which involve derivatives of a function $F(x,y,z,t)$ into conditions for the function F itself.

4.2 Let us apply these transformations to the equation div ϵ E = 0. We conclude first

$$\iiiint_D (\text{div } \epsilon\ E)\ dx\ dy\ dz\ dt\ =\ \iiint_\Gamma \epsilon\ (E_1 x_N\ +\ E_2 y_N\ +\ E_3 z_N)\ do\ . \tag{4.21}$$

The expression $E_1 x_N + E_2 y_N + E_3 z_N$ can be interpreted as the scalar product of the two three-dimensional vectors E and

$$M\ =\ (x_N, y_N, z_N) \tag{4.22}$$

The vector M is the projection of the four-dimensional unit vector N into the x,y,z space, i.e.,

$$M\ =\ (x_N, y_N, z_N, 0)\ .$$

By using the vector M we can write (4.21) in the form

$$\iiiint_D \text{div } \epsilon E \text{ dx dy dz dt} = \iiint_\Gamma \epsilon (E \cdot M) \text{ do} . \qquad (4.23)$$

This integral relation, of course, is nothing but the integral theorem of Gauss for four dimensions and applied to a vector ϵE for which the fourth component is zero.

Since div ϵE = div μH = 0, according to Maxwell's equations, we can formulate the statement:

The surface integrals

$$\iiint_\Gamma \epsilon (E \cdot M) \text{ do and } \iiint_\Gamma \mu (H \cdot M) \text{ do} \qquad (4.24)$$

are zero for any closed hypersurface Γ in the four-dimensional x,y,z,t space.

We have derived this result from Maxwell's equations under the assumption that ϵ, μ; E, H are continuous and have continuous derivatives. In this case the integral relations (4.24) are equivalent to the differential equations div ϵE = div μH = 0, as we can see easily. However the relations (4.24) can be applied directly to discontinuous functions as long as they are integrable. We will see presently that explicit conditions for discontinuities can be derived from (4.24). In view of this we consider the integral equations as the original source of the differential equations to which we have to go back in case of doubt.

4.3 We next apply our transformation to the equation

$$\text{curl } H - \frac{\epsilon}{c} E_t = 0 .$$

From curl H = i x H_x + j x H_y + k x H_z follows:

$$\iiiint_D \text{curl } H \text{ dx dy dz dt} = \iiint_\Gamma \left[(ix_N + jy_N + kz_N) \right] \text{ x H do}$$

$$= \iiint_\Gamma (M \text{ x } H) \text{ do} . \qquad (4.31)$$

Hence

$$\iiiint_D \left(\text{curl } H - \frac{\epsilon}{c} E_t \right) \text{dx dy dz dt} = \iiint_\Gamma \left(M \text{ x } H - \frac{\epsilon}{c} t_N E \right) \text{do} \qquad (4.32)$$

and similarly,

$$\iiiint_D \left(\text{curl } E + \frac{\mu}{c} H_t \right) dx \, dy \, dz \, dt = \iiint_\Gamma \left(M \times E + \frac{\mu}{c} t_N H \right) do \quad (4.33)$$

Hence: <u>The surface integrals</u>

$$\iiint_\Gamma \left(M \times H - \frac{\epsilon}{c} t_N E \right) do \quad \text{and} \quad \iiint_\Gamma \left(M \times E + \frac{\mu}{c} t_N H \right) do \quad (4.34)$$

<u>are zero for any closed hypersurface Γ in the four-dimensional x,y,z,t space.</u>

    Again we notice that these conditions involve only the vectors E and H and the functions ϵ and μ, and not their derivatives. They are equivalent to Maxwell's equations (1.20) if the derivatives exist. We require, however, that the integral relations (4.34) must be satisfied also by discontinuous electromagnetic fields.

§5.   GENERAL CONDITIONS FOR DISCONTINUITIES.

    5.1 We apply the integral equations (4.24) and (4.34) to the following problem. Let $\varphi(x,y,z,t) = 0$ represent a surface section on which ϵ, μ or E, H are discontinuous. What is the relation of the boundary values of E and H on the two sides of $\varphi = 0$ to each other? We consider a closed hypersurface Γ which is divided into two parts Γ_1 and Γ_2 by the hypersurface $\varphi = 0$. Let Γ_0 be the part of $\varphi = 0$ which lies inside of Γ. The normal of the surface $\varphi = 0$ is proportional to the vector $(\varphi_x, \varphi_y, \varphi_z, \varphi_t)$. Let us assume that on Γ_0 this vector points toward Γ_2. We denote the boundary values of (ϵ, μ, E, H) by $(\epsilon_1, \mu_1, E_1, H_1)$ if Γ_0 is approached from the domain D_1, and by $(\epsilon_2, \mu_2, E_2, H_2)$ if Γ_0 is approached from D_2.

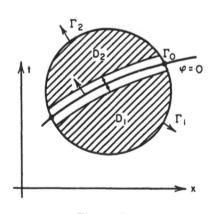

Figure 7

    5.2 We now apply the first equation (4.34) to the closed surface $\Gamma = \Gamma_1 + \Gamma_2$. We have

$$\iiint_\Gamma \left(M \times H - \frac{\epsilon}{c} t_N E \right) do = 0 . \quad (5.21)$$

However, this condition must also be satisfied if the closed surface $\Gamma_1 + \Gamma_0$ is chosen. On Γ_0 we have in this case

$$M = \frac{\text{grad } \varphi}{\sqrt{\varphi_x^2 + \varphi_y^2 + \varphi_z^2 + \varphi_t^2}} \quad ; \quad t_N = \frac{\varphi_t}{\sqrt{\varphi_x^2 + \varphi_y^2 + \varphi_z^2 + \varphi_t^2}} \tag{5.22}$$

and hence

$$\iiint\limits_{\Gamma_1} \left(M \times H - \frac{\epsilon}{c} t_N E \right) do$$

$$+ \iiint\limits_{\Gamma_0} \left(\text{grad } \varphi \times H_1 - \frac{\epsilon_1}{c} \varphi_t E_1 \right) \frac{do}{\sqrt{\varphi_x^2 + \varphi_y^2 + \varphi_z^2 + \varphi_t^2}} = 0 . \tag{5.23}$$

If the surface $\Gamma_2 + \Gamma_0$ is considered, in which case on Γ_0

$$M = - \frac{\text{grad } \varphi}{\sqrt{\varphi_x^2 + \varphi_y^2 + \varphi_z^2 + \varphi_t^2}} \quad ; \quad t_N = - \frac{\varphi_t}{\sqrt{\varphi_x^2 + \varphi_y^2 + \varphi_z^2 + \varphi_t^2}} , \tag{5.24}$$

we obtain

$$\iiint\limits_{\Gamma_2} \left(M \times H - \frac{\epsilon}{c} t_N E \right) do$$

$$- \iiint\limits_{\Gamma_0} \left(\text{grad } \varphi \times H_2 - \frac{\epsilon_2}{c} \varphi_t E_2 \right) \frac{do}{\sqrt{\varphi_x^2 + \varphi_y^2 + \varphi_z^2 + \varphi_t^2}} = 0 . \tag{5.25}$$

We finally subtract the equations (5.25) and (5.23) from (5.21). The result is

$$\iiint\limits_{\Gamma_0} \left\{ \text{grad } \varphi \times (H_2 - H_1) \right.$$

$$\left. - \frac{\varphi_t}{c} (\epsilon_2 E_2 - \epsilon_1 E_1) \right\} \frac{do}{\sqrt{\varphi_x^2 + \varphi_y^2 + \varphi_z^2 + \varphi_t^2}} = 0 . \tag{5.26}$$

This relation must be true for any part Γ_0 of the surface $\varphi = 0$. This is only possible if the integrand in (5.26) is zero; hence

$$\text{grad } \varphi \times [H] - \frac{\varphi_t}{c} [\epsilon E] = 0 \tag{5.261}$$

where

$$[H] = H_2 - H_1 \quad \text{and} \quad [\epsilon E] = \epsilon_2 E_2 - \epsilon_1 E_1 \tag{5.27}$$

denotes the size of the discontinuity of the quantity inside the bracket.

5.3 We next consider the first integral equation (4.24) and apply it to the three closed surfaces $\Gamma_1 + \Gamma_2$; $\Gamma_1 + \Gamma_0$; $\Gamma_2 + \Gamma_0$. We obtain

$$\iiint_{\Gamma_1} \epsilon\,(E\cdot M)do \; + \; \iiint_{\Gamma_2} \epsilon\,(E\cdot M)do \; = \; 0 \, , \tag{5.31}$$

$$\iiint_{\Gamma_1} \epsilon\,(E\cdot M)do \; + \; \iiint_{\Gamma_0} (\epsilon_1 E_1 \cdot grad\ \varphi)\frac{do}{\sqrt{\varphi_x{}^2 + \varphi_y{}^2 + \varphi_z{}^2 + \varphi_t{}^2}} \; = \; 0, \tag{5.32}$$

$$\iiint_{\Gamma_2} \epsilon\,(E\cdot M)do \; - \; \iiint_{\Gamma_0} (\epsilon_2 E_2 \cdot grad\ \varphi)\frac{do}{\sqrt{\varphi_x{}^2 + \varphi_y{}^2 + \varphi_z{}^2 + \varphi_t{}^2}} \; = \; 0, \tag{5.33}$$

and by subtraction

$$\iiint_{\Gamma_0} (\epsilon_2 E_2 \, - \, \epsilon_1 E_1)\cdot grad\ \varphi\ \frac{do}{\sqrt{\varphi_x{}^2 + \varphi_y{}^2 + \varphi_z{}^2 + \varphi_t{}^2}} \; = \; 0 \, . \tag{5.34}$$

This yields in the same way as above:

$$[\,\epsilon\,E\,]\cdot grad\ \varphi \; = \; 0 \, . \tag{5.35}$$

5.4 From (5.26) and (5.35) two more equations can be found by inter-changing the letters ϵ and μ, and E and $-$H. We summarize our results as follows:

An electromagnetic field which is discontinuous on a hypersurface $\varphi(x,y,z,t) = 0$ must satisfy the conditions:

$$grad\ \varphi\ x\ [\,H\,] - \frac{\varphi_t}{c}[\,\epsilon\,E\,] \; = \; 0, \quad [\,\epsilon\,E\,]\cdot grad\ \varphi \; = \; 0\,;$$

$$\tag{5.41}$$

$$grad\ \varphi\ x\ [\,E\,] + \frac{\varphi_t}{c}[\,\mu\,H\,] \; = \; 0, \quad [\,\mu\,H\,]\cdot grad\ \varphi \; = \; 0\, .$$

We notice that the second column of equations follows from the first column if $\varphi_t \neq 0$. The equations (5.41) may be considered as the counterpart of Maxwell's differential equations. They represent a system of linear difference equations which take the place of the differential equations (1.20) and (1.23).

§6. DISCONTINUITIES OF THE OPTICAL PROPERTIES.

We apply the general conditions (5.41) to the special case where discon-tinuities of E and H are introduced by discontinuities of the functions ϵ or μ. Any system of glass lenses gives an example for this case. Let $\psi(x,y,z) = 0$ represent a refracting surface on which $\epsilon\,(x,y,z)$ and $\mu(x,y,z)$ are discon-tinuous. We consider the hypersurface

$$\varphi(x,y,z,t) \; = \; \psi\,(x,y,z) \; = \; 0$$

in the four-dimensional x,y,z,t space. This is a cylindrical hypersurface the generating lines of which are parallel to the t-axis. In this case, since $\varphi_t = 0$, our conditions (5.41) become

$$\text{grad } \psi \times [H] = 0, \quad [\epsilon E] \cdot \text{grad } \psi = 0,$$

$$\text{grad } \psi \times [E] = 0, \quad [\mu H] \cdot \text{grad } \psi = 0.$$

(6.1)

The vectors $\dfrac{\text{grad } \psi \times H}{|\text{grad } \psi|}$ and $\dfrac{\text{grad } \psi \times E}{|\text{grad } \psi|}$ are linearly related to the tangential components of H and E. The quantities $\dfrac{\epsilon E \text{ grad } \psi}{|\text{grad } \psi|}$ and $\dfrac{\mu H \cdot \text{grad } \psi}{|\text{grad } \psi|}$ are the normal components of ϵE and μH. Therefore we may formulate the conditions (6.1) in the following customary way:

The tangential components of E and H and the normal components of ϵE and μH are continuous on a surface of discontinuity of ϵ and μ.

§7. PROPAGATION OF DISCONTINUITIES; WAVEFRONTS.

7.1 Discontinuities of the electromagnetic field can appear without being caused by a discontinuous distribution of substances. Let us, for example,

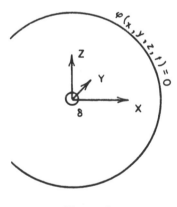

Figure 8

consider the case $\epsilon = \mu = 1$ and assume that at $t = 0$ the vectors $E(x,y,z,0)$ and $H(x,y,z,0)$ are different from zero only in a small sphere of radius δ around the origin. We expect, in analogy to other forms of wave motion, that this electromagnetic field expands with increasing time such that at a given time $t > 0$ the vectors E and H are different from zero in a larger sphere of radius $\delta + ct$. In other words we expect that the surface which separates the parts of the space which are still at rest from those penetrated by the original impulse travels over the space. A surface of this type is called a wavefront. In the above example the wave fronts are spherical and given by the equation

$$\varphi(x,y,z,t) = \sqrt{x^2 + y^2 + z^2} - \delta - ct = 0.$$

(7.11)

If the boundary values of $E(x,y,z,0)$ or $H(x,y,z,0)$ on the original sphere of radius δ are different from zero then this sphere is a surface on which the electromagnetic field is discontinuous. We must expect that at the time $t > 0$

the corresponding boundary values on the wavefront (7.11) are likewise different from zero so that the electromagnetic field is also discontinuous on the new wave front. This consideration leads us to define a wave front more generally as any surface in the x,y,z space on which, at a given time t, the electromagnetic field is discontinuous.

An observer at a point x,y,z will interpret such a discontinuity as a sudden signal which reaches him when the wave front goes through the point x,y,z.

Instead of illustrating the equation (7.11) by a set of surfaces in the three-dimensional x,y,z space depending on the parameter t, we can interpret such a relation $\varphi(x,y,z,t) = 0$ as a hypersurface in the four-dimensional space x,y,z,t. In our example this hypersurface is the cone

$$\sqrt{x^2 + y^2 + z^2} - ct = \delta$$

and the electromagnetic vectors are discontinuous on this cone. Its "contour lines", i.e., the cross sections of the hypercone, $\varphi(x,y,z,t) = 0$, with the hyperplanes t = const., then represent the above set of wave fronts in the x,y,z, space.

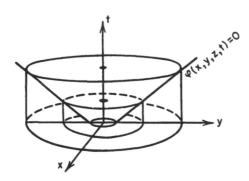

Figure 9

7.2 We may expect from the above example that the hypersurfaces $\varphi = 0$ which determine the propagation of discontinuities are not arbitrary but must fulfill certain conditions. We can derive these conditions easily with the aid of the general relations (5.41). Let us assume that $\varphi(x,y,z,t) = 0$ represents a hypersurface on which the vectors E and H are discontinuous. The functions $\epsilon(x,y,z)$ and $\mu(x,y,z)$ shall be continuous in the neighborhood of $\varphi = 0$. We introduce, on $\varphi = 0$, the vectors

$$U = [E] = E_2 - E_1,$$

$$V = [H] = H_2 - H_1,$$

$$(7.21)$$

which measure the discontinuity of E and H on $\varphi = 0$. It follows from (5.41):

$$\text{grad } \varphi \times V - \frac{\epsilon}{c} \varphi_t U = 0; \quad U \cdot \text{grad } \varphi = 0;$$

$$(7.22)$$

$$\text{grad } \varphi \times U + \frac{\mu}{c} \varphi_t V = 0; \quad V \cdot \text{grad } \varphi = 0.$$

The first column of these equations represents a system of six linear homogeneous equations for the six components $U_1, U_2, U_3; V_1, V_2, V_3$. This system can have non-trivial solutions, $U \neq 0$ and $V \neq 0$, only if the determinant is zero. This establishes the desired condition for the function $\varphi(x,y,z,t) = 0$. We can derive this condition as follows: We form the vector product of grad φ with one of the equations (7.22), for example, with the second equation:

$$\text{grad } \varphi \times (\text{grad } \varphi \times U) + \frac{\mu}{c} \varphi_t \text{ grad } \varphi \times V = 0$$

and introduce grad $\varphi \times V = \frac{\epsilon}{c} \varphi_t U$ from the first equation. It follows

$$\text{grad } \varphi \times (\text{grad } \varphi \times U) + \frac{\epsilon\mu}{c^2} \varphi_t^2 U = 0 . \tag{7.23}$$

If we apply the vector identity (Appendix: I.23) we obtain

$$(U \cdot \text{grad } \varphi) \text{ grad } \varphi - (\text{grad } \varphi)^2 U + \frac{\epsilon\mu}{c^2} \varphi_t^2 U = 0 ;$$

or, since $U \cdot \text{grad } \varphi = 0$,

$$\left((\text{grad } \varphi)^2 - \frac{\epsilon\mu}{c^2} \varphi_t^2\right) U = 0 . \tag{7.24}$$

In a similar way we find

$$\left((\text{grad } \varphi)^2 - \frac{\epsilon\mu}{c^2} \varphi_t^2\right) V = 0 \tag{7.25}$$

and conclude: If U and V are different from zero, i.e., if E and H are discontinuous on $\varphi = 0$, then $\varphi(x,y,z,t)$ must satisfy the equation

$$(\text{grad } \varphi)^2 = \varphi_x^2 + \varphi_y^2 + \varphi_z^2 = \frac{\epsilon\mu}{c^2} \varphi_t^2 . \tag{7.26}$$

This equation is called the <u>characteristic equation</u> of Maxwell's differential equation. Every function $\varphi(x,y,z,t)$ which, for $\varphi(x,y,z,t) = 0$, satisfies this equation (7.26) represents a hypersurface which is called a <u>characteristic surface</u> of the differential equations.

7.3 The characteristic equation (7.26) is not a true differential equation for $\varphi(x,y,z,t)$; indeed it does not have to be satisfied identically in x,y,z,t but only for those combinations x,y,z,t for which $\varphi(x,y,z,t) = 0$.

We can, however, assume, without loss of generality, that the characteristic surface is given in the form

$$\varphi = \psi(x,y,z) - ct = 0 ,$$

where $\psi(x,y,z)$ is independent of t. We obtain

$$\psi_x^2 + \psi_y^2 + \psi_z^2 = \epsilon \mu = n^2 \tag{7.31}$$

and this equation must be satisfied identically in x,y,z.

We may formulate our result as follows: <u>If E and H are discontinuous on a set of wave fronts $\psi(x,y,z) = ct$ then $\psi(x,y,z)$ must be a solution of the partial differential equation (7.31)</u>.

The equation (7.31) is called the <u>equation of the wave fronts</u>; in some literature it is known as the <u>Eiconal Equation</u>. It is the basic equation of Geometrical Optics; the greater part of this course is concerned with problems related to this equation.

If we introduce $\varphi = \psi - ct$ in the original equations (7.22) we obtain

$$\text{grad } \psi \times V + \epsilon U = 0 ,$$

$$\text{grad } \psi \times U - \mu V = 0 . \tag{7.32}$$

It is not necessary to add the other two equations (7.22) explicitly since both equations are a consequence of (7.32):

$$U \cdot \text{grad } \psi = 0 ,$$

$$V \cdot \text{grad } \psi = 0 . \tag{7.33}$$

We furthermore conclude

$$U \cdot V = 0 . \tag{7.34}$$

Hence: <u>The vectors U and V are tangential to the wave fronts and perpendicular to each other</u>.

If $\psi - ct = 0$ represents a set of wave fronts in the sense of our original definition, namely, boundaries of regions which have been penetrated by a light impulse, then U and V are equal to the vectors E and H on the wave front (because E = H = 0 on one side of the surface

$$\psi - ct = 0).$$

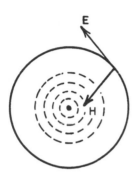

Figure 10

We find: The electromagnetic vectors on such a wave front are tangential to the wave front and perpendicular to each other.

§8. BICHARACTERISTICS; LIGHTRAYS.

8.1 The problem of integrating a partial differential equation of first order can be reduced to the problem of integration of a system of ordinary differential equations, the so-called characteristic differential equations. The integral curves of the characteristic equations are known as characteristics. The equation of the wave fronts

$$\psi_x^2 + \psi_y^2 + \psi_z^2 = n^2 \tag{8.11}$$

is itself a characteristic equation of Maxwell's differential equations. Therefore the characteristics of this first order equation are called Bicharacteristics of Maxwell's equations.

For our purpose it is not necessary to introduce these bicharacteristics by general considerations which would apply to any partial differential equation of first order. We would find that the bicharacteristics in our special case are nothing but the orthogonal trajectories of the wavefronts $\psi = ct$. Hence we prefer to introduce these bicharacteristics directly as orthogonal trajectories of a set of wavefronts $\psi = ct$. We call these trajectories the light rays of the optical medium and we will see in the following that this name is justified.

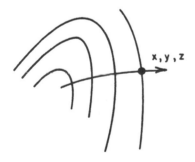

Figure 11

8.2 Let us consider a set of wavefronts $\psi(x,y,z) = $ const. An orthogonal trajectory of these surfaces at any point x,y,z is normal to the wavefront through this point. The complete manifold of orthogonal trajectories through the given set of wavefronts $\psi = $ const. thus must be identical with the solutions of the differential equations

$$\frac{dx}{d\sigma} = \lambda\psi_x$$

$$\frac{dy}{d\sigma} = \lambda\psi_y \qquad\qquad \frac{dz}{d\sigma} = \lambda\psi_z \tag{8.21}$$

where $\lambda = \lambda(x,y,z,\sigma)$ is an arbitrary factor. The choice of λ does not influence the geometrical form of the trajectories but only their parametric representation.

The orthogonal trajectories (8.21) depend of course upon the chosen set of wavefronts, i.e., on the particular solution ψ of (8.11). It is now significant that it is possible to determine light rays, i.e., orthogonal trajectories of surfaces ψ = const. without reference to a particular solution ψ of (8.11). Indeed, if we differentiate $\frac{1}{\lambda} \frac{dx}{d\sigma}$ with respect to σ we obtain

$$\frac{d}{d\sigma}\left(\frac{1}{\lambda}\frac{dx}{d\sigma}\right) = \psi_{xx}\frac{dx}{d\sigma} + \psi_{yx}\frac{dy}{d\sigma} + \psi_{zx}\frac{dz}{d\sigma} ,$$

$$= \lambda(\psi_{xx}\psi_x + \psi_{yx}\psi_y + \psi_{zx}\psi_z) ,$$

$$= \frac{1}{2}\lambda\frac{\partial}{\partial x}(\psi_x^2 + \psi_y^2 + \psi_z^2) ,$$

and hence, on account of (8.11)

$$\frac{1}{\lambda}\frac{d}{d\sigma}\left(\frac{1}{\lambda}\frac{dx}{d\sigma}\right) = \frac{1}{2}\frac{\partial n^2}{\partial x} . \tag{8.22}$$

By dealing similarly with the other equations (8.21) we find: The orthogonal trajectories (8.21) form a two-parameter manifold of solutions of the 2nd order equations:

$$\frac{1}{\lambda}\frac{d}{d\sigma}\left(\frac{1}{\lambda}\frac{dx}{d\sigma}\right) = \frac{1}{2}\frac{\partial}{\partial x}(n^2) ,$$

$$\frac{1}{\lambda}\frac{d}{d\sigma}\left(\frac{1}{\lambda}\frac{dy}{d\sigma}\right) = \frac{1}{2}\frac{\partial}{\partial y}(n^2) , \tag{8.23}$$

$$\frac{1}{\lambda}\frac{d}{d\sigma}\left(\frac{1}{\lambda}\frac{dz}{d\sigma}\right) = \frac{1}{2}\frac{\partial}{\partial z}(n^2) .$$

We remark again that the choice of λ does not affect the geometric form of the integral curves of (8.23). This can be seen from the following fact: Any particular solution $x(\sigma)$, $y(\sigma)$, $z(\sigma)$ of (8.23) can be transformed by a transformation of the parameter σ into a solution $x(\sigma')$, $y(\sigma')$, $z(\sigma')$ of the equations

$$\frac{d^2x}{d\sigma'^2} = \frac{1}{2}\frac{\partial}{\partial x}(n^2) , \quad \frac{d^2y}{d\sigma'^2} = \frac{1}{2}\frac{\partial}{\partial y}(n^2) , \quad \frac{d^2z}{d\sigma'^2} = \frac{1}{2}\frac{\partial}{\partial z}(n^2) , \tag{8.24}$$

where $\lambda = 1$. Such a transformation, however, does not affect the geometric shape of the curve.

8.3 We choose first $\lambda = 1$ and denote the parameter σ by τ. The equations (8.21) become

$$\frac{dx}{d\tau} = \psi_x \; ; \quad \frac{dy}{d\tau} = \psi_y \; ; \quad \frac{dz}{d\tau} = \psi_z \; , \tag{8.31}$$

and the equations (8.23) become:

$$\frac{d^2x}{d\tau^2} = \frac{\partial}{\partial x}\left(\frac{1}{2}\,n^2\right) \; ,$$

$$\frac{d^2y}{d\tau^2} = \frac{\partial}{\partial y}\left(\frac{1}{2}\,n^2\right) \; , \tag{8.32}$$

$$\frac{d^2z}{d\tau^2} = \frac{\partial}{\partial z}\left(\frac{1}{2}\,n^2\right) \; .$$

The orthogonal trajectories (8.31) thus form a two-parameter manifold of solutions of the equations (8.32). According to (8.31) we have

$$\left(\frac{dx}{d\tau}\right)^2 + \left(\frac{dy}{d\tau}\right)^2 + \left(\frac{dz}{d\tau}\right)^2 = n^2 \; . \tag{8.33}$$

The analogy of (8.32) to the equations of mechanics is obvious. If we interpret $-\frac{1}{2}\,n^2$ as a potential field our light rays can be regarded as paths of particles moving in this field with energy $\frac{1}{2}(\dot{x}^2 + \dot{y}^2 + \dot{z}^2) - \frac{n^2}{2} = 0$.

8.4 We choose next $\lambda = 1/n$ and denote the parameter σ by s. It follows:

$$n\,\frac{dx}{ds} = \psi_x \; ; \quad n\,\frac{dy}{ds} = \psi_y \; ; \quad n\,\frac{dz}{ds} = \psi_z \tag{8.41}$$

and hence $\left(\frac{dx}{ds}\right)^2 + \left(\frac{dy}{ds}\right)^2 + \left(\frac{dz}{ds}\right)^2 = 1$, i.e., the parameter s measures the length along the light rays. Equations (8.23) become

$$\frac{d}{ds}\left(n\,\frac{dx}{ds}\right) = \frac{\partial n}{\partial x} \; ,$$

$$\frac{d}{ds}\left(n\,\frac{dy}{ds}\right) = \frac{\partial n}{\partial y} \; , \tag{8.42}$$

$$\frac{d}{ds}\left(n\,\frac{dz}{ds}\right) = \frac{\partial n}{\partial z} \; .$$

These equations can easily be recognized as the Euler equations of the variation problem

$$V = \int_{P_0}^{P_1} n \, ds \text{ is an extremum.} \tag{8.43}$$

Let us consider two points P_0 and P_1 in the x,y,z space. Let x(s), y(s), z(s) be a continuous curve between P_0 and P_1 which also shall have a continuous tangent. We define the optical length of this curve by the integral

$$V = \int_{P_0}^{P_1} n \, ds. \tag{8.44}$$

In case n = 1 this optical length coincides with the geometrical length.

The problem is to find the curve for which the optical length is a minimum. Let us assume that a solution exists and is given in the form $x = x(\sigma)$, $y = y(\sigma)$, $z = z(\sigma)$ where σ is a parameter, such that $\sigma = 0$ at P_0 and $\sigma = 1$ at P_1; hence

$$V =$$

$$\int_0^1 n(x,y,z) \sqrt{x'^2 + y'^2 + z'^2} \, d\sigma \tag{8.45}$$

is a minimum.

Figure 12

The necessary conditions which the solution must satisfy are Euler's differential equations, i.e., in case of (8.45):

$$\frac{d}{d\sigma} \left(\frac{nx'}{\sqrt{x'^2 + y'^2 + z'^2}} \right) - n_x \sqrt{x'^2 + y'^2 + z'^2} = 0,$$

$$\frac{d}{d\sigma} \left(\frac{ny'}{\sqrt{x'^2 + y'^2 + z'^2}} \right) - n_y \sqrt{x'^2 + y'^2 + z'^2} = 0, \tag{8.46}$$

$$\frac{d}{d\sigma} \left(\frac{nz'}{\sqrt{x'^2 + y'^2 + z'^2}} \right) - n_z \sqrt{x'^2 + y'^2 + z'^2} = 0.$$

If we introduce in these equations the geometric length s of the solution as a parameter we obtain the differential equations (8.42).

The light ray between two points P_0 and P_1 is the curve for which the optical path attains an extreme value.

This result is known as Fermat's principle of geometrical optics.

If the index of refraction is interpreted as the ratio c/v of the velocity of light in a vacuum to the velocity $v(x,y,z)$ in the medium, the optical path

$$V = \int n \, ds = c \int \frac{ds}{v} = c \int dt$$

becomes proportional to the time needed to travel from P_0 to P_1. The principle of Fermat states that the light ray is a curve on which this time is a minimum, or at least an extremum.

§9. CONSTRUCTION OF WAVE FRONTS WITH THE AID OF LIGHT RAYS.

9.1 Every solution $\psi(x,y,z)$ of the equation

$$\psi_x^2 + \psi_y^2 + \psi_z^2 = n^2 \tag{9.11}$$

determines a two-parameter manifold of light rays, i.e., of orthogonal trajectories. We have seen that these light rays satisfy a system of ordinary differential equations

$$\ddot{x} = n \, n_x$$
$$\ddot{y} = n \, n_y \tag{9.12}$$
$$\ddot{z} = n \, n_z$$

and the condition

$$\dot{x}^2 + \dot{y}^2 + \dot{z}^2 = n^2 . \tag{9.13}$$

The dot means differentiation with regard to the parameter t.

Our aim in the following is to show that the two problems of integrating the partial differential equation (9.11) or the system of ordinary differential equations (9.12) are equivalent.

Let us first assume that the solutions of (9.12) are known. We show that it is possible then to solve the following problem simply by quadratures and eliminations.

Let Γ be an arbitrary surface section given in parametric form

$$x = f(\xi, \eta)$$

$$y = g(\xi, \eta) \tag{9.14}$$

$$z = h(\xi, \eta)$$

To find a solution $\psi(x,y,z)$ of (9.11) which, on Γ, has given values $\psi = F(\xi, \eta)$.

9.2 We know, if ψ is the desired solution, that the orthogonal trajectories of the surfaces $\psi = $ const. are solutions of (9.12) and (9.13) and the optical length

$$V(P_0, P_1) = \int_{\tau_0}^{\tau_1} n^2 \, d\tau$$

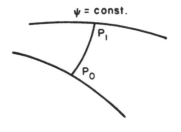

$\psi = $ const.

Figure 13

between two points P_0 and P_1 of such a trajectory is given by the difference

$$\psi(P_1) - \psi(P_0) = \int_{\tau_0}^{\tau} n^2 \, d\tau . \tag{9.21}$$

This leads to the following attempt to solve the above problem. We determine through every point ξ, η of Γ a light ray, i.e., a solution

$$x = x(\xi, \eta; \tau) ,$$

$$y = y(\xi, \eta; \tau) , \tag{9.22}$$

$$z = z(\xi, \eta; \tau) ,$$

of (9.12) and (9.13) which satisfies the boundary conditions

$$x(\xi, \eta, 0) = f(\xi, \eta) , \qquad \dot{x}(\xi, \eta, 0) = a(\xi, \eta) ,$$

$$y(\xi, \eta, 0) = g(\xi, \eta) , \qquad \dot{y}(\xi, \eta, 0) = b(\xi, \eta) , \tag{9.23}$$

$$z(\xi, \eta, 0) = h(\xi, \eta) , \qquad \dot{z}(\xi, \eta, 0) = c(\xi, \eta) .$$

The functions a,b,c must obey the condition

$$a^2 + b^2 + c^2 = n^2 (\xi, \eta) \tag{9.24}$$

in order to insure that the functions (9.22) satisfy the condition (9.13) but are otherwise arbitrary.

We now consider the expression

$$\psi(\xi, \eta, \tau) = F(\xi, \eta) + \int_0^\tau n^2 \, d\tau \tag{9.25}$$

and expect that the solution of our problem can be obtained in this form when ξ, η, τ are expressed as functions of x,y,z with the aid of (9.22). We assume that the Jacobian $\dfrac{\partial(x,y,z)}{\partial(\xi,\eta,\tau)}$ is not zero in the neighborhood of Γ in order to be able to carry out this elimination.

9.3 It is clear that $\psi(x,y,z)$ has the correct boundary values on Γ; for $\tau = 0$, we have $\psi = F(\xi, \eta)$. We show next that ψ satisfies the equation (9.11) if the functions a,b,c are chosen suitably. We determine the derivatives of $\psi(\xi, \eta, \tau)$. First,

$$\frac{\partial \psi}{\partial \tau} = n^2 = \dot{x}^2 + \dot{y}^2 + \dot{z}^2 \tag{9.31}$$

Then, we write (9.25) in the form

$$\psi = F(\xi, \eta) + \frac{1}{2} \int_0^\tau (n^2 + \dot{x}^2 + \dot{y}^2 + \dot{z}^2) \, d\tau \tag{9.32}$$

and obtain

$$\frac{\partial \psi}{\partial \xi} = F_\xi + \int_0^\tau n\left(n_x \frac{\partial x}{\partial \xi} + n_y \frac{\partial y}{\partial \xi} + n_z \frac{\partial z}{\partial \xi}\right) d\tau$$

$$+ \int_0^\tau \left(\dot{x} \frac{\partial \dot{x}}{\partial \xi} + \dot{y} \frac{\partial \dot{y}}{\partial \xi} + \dot{z} \frac{\partial \dot{z}}{\partial \xi}\right) d\tau \; .$$

We introduce $n \, n_x = \ddot{x}$, $n \, n_y = \ddot{y}$, $n \, n_z = \ddot{z}$ and find

$$\psi_\xi = F_\xi + \int_0^\tau \frac{d}{d\tau} \left(\dot{x} \frac{\partial x}{\partial \xi} + \dot{y} \frac{\partial y}{\partial \xi} + \dot{z} \frac{\partial z}{\partial \xi}\right) d\tau$$

$$\psi_\xi = F_\xi - \left(a \frac{\partial f}{\partial \xi} + b \frac{\partial g}{\partial \xi} + c \frac{\partial h}{\partial \xi}\right) + \dot{x} \frac{\partial x}{\partial \xi} + \dot{y} \frac{\partial y}{\partial \xi} + \dot{z} \frac{\partial z}{\partial \xi} \; , \tag{9.33}$$

and similarly

$$\psi_\eta = F_\eta - \left(a \frac{\partial f}{\partial \eta} + b \frac{\partial g}{\partial \eta} + c \frac{\partial h}{\partial \eta}\right) + \dot{x} \frac{\partial x}{\partial \eta} + \dot{y} \frac{\partial y}{\partial \eta} + \dot{z} \frac{\partial z}{\partial \eta} \; . \tag{9.34}$$

We now assume that the functions a,b,c satisfy the conditions

$$a \frac{\partial f}{\partial \xi} + b \frac{\partial g}{\partial \xi} + c \frac{\partial h}{\partial \xi} = F_\xi \, ,$$

$$a \frac{\partial f}{\partial \eta} + b \frac{\partial g}{\partial \eta} + c \frac{\partial h}{\partial \eta} = F_\eta \, , \qquad (9.35)$$

$$a^2 + b^2 + c^2 = n^2 \, .$$

These equations have two solutions (a,b,c) provided that not all of the sub-determinants of the matrix

$$\begin{pmatrix} f_\xi & g_\xi & h_\xi \\ f_\eta & g_\eta & h_\eta \end{pmatrix} \qquad (9.36)$$

are zero. Let (a,b,c) be one of these two solutions. With this choice of a,b,c we obtain

$$\psi_\xi = \dot{x} \frac{\partial x}{\partial \xi} + \dot{y} \frac{\partial y}{\partial \xi} + \dot{z} \frac{\partial z}{\partial \xi} \, ,$$

$$\psi_\eta = \dot{x} \frac{\partial x}{\partial \eta} + \dot{y} \frac{\partial y}{\partial \eta} + \dot{z} \frac{\partial z}{\partial \eta} \, , \qquad (9.37)$$

$$\psi_\tau = \dot{x} \frac{\partial x}{\partial \tau} + \dot{y} \frac{\partial y}{\partial \tau} + \dot{z} \frac{\partial z}{\partial \tau} \, .$$

On the other hand

$$\psi_\xi = \psi_x \frac{\partial x}{\partial \xi} + \psi_y \frac{\partial y}{\partial \xi} + \psi_z \frac{\partial z}{\partial \xi} \, ,$$

$$\psi_\eta = \psi_x \frac{\partial x}{\partial \eta} + \psi_y \frac{\partial y}{\partial \eta} + \psi_z \frac{\partial z}{\partial \eta} \, , \qquad (9.38)$$

$$\psi_\tau = \psi_x \frac{\partial x}{\partial \tau} + \psi_y \frac{\partial y}{\partial \tau} + \psi_z \frac{\partial z}{\partial \tau} \, .$$

Since the determinant $\frac{\partial(x,y,z)}{\partial(\xi,\eta,\tau)}$ is different from zero it follows by comparing (9.37) and (9.38):

$$\psi_x = \dot{x}, \quad \psi_y = \dot{y}, \quad \psi_z = \dot{z} \, . \qquad (9.39)$$

Therefore, on account of (9.13)

$$\psi_x^2 + \psi_y^2 + \psi_z^2 = n^2 ,$$

i.e., ψ is a solution of (9.11).

Actually, our method allows us to find two solutions of the above problem as we expect from the quadratic nature of (9.11).

9.4 We apply our result to the case $F(\xi,\eta) = 0$. The surface Γ itself is thus a wave front, namely, the wave front at the time $t = 0$. The problem is to find the position of the wave front at the time t. We calculate the two-parameter set of solutions of (9.12) which intersect the surface Γ with directions $a(\xi,\eta)$, $b(\xi,\eta)$, $c(\xi,\eta)$. The quantities a,b,c can be found from (9.35), i.e., from

$$a \frac{\partial f}{\partial \xi} + b \frac{\partial g}{\partial \xi} + c \frac{\partial h}{\partial \xi} = 0 ,$$

$$a \frac{\partial f}{\partial \eta} + b \frac{\partial g}{\partial \eta} + c \frac{\partial h}{\partial \eta} = 0 , \qquad (9.41)$$

$$a^2 + b^2 + c^2 = n^2 .$$

From the first two of the above equations it follows that the light rays must be normal to the surface Γ. When these rays have been found:

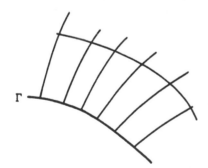

$$x = x(\xi,\eta,\tau) ,$$
$$y = y(\xi,\eta,\tau) , \qquad (9.42)$$
$$z = z(\xi,\eta,\tau) ,$$

we obtain the solution ψ by the integral

$$\psi(\xi,\eta,\tau) = \int_0^\tau n^2 (x,y,z) \, d\tau , \qquad (9.43)$$

Figure 14

in which ξ,η,τ have to be expressed in terms of x,y,z with the aid of (9.42).

9.5 The original wave surface Γ of the preceding section may degenerate into a point, (x_0, y_0, z_0). The functions f, g, h are constant in this case, so that the conditions (9.41) reduce to only one condition:

$$a^2 + b^2 + c^2 = n^2 (x_0,y_0,z_0) . \qquad (9.51)$$

Let us determine the two-parameter set of light rays through the point (x_0, y_0, z_0), i.e., solutions

$$x = x(a,b,c,\tau) ,$$

$$y = y(a,b,c,\tau) , \qquad (9.52)$$

$$z = z(a,b,c,\tau) ,$$

of (9.12) which satisfy the boundary conditions

$$x(a,b,c,0) = x_0 , \qquad \dot{x}(a,b,c,0) = a ,$$

$$y(a,b,c,0) = y_0 , \qquad \dot{y}(a,b,c,0) = b , \qquad (9.53)$$

$$z(a,b,c,0) = z_0 , \qquad \dot{z}(a,b,c,0) = c .$$

We obtain a solution $\psi(x,y,z)$ of (9.11) in the form of the integral

$$\psi(a,b,c,\tau) = \int_0^\tau n^2(x,y,z) \, d\tau \qquad (9.54)$$

after a,b,c,τ have been expressed by x,y,z with the aid of the relations (9.51) and (9.52).

These special solutions are called "Spherical" Waves or simply wavelets. If x_0, y_0, z_0 are considered as variable parameters ψ becomes a function of two points

$$V(x_0, y_0, z_0; x, y, z) = \int_0^\tau n^2 \, d\tau \qquad (9.55)$$

It determines the optical distance of the two points (x_0, y_0, z_0) and (x,y,z). The spherical wave fronts around a point (x_0, y_0, z_0) then are given by the surfaces

$$V(x_0, y_0, z_0; x, y, z) - ct = 0. \qquad (9.56)$$

9.6 Huyghens' Construction. With the aid of the wavelet $V(x_0, y_0, z_0; x, y, z)$ another method can be obtained to determine the wave fronts belonging to a given single wave front Γ. This method is of prime importance and is known as Huyghens' construction. We consider the wavelet functions $V(\xi, \eta: x,y,z)$ which belong to the points (ξ,η) of the surface Γ. At the time t a two-parameter set,

$$V(\xi, \eta; x,y,z) - ct = 0 \qquad (9.61)$$

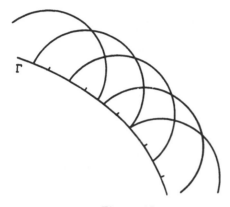

Figure 15

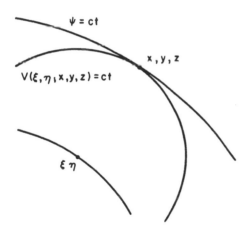

Figure 16

of spherical wave fronts is obtained; we show that the envelope of these wave fronts is the wave front

$$\psi(x,y,z) - ct = 0$$

which, at $t = 0$, coincides with the given surface Γ.

We find the envelope of the surfaces (9.61) by eliminating the parameters (ξ, η) from the three equations,

$$V_\xi(\xi, \eta; x,y,z) = 0 ,$$

$$V_\eta(\xi, \eta; x,y,z) = 0 , \qquad (9.62)$$

$$V(\xi, \eta; x,y,z) - ct = 0 .$$

Let

$$\xi = A(x,y,z) , \qquad (9.63)$$
$$\eta = B(x,y,z)$$

be the result of calculating ξ, η from the first two equations (9.62). We introduce ξ and η in $V(\xi, \eta; x,y,z)$ and show that

$$\psi(x,y,z) =$$

$$V\Big(A(x,y,z), B(x,y,z); x,y,z\Big) \qquad (9.64)$$

is a solution of (9.11). Indeed,

$$\psi_x = V_\xi A_x + V_\eta B_x + V_x$$

or, on account of (9.62):

$$\psi_x = V_x ,$$

and similarly,

$$\psi_y = V_y , \qquad \psi_z = V_z . \qquad (9.65)$$

These equations state that, at the point (x,y,z) the surface ψ - ct = 0 has the same tangential plane as the wave $V(\xi, \eta; x,y,z)$ - ct from the point (ξ, η), by (9.63). Since V satisfies the equation (9.11), the same is true for the function (9.64).

The function V(A,B;x,y,z) determines the optical distance of a point (x,y,z) from the corresponding point (ξ, η) defined by (9.63). If the point (x,y,z) approaches the surface Γ the corresponding point (ξ, η) on Γ approaches the same limiting point. Instead of proving this analytically, we will refer to the geometric evidence. Hence $\psi = V(\xi, \eta, x,y,z) \rightarrow 0$ if (x,y,z) approaches the surface Γ. The function (9.64) thus is the desired solution.

§10. JACOBI'S THEOREM.

10.1 In this section we shall be concerned with the inverse problem. Suppose we are in a position to integrate the partial differential equation

$$\psi_x^2 + \psi_y^2 + \psi_z^2 = n^2 .$$

(10.11)

To find the general solution of the differential equations of the light rays.

The answer is given in a general theorem of Jacobi. This theorem, applied to the differential equations (10.11) states: Let $\psi(x,y,z; a,b)$ be a complete integral of the equation (10.11). A complete integral is defined as a set of solutions which depend on two arbitrary parameters a and b such that not all of the subdeterminants of the matrix

$$\begin{pmatrix} \psi_{xa}, & \psi_{ya}, & \psi_{za}, \\ \psi_{xb}, & \psi_{yb}, & \psi_{zb}, \end{pmatrix}$$

(10.12)

are zero. Then the light rays of the medium of refractive index n(x,y,z) are given by the equations

$$\frac{\partial}{\partial a} \psi(x,y,z; a,b) = \alpha ,$$

$$\frac{\partial}{\partial b} \psi(x,y,z; a,b) = \beta ,$$

(10.13)

where α and β are arbitrary constants.

If, for example, the determinant

$$\begin{vmatrix} \psi_{ya} & \psi_{za} \\ \psi_{yb} & \psi_{zb} \end{vmatrix} \neq 0 ,$$

then we may calculate y and z as functions of x:

$$y = y(x; a,b,\alpha,\beta) ,$$

$$z = z(x; a,b,\alpha,\beta) .$$

(10.14)

These functions represent a four-parameter set of curves which, according to Jacobi's theorem, are the light rays of the medium.

10.2 For the proof of Jacobi's theorem let us assume that the curves (10.13) are given in parametric representation:

$$x = x(\sigma; a,b,\alpha,\beta) ,$$

$$y = y(\sigma; a,b,\alpha,\beta) ,$$

$$z = z(\sigma; a,b,\alpha,\beta) .$$

(10.21)

By introducing these functions in (10.13) we obtain identities in σ,a,b,α,β. Hence by differentiation with respect to σ:

$$\psi_{ax} \dot{x} + \psi_{ay} \dot{y} + \psi_{az} \dot{z} = 0 ,$$

$$\psi_{bx} \dot{x} + \psi_{by} \dot{y} + \psi_{bz} \dot{z} = 0 .$$

(10.22)

The six quantities

$$\psi_{ax}, \qquad \psi_{ay}, \qquad \psi_{az},$$

$$\psi_{bx}, \qquad \psi_{by}, \qquad \psi_{bz},$$

(10.23)

can be interpreted as two vectors which, on account of (10.12), are not linearly dependent and thus determine a plane. The equations (10.22) state that the vector $(\dot{x},\dot{y},\dot{z})$ is perpendicular to this plane. From (10.11) we have by differentiation with respect to a and b:

$$\psi_x\psi_{ax} + \psi_y\psi_{ay} + \psi_z\psi_{az} = 0 ,$$

$$\psi_x\psi_{bx} + \psi_y\psi_{by} + \psi_z\psi_{bz} = 0 ,$$

(10.24)

i.e., the vector (ψ_x, ψ_y, ψ_z) is also normal to the above plane. Hence,

$$\dot{x} = \lambda\psi_x, \quad \dot{y} = \lambda\psi_y, \quad \dot{z} = \lambda\psi_z, .$$

(10.25)

By differentiating these last equations with respect to σ we have as in section 8.2:

$$\frac{1}{\lambda}\frac{d}{d\sigma}\left(\frac{\dot{x}}{\lambda}\right) = nn_x ,$$

$$\frac{1}{\lambda}\frac{d}{d\sigma}\left(\frac{\dot{y}}{\lambda}\right) = nn_y , \qquad (10.26)$$

$$\frac{1}{\lambda}\frac{d}{d\sigma}\left(\frac{\dot{z}}{\lambda}\right) = nn_z ,$$

which shows that the curves (10.21) are light rays.

10.3 **Example.** Let us consider the case of a stratified medium where $n = n(z)$. We verify easily that

$$\psi = ax + by + \int_0^z \sqrt{n^2 - a^2 - b^2}\ d\zeta \qquad (10.31)$$

is a complete integral of the equation

$$\psi_x^2 + \psi_y^2 + \psi_z^2 = n^2(z) .$$

The light rays in such a medium thus are given by

$$\frac{\partial\psi}{\partial a} = x - a \int_0^z \frac{d\zeta}{\sqrt{n^2(\zeta) - a^2 - b^2}} = \alpha ,$$

$$\frac{\partial\psi}{\partial b} = y - b \int_0^z \frac{d\zeta}{\sqrt{n^2(\zeta) - a^2 - b^2}} = \beta \qquad (10.32)$$

or

$$x = \alpha + a \int_0^z \frac{d\zeta}{\sqrt{n^2(\zeta) - a^2 - b^2}}$$

$$y = \beta + b \int_0^z \frac{d\zeta}{\sqrt{n^2(\zeta) - a^2 - b^2}} . \qquad (10.33)$$

§11. TRANSPORT EQUATIONS FOR DISCONTINUITIES IN CONTINUOUS
OPTICAL MEDIA.

We shall derive certain differential relations in this section which allow
us to calculate the discontinuities of an electromagnetic field along a given
light ray if the discontinuity is known at one point of the ray. We assume
explicitly that the functions ϵ $=$ $\epsilon(x,y,z)$ and μ $=$ $\mu(x,y,z)$ are continuous
functions. The case of discontinuous optical media will be studied later and
a principal difference between both cases will be found.

11.1 Differentiation along a light ray. Let $F(x,y,z)$ be a differentiable
function. Along a given light ray x = $x(\tau)$, y = $y(\tau)$, z = $z(\tau)$, a function

$$F(\tau) = F\big(x(\tau), y(\tau), z(\tau)\big) ,$$

is obtained whose differential quotient is

$$\frac{dF}{d\tau} = F_x \dot{x} + F_y \dot{y} + F_z \dot{z} .$$

On account of the relations $\dot{x}$ = ψ_x, etc. this becomes

$$\frac{\partial F}{\partial \tau} = \frac{d}{d\tau} F\big(x(\tau), y(\tau), z(\tau)\big) = F_x \psi_x + F_y \psi_y + F_z \psi_z . \qquad (11.11)$$

Hence the differential operator

$$\frac{\partial}{\partial \tau} = \psi_x \frac{\partial}{\partial x} + \psi_y \frac{\partial}{\partial y} + \psi_z \frac{\partial}{\partial z} \qquad (11.12)$$

can be interpreted as differentiation along a light ray provided that $\psi(x,y,z)$
is a solution of the equation

$$\psi_x^2 + \psi_y^2 + \psi_z^2 = n^2 .$$

Let $F(x,y,z)$ = $\psi(x,y,z)$, for example. It follows

$$\frac{\partial \psi}{\partial \tau} = \psi_x^2 + \psi_y^2 + \psi_z^2 = n^2 . \qquad (11.14)$$

The operator (11.12) can also be applied to a vector field; for example

$$\frac{\partial}{\partial \tau}(\text{grad } \psi) = \left(\frac{\partial}{\partial \tau}\psi_x, \frac{\partial}{\partial \tau}\psi_y, \frac{\partial}{\partial \tau}\psi_z\right) = \frac{1}{2}\text{grad } n^2 . \qquad (11.15)$$

11.2 We consider an electromagnetic field which is discontinuous on the
hypersurface

$$\varphi(x,y,z,t) = \psi(x,y,z) - ct = 0 . \qquad (11.21)$$

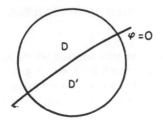

Let D and D' be two domains of the x,y,z,t space which are separated by the hypersurface $\varphi = 0$. We assume that E and H are continuous functions with continuous derivatives in the individual domains D and D'. The boundary values on $\varphi = 0$ which are assumed if this surface is approached from D and D' are denoted by

Figure 17

$$E, H, E_x, H_x, \ldots,$$

$$E', H', E_x', H_x', \ldots, \text{ respectively.}$$

The discontinuities U and V of E and H are then given by

$$U = E' - E,$$

$$V = H' - H.$$

First we consider the boundary values of E,H in the domain D. Both E and $H(x,y,z,t)$ become functions of x,y,z on $\varphi = 0 = \psi - ct$. We denote these vectors by

$$E^*(x,y,z) = E\left(x,y,z; \frac{1}{c}\psi(x,y,z)\right),$$

$$H^*(x,y,z) = H\left(x,y,z; \frac{1}{c}\psi(x,y,z)\right). \tag{11.22}$$

The derivatives, for example of E^*, are given by

$$E_x^* = E_x + \frac{1}{c}\psi_x E_t,$$

$$E_y^* = E_y + \frac{1}{c}\psi_y E_t, \tag{11.23}$$

$$E_z^* = E_z + \frac{1}{c}\psi_z E_t.$$

Therefore

$$\text{curl } E^* = \text{curl } E + \frac{1}{c}\text{grad }\psi \times E_t,$$

$$\tag{11.23}$$

and similarly $\text{curl } H^* = \text{curl } H + \frac{1}{c}\text{grad }\psi \times H_t.$

From Maxwell's equations: $\text{curl } E = -\dfrac{\mu}{c} H_t$ and $\text{curl } H = \dfrac{\epsilon}{c} E_t$;

hence:

$$c \text{ curl } E^* = -\mu H_t + \text{grad } \psi \times E_t \ ,$$

$$c \text{ curl } H^* = \epsilon E_t + \text{grad } \psi \times H_t \ .$$

(11.24)

These equations can be considered as a system of six linear equations for the six components of the vectors E_t and H_t. The matrix of these equations is the same as the matrix of the homogeneous equations for the discontinuities U and V on $\varphi = 0$:

$$\text{grad } \psi \times V + \epsilon U = 0 \ ,$$

$$\text{grad } \psi \times U - \mu V = 0 \ ,$$

(11.25)

and we know that the determinant of this matrix is zero on $\psi - ct = 0$. We conclude that the equations (11.24) are possible only if the left sides satisfy certain conditions. These conditions will now be derived.

We form the vector product of $\text{grad } \psi$ with the second equation (11.24):

$$c \text{ grad } \psi \times \text{curl } H^* = \epsilon \text{ grad } \psi \times E_t + \text{grad } \psi \times (\text{grad } \psi \times H_t)$$

or on account of the vector identity (Appendix I.23)

$$c \text{ grad } \psi \times \text{curl } H^* = \epsilon(\text{grad } \psi \times E_t - \mu H_t) + (H_t \cdot \text{grad } \psi) \text{ grad } \psi \ .$$

Hence with the aid of the first equation (11.24):

$$\text{grad } \psi \times \text{curl } H^* - \epsilon \text{ curl } E^* = \frac{1}{c} (H_t \cdot \text{grad } \psi) \text{ grad } \psi \ . \qquad (11.26)$$

This equation states: <u>The vector $\text{grad } \psi \times \text{curl } H^* - \epsilon \text{ curl } E^*$ has the direction of $\text{grad } \psi$</u>, i.e., is normal to the wave front $\psi = ct$.

The same considerations can be applied to the boundary values

$$E'^* = E'(x,y,z,\frac{1}{c}\psi) \ ,$$

$$H'^* = H'(x,y,z,\frac{1}{c}\psi) \ .$$

(11.27)

If $\psi - ct = 0$ is approached from the domain D', we find that the vector

$$\text{grad } \psi \times \text{curl } H'^* - \epsilon \text{ curl } E'^*$$

is normal to the wave front ψ = ct. Finally, by considering the differences U = E'* - E* and V = H'* - H* we have: <u>The vector grad ψ x curl V - ϵ curl U is normal to the wave front ψ = ct.</u>

We formulate this statement in the equation

$$\epsilon \text{ curl } U - \text{grad } \psi \times \text{curl } V = R \text{ grad } \psi \qquad (11.28)$$

where R is a certain scalar function of x,y,z. We can determine R explicitly by forming the scalar product of grad ψ with equation (11.28). It follows

$$R = \frac{1}{\mu} (\text{grad } \psi \cdot \text{curl } U) . \qquad (11.281)$$

Let us, finally, introduce $U = -\frac{1}{\epsilon} (\text{grad } \psi \times V)$ with the aid of (11.25). The result is

$$\text{curl} \left(\frac{1}{\epsilon} \text{ grad } \psi \times V \right) + \frac{1}{\epsilon} \text{ grad } \psi \times \text{curl } V = -\frac{R}{\epsilon} \text{ grad } \psi \qquad (11.29)$$

which is a differential equation of first order in the discontinuity V.

11.3 We can transform the equation (11.29) into a much simpler form. We remark that the vector $V \times \text{curl} \left(\frac{1}{\epsilon} \text{ grad } \psi \right)$ has the direction of grad ψ. Indeed, on account of the vector formula (Appendix I.412), we have

$$\text{curl} \left(\frac{1}{\epsilon} \text{ grad } \psi \right) = \frac{1}{\epsilon} \text{ curl grad } \psi + \left(\text{grad } \frac{1}{\epsilon} \right) \times \text{grad } \psi$$

$$= \left(\text{grad } \frac{1}{\epsilon} \right) \times \text{grad } \psi .$$

Hence,

$$V \times \text{curl} \left(\frac{1}{\epsilon} \text{ grad } \psi \right) = (V \cdot \text{grad } \psi) \text{ grad } \frac{1}{\epsilon} - \left(V \cdot \text{grad } \frac{1}{\epsilon} \right) \text{ grad } \psi$$

which proves our statement, for $V \cdot \text{grad } \psi = 0$. Consequently, we can write (11.29) in the form

$$\text{curl} \left(\frac{1}{\epsilon} \text{grad } \psi \times V \right) + \frac{1}{\epsilon} \text{ grad } \psi \times \text{curl } V + V \times \text{curl} \left(\frac{1}{\epsilon} \text{grad } \psi \right)$$

$$= R' \text{ grad } \psi \qquad (11.31)$$

where R' is a certain scalar function. The left side can be transformed with the aid of the vector identity (Appendix I.48) by introducing

$$A = \frac{1}{\epsilon} \text{ grad } \psi , \quad B = V$$

and   (11.32)

$$\frac{\partial}{\partial \alpha} = \frac{1}{\epsilon} \left(\psi_x \frac{\partial}{\partial x} + \psi_y \frac{\partial}{\partial y} + \psi_z \frac{\partial}{\partial z} \right) = \frac{1}{\epsilon} \frac{\partial}{\partial \tau}$$

Using $A \cdot B = 0$, we obtain

$$- \frac{2}{\epsilon} \frac{\partial V}{\partial \tau} - V \text{ div} \left(\frac{1}{\epsilon} \text{ grad } \psi \right) = - R^* \text{ grad } \psi ,$$ (11.33)

R^* being a new factor. We find R^* by forming the scalar product of (11.33) with grad ψ. We obtain

$$n^2 R^* = \frac{2}{\epsilon} \frac{\partial V}{\partial \tau} \cdot \text{grad } \psi = - \frac{2}{\epsilon} V \cdot \frac{\partial}{\partial \tau} \text{ grad } \psi$$

or, by (11.15)

$$n^2 R^* = - \frac{1}{\epsilon} V \cdot \text{grad } n^2 ,$$

i.e.,

$$R^* = - \frac{2}{\epsilon} \frac{V \cdot \text{grad } n}{n} .$$ (11.34)

Equation (11.33) becomes

$$\frac{\partial V}{\partial \tau} + \frac{1}{2} \epsilon \text{ div} \left(\frac{1}{\epsilon} \text{ grad } \psi \right) V + \left(V \cdot \frac{\text{grad } n}{n} \right) \text{grad } \psi = 0.$$ (11.35)

Finally we introduce the notation

$$\Delta_\epsilon \psi = \epsilon \text{ div} \left(\frac{1}{\epsilon} \text{ grad } \psi \right) ,$$

i.e.,

$$\Delta_\epsilon \psi = \epsilon \left[\left(\frac{\psi_x}{\epsilon} \right)_x + \left(\frac{\psi_y}{\epsilon} \right)_y + \left(\frac{\psi_z}{\epsilon} \right)_z \right]$$ (11.36)

and we get the equation

$$\frac{\partial V}{\partial \tau} + \frac{1}{2} \Delta_\epsilon \psi V + \frac{1}{n} (V \cdot \text{grad } n) \text{grad } \psi = 0. \tag{11.37}$$

A similar relation can be found for the discontinuity U by replacing ϵ and V by μ and $-$U in (11.37).

Thus our complete result is: The discontinuities U and V satisfy the differential equations:

$$\frac{\partial U}{\partial \tau} + \frac{1}{2} \Delta_\mu \psi U + \frac{1}{n} (U \cdot \text{grad } n) \text{grad } \psi = 0,$$

$$\tag{11.38}$$

$$\frac{\partial V}{\partial \tau} + \frac{1}{2} \Delta_\epsilon \psi V + \frac{1}{n} (V \cdot \text{grad } n) \text{grad } \psi = 0$$

where the differential operators $\Delta_\epsilon \psi$ and $\Delta_\mu \psi$ are defined by

$$\Delta_\epsilon \psi = \epsilon \left[\left(\frac{1}{\epsilon} \psi_x \right)_x + \left(\frac{1}{\epsilon} \psi_y \right)_y + \left(\frac{1}{\epsilon} \psi_z \right)_z \right],$$

$$\tag{11.39}$$

$$\Delta_\mu \psi = \mu \left[\left(\frac{1}{\mu} \psi_x \right)_x + \left(\frac{1}{\mu} \psi_y \right)_y + \left(\frac{1}{\mu} \psi_z \right)_z \right].$$

§12. TRANSPORT OF DISCONTINUITIES. (CONTINUED).

12.1 On a given light ray the equations (11.38) represent a system of ordinary differential equations. Indeed, we have shown in (11.1) that the differential operator $\partial/\partial \tau$ differentiates a function in the direction of a light ray. Let us introduce, in (11.38), instead of U and V, the vectors

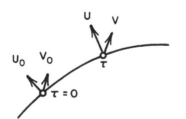

$$P = U \, e^{\frac{1}{2} \int_0^\tau \Delta_\mu \psi \, d\tau}$$

$$\tag{12.11}$$

$$Q = V \, e^{\frac{1}{2} \int_0^\tau \Delta_\epsilon \psi \, d\tau}$$

Figure 18

which have the same directions respectively as U and V, but different lengths. The differential equations (11.38) then assume an even simpler form:

$$\frac{dP}{d\tau} + \frac{1}{n} (P \cdot grad\ n) grad\ \psi = 0 ,$$

$$\frac{dQ}{d\tau} + \frac{1}{n} (Q \cdot grad\ n) grad\ \psi = 0 .$$

(12.12)

Since U and V are orthogonal to grad ψ, the same is true for P and Q. From $U \cdot V = 0$ it follows that $P \cdot Q = 0$.

By forming the scalar product of P and Q with the equations (12.12) it follows that

$$P \cdot \frac{dP}{d\tau} = 0 , \qquad Q \cdot \frac{dQ}{d\tau} = 0 .$$

(12.13)

This shows:

The lengths of the vectors P and Q are not changed on a given light ray. Thus without loss of generality we can assume

$$|P| = |Q| = 1$$

and interpret P and Q as unit vectors which determine the directions of the vectors U and V.

If P and Q have been found as solutions of (12.12), we obtain U and V from

$$U = |U_0|\ P\ e^{-\frac{1}{2} \int_0^\tau \Delta_\mu \psi\, d\tau} ,$$

(12.14)

$$V = |V_0|\ Q\ e^{-\frac{1}{2} \int_0^\tau \Delta_\epsilon \psi\, d\tau}$$

These equations make it evident that U and V are zero on the whole light ray if they are zero on one particular point, $\tau = 0$, of the ray. The light rays

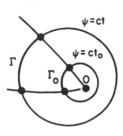

thus determine the region of the space where directed signals can be seen. Let us assume that from the point 0 a light signal is released at the time $t = 0$. Let us furthermore assume that the discontinuities U and V which represent the signal are different from zero only on a section Γ_0 of the wave front $\psi = ct_0$. From (12.14) it follows that, at a time $t > t_0$, only on the corresponding section Γ of the wave front $\psi = ct$ will discontinuities U,V be observed. This section is determined by the light rays

Figure 19

through Γ_0. In other words, the light signal will be observed only in the part of the space which is covered by the light rays through Γ_0.

This does not exclude the possibility that light penetrates into other regions of the space. However, this "diffracted" excitation does not have a sudden discontinuous beginning.

12.2 The exponential factors in (12.14) have a simple geometric meaning. We have

$$\Delta_\epsilon \psi = \Delta \psi - \frac{1}{\epsilon} (\epsilon_x \psi_x + \epsilon_y \psi_y + \epsilon_z \psi_z)$$

or

$$\Delta_\epsilon \psi = \Delta \psi - \frac{\partial}{\partial \tau} (\log \epsilon) .$$

(12.21)

Similarly

$$\Delta_\mu \psi = \Delta \psi - \frac{\partial}{\partial \tau} (\log \mu) .$$

Let us now consider a "tube" of light rays, i.e., a domain D of the x,y,z space which is enclosed by a surface Γ consisting of two sections Γ_1 and Γ_2 of the wave fronts

$$\psi = \rho_1 \text{ and } \psi = \rho_2$$

and the cylindrical wall Γ_3 formed by the light rays through the circumference of Γ_2 and Γ_1. We apply the theorem of Gauss to this domain D:

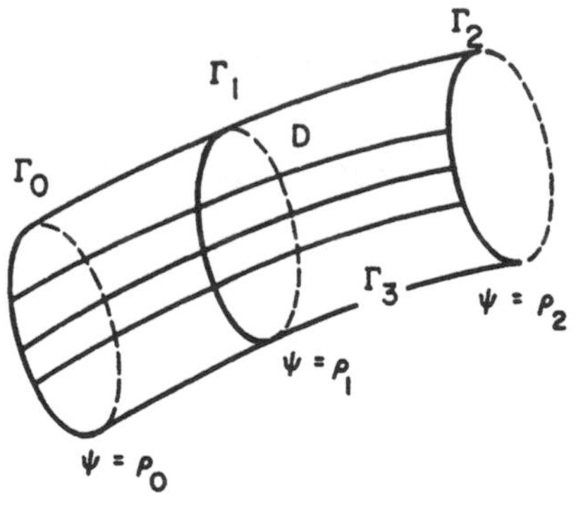

Figure 20

$$\iiint_D \Delta \psi \; dx \; dy \; dz = \iint_\Gamma \frac{\partial \psi}{\partial \nu} \; do$$

(12.22)

where $\frac{\partial \psi}{\partial \nu}$ is the derivative of ψ in direction of the outside normals. However,

$$\frac{\partial \psi}{\partial \nu} = 0 \text{ on } \Gamma_3 ,$$

$$\frac{\partial \psi}{\partial \nu} = n_2 \text{ on } \Gamma_2 ,$$

$$\frac{\partial \psi}{\partial \nu} = - n_1 \text{ on } \Gamma_1 .$$

Hence,

$$\iiint_D \Delta \psi \ dx \ dy \ dz \ = \ \iint_{\Gamma_2} n_2 \ do_2 \ - \ \iint_{\Gamma_1} n_1 \ do_1 \ . \tag{12.23}$$

We now express the surface elements do_2 and do_1 by the corresponding surface elements do of an arbitrarily chosen wavefront $\psi = ct_0$. We write

$$do_2 = K_2 \ do \ ,$$
$$do_1 = K_1 \ do \ . \tag{12.24}$$

The factor K measures the expansion of an infinitesimally narrow tube of light rays. It follows that

$$\iiint_D \Delta \psi \ dx \ dy \ dz \ = \ \iint_{\Gamma_0} (n_2 K_2 \ - \ n_1 K_1) do \ = \ \iiint_{\Gamma_0 \tau_1}^{\tau_2} \frac{d(nK)}{d\tau} \ d\tau \ do \ .$$

The volume element $dx \ dy \ dz$ can be expressed as follows:

$$dx \ dy \ dz \ = \ K \ do \ ds \ = \ nK \ do \ d\tau \ .$$

Hence,

$$\iiint_D \Delta \psi \ dx \ dy \ dz \ = \ \iiint_D \frac{1}{nK} \ \frac{\partial(nK)}{\partial \tau} \ dx \ dy \ dz \ .$$

Since D is of arbitrary size, we find

$$\Delta \psi \ = \ \frac{1}{nK} \ \frac{\partial}{\partial \tau} \ (nK) \ = \ \frac{\partial}{\partial \tau} (\log nK) \tag{12.25}$$

and hence

$$\Delta_\epsilon \psi \ = \ \frac{\partial}{\partial \tau} \log \frac{nK}{\epsilon} \ ,$$
$$\Delta_\mu \psi \ = \ \frac{\partial}{\partial \tau} \log \frac{nK}{\mu} \ . \tag{12.26}$$

The exponential factors in equation (12.14) become

$$e^{-\frac{1}{2}\int_0^\tau \Delta_\mu \psi \, d\tau} = \sqrt{\frac{n_0 K_0}{\mu_0}} \Big/ \sqrt{\frac{nK}{\mu}} \, ,$$

$$e^{-\frac{1}{2}\int_0^\tau \Delta_\epsilon \psi \, d\tau} = \sqrt{\frac{n_0 K_0}{\epsilon_0}} \Big/ \sqrt{\frac{nK}{\epsilon}} \, ,$$

(12.27)

and we conclude from (12.14):

$$\frac{K}{n}\epsilon |U|^2 = \frac{K_0}{n_0}\epsilon_0 |U_0|^2 \, ,$$

$$\frac{K}{n}\mu |V|^2 = \frac{K_0}{n_0}\mu_0 |V_0|^2 \, .$$

(12.28)

That is, the quantities $\frac{K}{n}\epsilon |U|^2$ and $\frac{K}{n}\mu |V|^2$ are constant along a light ray.

This result allows us to determine the lengths of U and V along the light rays of a given set of wave fronts ψ = ct without integration, simply by calculating the ratio $\frac{K}{K_0} = \frac{do}{do_0}$ of corresponding surface elements of the wave fronts.

12.3 Energy and flux on a wave front. Let us assume that the electromagnetic field is zero on one side of the wave fronts ψ - ct = 0. In this case E = U and H = V; hence

$$W = \frac{1}{8\pi}(\epsilon U^2 + \mu V^2), \quad S = \frac{c}{4\pi}(U \times V) \, .$$

(12.31)

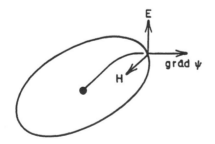

Figure 21

It follows from (11.25) that

$$\epsilon U^2 = U \cdot (V \times \text{grad } \psi) \, ,$$

$$\mu V^2 = (\text{grad } \psi \times U) \cdot V \, ,$$

and thus

$$\epsilon U^2 = \mu V^2$$

(12.32)

i.e. Electric and magnetic energies are equal on a wave front. From (11.25) it follows further that

$$U \times V = \frac{1}{\epsilon}(V \times \text{grad } \psi) \times V = \frac{1}{n^2}\mu V^2\text{grad } \psi$$

$$= \frac{4\pi}{n^2} W \text{ grad } \psi .$$

Hence,

$$S = \frac{c}{n^2} W \text{ grad } \psi . \qquad (12.33)$$

This yields for the absolute value of the Poynting vector:

$$|S| = \frac{c}{n} W . \qquad (12.34)$$

This result allows us to interpret the equations (12.28). If we add both equations we find $\frac{K}{n}W = \frac{K_0}{n_0} W_0$ and, introducing $\frac{K}{K_0} = \frac{do}{do_0}$, we obtain $\frac{1}{n} W \text{ do} = \frac{1}{n_0} W \text{ do}_0$. Finally, on account of (12.34):

$$|S| \text{ do} = |S_0| \text{ do}_0 . \qquad (12.35)$$

The flux through corresponding surface elements of a set of wave fronts is constant.

12.4 We continue the investigation of the differential relations (11.38) for U and V. The preceding results show that it is sufficient to consider the equations (12.12) for the directions P and Q of the vectors U and V. Let us represent the light ray in the vectorial form

$$X(\tau) = \bigl(x(\tau), y(\tau), z(\tau)\bigr) .$$

We have

$$\dot{X} = \text{grad } \psi ,$$
$$\ddot{X} = n \text{ grad } n \qquad (12.41)$$

and we can write (12.12) in the form

$$\dot{P} + \frac{1}{n^2}(P \cdot \ddot{X})\dot{X} = 0 ,$$
$$\dot{Q} + \frac{1}{n^2}(Q \cdot \ddot{X})\dot{X} = 0 . \qquad (12.42)$$

Instead of the parameter τ we introduce the geometrical length s of the light ray, by using the relation $\dfrac{d\tau}{ds} = \dfrac{1}{n}$. We have

$$\dot{X} = n\frac{dX}{ds} = nX' ,$$

$$\ddot{X} = n(nX')' ,$$

$$\dot{P} = nP' ,$$

and hence,

$$P' + \frac{1}{n}\left(P \cdot (nX')'\right) X' = 0 . \tag{12.43}$$

However,

$$P \cdot (nX')' = P \cdot (nX'' + n'X') = n(P \cdot X'')$$

on account of $P \cdot X' = 0$. This yields

$$P' + (P \cdot X'')X' = 0 ,$$

and similarly $\qquad\qquad Q' + (Q \cdot X'')X' = 0 .$

$$\tag{12.44}$$

The equations (12.44) demonstrate that the two unit vectors P and Q are determined by the light ray alone. The same light ray, of course, can be an orthogonal trajectory to many different sets of wave fronts. For example, in case n = 1, a given straight line can be orthogonal to systems of spherical wave fronts or to a system of plane wave fronts. Equations (12.44) however, state that the vectors U and V are submitted to the same rotation around the light ray no matter to which type of wave fronts the light ray is orthogonal. The wave fronts influence only the size of the discontinuities U and V.

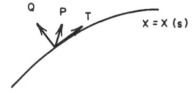

Figure 22

12.5 The tangential vector T = X' of the light ray and the vectors P and Q define an orthogonal system of unit vectors which travel along the light ray X = X(s). In general the change of such a system along a curve is determined by three kinematic formulae:

$$T' = \qquad * \qquad cP \quad +bQ$$

$$P' = -cT \qquad * \qquad +aQ \tag{12.51}$$

$$Q' = -bT \quad -aP \qquad *$$

The coefficients a,b,c are functions of s. In our special case these formulae reduce to

$$T' = \quad * \quad\quad cP \quad +bQ$$

$$P' = -cT \quad\quad * \quad\quad * \quad\quad\quad (12.52)$$

$$Q' = -bT \quad\quad * \quad\quad *$$

as is shown by (12.44). We have, incidentally,

$$a = 0, \quad b = (Q \cdot X''), \quad c = (P \cdot X'') .$$

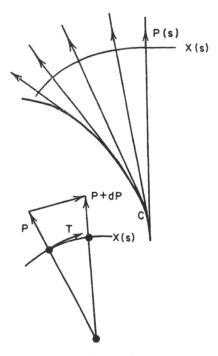

Figure 23

The fact that a = 0 for our system T,P,Q has a simple geometrical meaning. Let us consider the ruled surface which is formed by the straight lines through the vectors P on X(τ). From

$$dP = -cT \, ds$$

it follows that

$$P + dP = P - c \, ds \, T$$

lies in the plane formed by the vectors P and T. This means that two neighboring unit vectors P and P + dP are not skew but intersect each other in a point of the plane of P and T. The total manifold of straight lines through the vectors P(s) thus envelopes a certain curve C in space and can be interpreted as the manifold of tangents of this curve. A ruled surface which consists of the tangents of a curve in space is called an applicable surface, since it is possible to apply it to a plane by bending without strain. The same consideration applies to the ruled surface of the vectors Q(s). Hence we can formulate the statement:
The vector discontinuities U and V along a light ray determine a ruled surface which is applicable.

12.6 We can interpret the light rays as the geodetic lines in a space whose line element has the form

$$d\sigma^2 = n^2 (dx^2 + dy^2 + dz^2) \quad\quad\quad (12.61)$$

By introducing, on a light ray $X(\sigma)$ the parameter $\sigma = \int n \; ds = \int n^2 \; d\tau$ which measures the optical length on the ray we have the relations

$$\frac{dX}{d\sigma} = \frac{1}{n^2} \text{ grad } \psi \; ,$$

$$\frac{dP}{d\sigma} = \frac{1}{n^2} \frac{dP}{d\tau} \; .$$

With this choice of the parameter the equations (12.12) become

$$\frac{dP}{d\sigma} + \frac{1}{n} (P \cdot \text{grad } n) \frac{dX}{d\sigma} = 0 \; ,$$

$$\frac{dQ}{d\sigma} + \frac{1}{n} (Q \cdot \text{grad } n) \frac{dX}{d\sigma} = 0 \; . \tag{12.62}$$

We shall see that these equations characterize the vectors P or Q as being "parallel" along the light ray; parallelism being defined for the line element (12.61) in accordance with a definition which was introduced by Levy-Civita. The equations of the geodetic lines, i.e., of the light rays with σ as parameter, follow from (8.23) letting $\lambda = 1/n^2$ and hence

$$\left(\frac{dx}{d\sigma}\right)^2 + \left(\frac{dy}{d\sigma}\right)^2 + \left(\frac{dz}{d\sigma}\right)^2 = \frac{1}{n^2} \; .$$

We find

$$n^2 \frac{d}{d\sigma} \left(n^2 \frac{dx}{d\sigma}\right) = nn_x \; ,$$

$$n^2 \frac{d}{d\sigma} \left(n^2 \frac{dy}{d\sigma}\right) = nn_y \; , \tag{12.63}$$

$$n^2 \frac{d}{d\sigma} \left(n^2 \frac{dz}{d\sigma}\right) = nn_z \; .$$

Let us denote temporarily the components of the vector $X(\sigma) = (x,y,z)$ by $X_1(\sigma)$, $X_2(\sigma)$, $X_3(\sigma)$, and the partial derivatives of n by

$$n_1 = \frac{\partial n}{\partial x} \; , \quad n_2 = \frac{\partial n}{\partial y} \; , \quad n_3 = \frac{\partial n}{\partial z} \; .$$

Then (12.63) assumes the form

$$n^2 \frac{d}{d\sigma} \left(n^2 \frac{dX_\alpha}{d\sigma}\right) = nn_\alpha \tag{12.64}$$

or

$$n^2 \frac{d^2 X_\alpha}{d\sigma^2} + 2n \frac{dX_\alpha}{d\sigma} \left(\Sigma\, n_i \frac{dX_i}{d\sigma} \right) = \frac{n_\alpha}{n}$$

$$\frac{d^2 X_\alpha}{d\sigma^2} + 2 \frac{dX_\alpha}{d\sigma} \left(\Sigma\, \frac{n_i}{n} \frac{dX_i}{d\sigma} \right) = \frac{n_\alpha}{n} \frac{1}{n^2} \,.$$

By introducing on the right side $\frac{1}{n^2} = \Sigma \left(\frac{dX_i}{d\sigma} \right)^2$, we obtain the differential equations of the geodetic lines in the form,

$$\frac{d^2 X_\alpha}{d\sigma^2} + 2 \frac{dX_\alpha}{d\sigma} \left(\Sigma \frac{n_i}{n} \frac{dX_i}{d\sigma} \right) - \frac{n_\alpha}{n} \Sigma \left(\frac{dX_i}{d\sigma} \right)^2 = 0 \,. \qquad (12.65)$$

Thus, the second derivative $\dfrac{d^2 X_\alpha}{d\sigma^2}$ of each component is equal to a quadratic form of the first derivatives $\dfrac{dX_i}{d\sigma}$; we write

$$\frac{d^2 X_\alpha}{d\sigma^2} + \Sigma_{i,\,k} \Gamma_{ik}^\alpha \frac{dX_i}{d\sigma} \frac{dX_k}{d\sigma} = 0 \,, \qquad (12.66)$$

where

$$\Gamma_{ik}^\alpha = \frac{1}{n} \left(\delta_{\alpha i} \frac{\partial n}{\partial x_k} + \delta_{\alpha k} \frac{\partial n}{\partial x_i} - \delta_{ik} \frac{\partial n}{\partial x_\alpha} \right), \qquad (12.67)$$

as one can easily verify. The symbols δ_{ik} are Kronecker symbols

$$\delta_{ik} = 0 \,, \quad i \neq k$$

$$\delta_{ik} = 1 \,, \quad i = k \,. \qquad (12.671)$$

In general the equations of the geodetic lines $X(\sigma)$ of any line element

$$d\sigma^2 = \Sigma_{i,\,k} g_{ik}\, dx_i\, dx_k \qquad (12.68)$$

can be written in the above form (12.66). The matrices Γ_{ik}^α are given by the coefficient g_{ik} and are called Christoffel's symbols. In the special case of optics, i.e., for

$$d\sigma^2 = n^2 \Sigma_i dx_i^2$$

these coefficients are given by (12.67).

The theory of curvature of a space with the line element (12.68) can be developed in a similar way as the curvature of curves or surfaces, if one can compare the directions of vectors which are not attached to the same point. Obviously this involves a definition of parallelism in a non-Euclidean space, i.e., a criterion for the parallelism of two vectors at different points in space. Levy-Civita's definition of parallelism is as follows: A vector $A(\sigma)$ is moved parallel on a given curve $X(\sigma)$ if it satisfies the differential relations

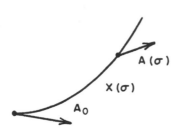

Figure 24

$$\frac{dA_\alpha}{d\sigma} + \sum_{i,k} \Gamma^\alpha_{i,k}\, A_i\, \frac{dX_k}{d\sigma} = 0. \qquad (12.681)$$

According to this definition, for example, the tangential vectors $\dfrac{dX_\alpha}{d\sigma}$ of a geodetic line are parallel.

In case of the optical line element (12.61) we find by using (12.67):

$$\frac{dA_\alpha}{d\sigma} + \frac{1}{n} A_\alpha \left(\sum_k \frac{\partial n}{\partial x_k} A_k \right) + \frac{1}{n} \frac{dX_\alpha}{d\sigma} \left(\sum_k \frac{\partial n}{\partial x_k} A_k \right) - \frac{1}{n} \frac{\partial n}{\partial x_\alpha} \left(\sum_k A_k \frac{dX_k}{d\sigma} \right) = 0.$$

$$(12.682)$$

or

$$\frac{d(nA_\alpha)}{d\sigma} + \frac{1}{n} \frac{dX_\alpha}{d\sigma} \left(\sum_k nA_k \frac{\partial n}{\partial x_k} \right) - \frac{1}{n} \frac{\partial n}{\partial x_\alpha} \left(\sum_k nA_k \frac{dX_k}{d\sigma} \right) = 0.$$

In vector notation:

$$\frac{d(nA)}{d\sigma} + \frac{1}{n} (nA \cdot \text{grad } n) \frac{dX}{d\sigma} - \left(nA \cdot \frac{dX}{d\sigma} \right) \frac{\text{grad } n}{n} = 0. \qquad (12.683)$$

which is the condition of parallelism in our optical medium.

Let us now consider a vector $A(\sigma)$ along a light ray; we have

$$\frac{d(nA)}{d\sigma} + \frac{1}{n} (nA \cdot \text{grad } n) \frac{dX}{d\sigma} - \left((nA) \cdot \frac{dX}{d\sigma} \right) \frac{\text{grad } n}{n} = 0$$

and

$$\frac{d}{d\sigma} \left(n \frac{dX}{d\sigma} \right) + \frac{1}{n} \left(n \frac{dX}{d\sigma} \cdot \text{grad } n \right) \frac{dX}{d\sigma} - n \left(\frac{dX}{d\sigma} \right)^2 \frac{\text{grad } n}{n} = 0.$$

We multiply the first equation by $n \dfrac{dX}{d\sigma}$ and the second equation by nA. It follows, by adding the results that

$$\frac{d}{d\sigma}\left(n^2 A \cdot \frac{dX}{d\sigma}\right) = 0 . \qquad (12.684)$$

Hence: If $A \cdot \dfrac{dX}{d\sigma} = 0$ at one point of the ray, then it is zero at all points of the ray. If a vector is normal to the ray then its parallel vectors on the ray are also normal to the ray.

Such normal vectors thus obey the condition

$$\frac{d(nA)}{d\sigma} + \frac{1}{n}(nA \cdot \operatorname{grad} n)\frac{dX}{d\sigma} = 0 . \qquad (12.69)$$

By comparing this with (12.62), we find that the vectors $\dfrac{1}{n}P$ and $\dfrac{1}{n}Q$ satisfy the above condition and thus demonstrate the parallelism of the directions P and Q on the light ray.

12.7 <u>Integration of the transport equations</u>. We introduce the following orthogonal system of unit vectors on the light ray, $X = X(s)$:

Tangential vector: $\qquad T = X'$.

Principal normal: $\qquad N = \dfrac{1}{\sqrt{X''^2}} X''$. $\qquad (12.71)$

Binormal: $\qquad S = T \times N$.

The derivatives of these vectors, and the vectors themselves, are related by a system of formulae of the type (12.51) which, in this case, are known as Frenet's formulae:

$$T' = \frac{1}{\rho}N$$

$$N' = -\frac{1}{\rho}T \qquad +\frac{1}{\tau}S \qquad (12.72)$$

$$S' = \qquad -\frac{1}{\tau}N$$

$\dfrac{1}{\rho}$ is the principal curvature, and $\dfrac{1}{\tau}$ is the torsion of the ray.

Let us consider the equation

$$P' + (X'' \cdot P)X' = 0 \tag{12.73}$$

for the vector P. By introducing the notation (12.71), we obtain

$$P' + (T' \cdot P)T = 0 ,$$

or on account of (12.72):

$$P' + \frac{1}{\rho} (N \cdot P)T = 0 . \tag{12.74}$$

Since P is normal to T, we can express it as a linear combination of N and S. We introduce

$$P = \alpha N + \beta S \tag{12.75}$$

in (12.74). This yields

$$\alpha 'N + \alpha N' + \beta 'S + \beta S' + \frac{\alpha}{\rho} T = 0 ,$$

or, on account of (12.72):

$$\alpha 'N + \alpha \left(-\frac{1}{\rho} T + \frac{1}{\tau} S \right) + \beta 'S - \frac{\beta}{\tau} N + \frac{\alpha}{\rho} T = 0 .$$

It follows that

$$\left(\alpha ' - \frac{\beta}{\tau} \right) N + \left(\beta ' + \frac{\alpha}{\tau} \right) S = 0 ,$$

whence

$$\alpha ' - \frac{\beta}{\tau} = 0 ,$$

$$\tag{12.76}$$

$$\beta ' + \frac{\alpha}{\tau} = 0 .$$

These two differential equations can be written in the form

$$\frac{d}{ds} (\alpha + i\beta) + \frac{i}{\tau} (\alpha + i\beta) = 0 . \tag{12.77}$$

and have a solution in the form

$$\alpha + i\beta = (\alpha_0 + i\beta_0)e^{-i\int_0^s \frac{ds}{\tau}} .$$

(12.78)

We introduce

$$\theta = \int_0^s \frac{ds}{\tau}$$

(12.79)

and

$$\alpha = \cos \vartheta , \qquad \alpha_0 = \cos \vartheta_0 ,$$

$$\beta = \sin \vartheta , \qquad \beta_0 = \sin \vartheta_0 .$$

Equations (12.78) become

$$e^{i\vartheta} = e^{i(\vartheta_0 - \theta)} ,$$

whence,

$$\vartheta = \vartheta_0 - \theta .$$

(12.791)

The vector P thus is given by

$$P = N \cos(\vartheta_0 - \theta) + S \sin(\vartheta_0 - \theta) .$$

(12.792)

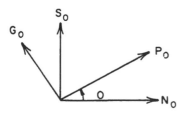

P changes its position relative to the principal normal and binormal of the ray; the angle of rotation with respect to N being given by

$$-\theta = -\int_0^S \frac{dS}{\tau} .$$

In case $\frac{1}{\tau} = 0$, i.e., if the light ray remains in one plane, then the vector P remains unchanged relative to the vectors N and S.

Finally we determine the vector Q by

$$Q = T \times P = - N \sin (\vartheta_0 - \theta)$$
$$+ S \cos (\vartheta_0 - \theta) .$$

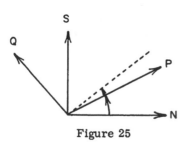

Figure 25

§13. SPHERICAL WAVES IN A HOMOGENEOUS MEDIUM.

We illustrate the former results with the example of the spherical wave which represents the electromagnetic field of a dipole. We assume the medium is homogeneous. Without loss of generality we let $\mu = \epsilon = 1$. In optics we may consider this electromagnetic field as the simplest mathematical representation of the light wave which is radiated from a point source.

13.1 Maxwell's equations in vacuum are

$$\text{curl } H - \frac{1}{c} E_t = 0 ,$$

$$\text{curl } E + \frac{1}{c} H_t = 0 .$$

(13.11)

The vectors E and H satisfy the second order equations

$$\frac{1}{c^2} E_{tt} - \Delta E = 0 ,$$

$$\frac{1}{c^2} H_{tt} - \Delta H = 0 .$$

(13.12)

However, only such vectors E or H are permitted for which

$$\text{div } E = \text{div } H = 0 .$$

In the case of the wave equation $\frac{1}{c^2} u_{tt} - \Delta u = 0$ for a scalar function $u(x,y,z,t)$ it is easy to find spherical waves. We simply ask for solutions u which depend only on $r = \sqrt{x^2 + y^2 + z^2}$ and on t. We find the equation

$$\frac{1}{c^2} (ru)_{tt} - (ru)_{rr} = 0 ,$$

(13.13)

which possess the solution

$$u = \frac{1}{r} \left(f(r - ct) + g(r + ct) \right) ,$$

(13.14)

where f and g are arbitrary functions. Outgoing spherical waves are obtained if $g = 0$, i.e.,

$$u = \frac{1}{r} f(r - ct) .$$

(13.15)

From this it follows that the first one of the equations (13.12) can be satisfied by a vector

$$E = \frac{1}{r} F(r - ct) \qquad (13.16)$$

where $F = (F_1, F_2, F_3)$ is a vector whose components are functions of $(r - ct)$. However, from

$$\text{div } E = (\text{grad } r) \cdot \frac{\partial}{\partial r} \left(\frac{F(r - ct)}{r} \right)$$

$$= \frac{1}{r^3} \left[x(rF_1' - F_1) + y(rF_2' - F_2) + z(rF_3' - F_3) \right] \qquad (13.17)$$

it follows that div $E = 0$ is possible only in the case $F_1 = F_2 = F_3 = 0$. Physically, this means that electric fields are not possible in which the electric vector is constant for points on the same sphere at a given time, t.

13.2 In the case of the scalar wave equation

$$\frac{1}{c^2} u_{tt} - \Delta u = 0$$

we can find other spherical waves from (13.15) by differentiation. Let L be the differential operator $L = a \frac{\partial}{\partial x} + b \frac{\partial}{\partial y} + c \frac{\partial}{\partial z}$, where a,b,c are constants. Then

$$u = L \left\{ \frac{1}{r} f \right\} \qquad (13.21)$$

is a solution of the wave equations. It can be interpreted as the wave which is radiated from a dipole with an axis (a,b,c).

More general solutions can be found by repeated differentiation. Let $L_\nu = a_\nu \frac{\partial}{\partial x} + b_\nu \frac{\partial}{\partial y} + c_\nu \frac{\partial}{\partial z}$, then

$$u = L_1 L_2 \ldots L_k \left\{ \frac{1}{r} f(r - ct) \right\} \qquad (13.22)$$

represents a wave from a "multipole" with K axes (a_ν, b_ν, c_ν).

13.3 We proceed in a similar way for vector waves. Let

$$L = A \frac{\partial}{\partial x} + B \frac{\partial}{\partial y} + C \frac{\partial}{\partial z} \qquad (13.31)$$

be a differential operator in which A, B, C are matrices with constant elements A_{ik}, B_{ik}, C_{ik}. We then obtain a variety of new solutions of $\frac{1}{c^2} E_{tt} - \Delta E$ in the form

$$E = L \left\{ \frac{1}{r} F(r - ct) \right\} ; \tag{13.32}$$

or explicitly

$$E_i = \sum_{k=1}^{3} \left(A_{ik} \frac{\partial}{\partial x} + B_{ik} \frac{\partial}{\partial y} + C_{ik} \frac{\partial}{\partial z} \right) \frac{1}{r} F_k (r - ct) . \tag{13.33}$$

More complicated "multipole" waves can be found by repeated differentiation:

$$E = L_1 L_2 \ldots L_k \left\{ \frac{1}{r} F(r - ct) \right\} \tag{13.34}$$

of a vector $\frac{1}{r} F(r - ct)$.

13.4 All these solutions satisfy the equation $\frac{1}{c^2} E_{tt} - \Delta E = 0$ and the remaining problem is to find operators L_ν such that div $E = 0$. For the dipole wave (13.32) this problem is solved by the operator $L = $ curl. In fact, this operator is of the type (13.31), namely

$$A = \begin{pmatrix} 0 & 0 & 0 \\ 0 & 0 & -1 \\ 0 & 1 & 0 \end{pmatrix}, \quad B = \begin{pmatrix} 0 & 0 & 1 \\ 0 & 0 & 0 \\ -1 & 0 & 0 \end{pmatrix}, \quad C = \begin{pmatrix} 0 & -1 & 0 \\ 1 & 0 & 0 \\ 0 & 0 & 0 \end{pmatrix}.$$

$$\tag{13.41}$$

Furthermore, div $L \left(\frac{1}{r} F \right)$ = div curl $\left(\frac{1}{r} F \right)$ = 0 .

In order to obtain our solution in a suitable form, let us write the vector F as the derivative of a vector $M = (M_1, M_2, M_3)$, i.e.,

$$F(r - ct) = M'(r - ct) . \tag{13.42}$$

We then know that

$$E = - \text{curl} \frac{1}{r} M'(r - ct) \tag{13.43}$$

satisfies the equations $\frac{1}{c^2} E_{tt} - \Delta E = 0$; div $E = 0$.

Now we have to construct a vector H such that both equations (13.11) are satisfied. From the equation curl E $+ \frac{1}{c}$ H$_t$ = 0, it follows that

$$H_t = c \text{ curl curl } \frac{1}{r} M'(r - ct) .$$

Hence, if

$$H = - \text{ curl curl } \frac{1}{r} M(r - ct) , \qquad (13.44)$$

we know that the second equation (13.11) is satisfied. Clearly, div H = 0. In order to show that the first equation (13.11) is also satisfied, we write with the aid of the vector identity (I.43)

$$H = \Delta \frac{1}{r} M(r - ct) - \text{grad div } \frac{1}{r} M(r - ct)$$

$$= \frac{1}{c^2} \frac{\partial^2}{\partial t^2} \left(\frac{1}{r} M \right) - \text{grad div } \left(\frac{1}{r} M \right)$$

$$= \frac{1}{r} M'' (r - ct) - \text{grad div } \left(\frac{1}{r} M \right) .$$

It follows that

$$\text{curl } H = \text{curl } \left(\frac{1}{r} M''(r - ct) \right) . \qquad (13.45)$$

On the other hand, from (13.43):

$$E_t = c \text{ curl } \left(\frac{1}{r} M''(r - ct) \right) , \qquad (13.46)$$

and hence

$$\text{curl } H - \frac{1}{c} E_t = 0 .$$

Our result is:

Let M(r - ct) be an arbitrary vector function of φ = r - ct with continuous derivatives to the third order, M', M'', M'''. Then a solution of Maxwell's equations is given by the vectors

$$E = - \text{ curl } \left(\frac{1}{r} M'(r - ct) \right)$$

$$\qquad (13.47)$$

$$H = - \text{ curl curl } \left(\frac{1}{r} M(r - ct) \right) .$$

This electromagnetic field can be regarded as the field of a dipole at the point $r = 0$ whose momentum $M = M(-ct)$ is given as a function of t.

Of course, a second solution can be obtained by replacing E by $+H$ and H by $-E$. This yields

$$E = \text{curl curl} \left(\frac{1}{r} M(r - ct) \right)$$

$$H = - \text{curl} \left(\frac{1}{r} M'(r - ct) \right) . \tag{13.48}$$

From a mathematical point of view either one of these solutions may represent the radiation of a point source.

13.5 Let us consider, in the following, the solution (13.47). We have

$$E = - \frac{1}{r} (ix + jy + zk) \times \frac{\partial}{\partial r} \left(\frac{1}{r} M' \right)$$

$$= \frac{1}{r} \left(M'' - \frac{M'}{r} \right) \times \rho , \tag{13.51}$$

where ρ is the unit vector

$$\rho = \left(\frac{x}{r}, \frac{y}{r}, \frac{z}{r} \right) . \tag{13.52}$$

The expression for H can be transformed as follows:

$$\text{curl} \frac{M}{r} = \vec{r} \times \frac{1}{r^2} \left(M' - \frac{M}{r} \right) .$$

Hence

$$H = \text{curl} \left[\left(\frac{M'}{r^2} - \frac{M}{r^3} \right) \times \vec{r} \right] ,$$

$$= \vec{r} \times \left[\left(\frac{M''}{r^3} - 3 \frac{M'}{r^4} + 3 \frac{M}{r^5} \right) \times \vec{r} \right] + 2 \left(\frac{M'}{r^2} - \frac{M}{r^3} \right) ,$$

$$= \rho \times \left[\left(\frac{M''}{r} - 3 \frac{M'}{r^2} + 3 \frac{M}{r^3} \right) \times \rho \right] + 2 \left(\frac{M'}{r^2} - \frac{M}{r^3} \right) .$$

This however is equal to

$$H = \rho \times E + \frac{M}{r^3} + \frac{1}{r^2} \left(2(M' \cdot \rho) - \frac{3}{r} (M \cdot \rho) \right) \rho . \tag{13.53}$$

Hence we get the final result:

$$E = \left(\frac{M''}{r} - \frac{M'}{r^2}\right) \times \rho ,$$

$$H = \rho \times E + \frac{M}{r^3} + \frac{1}{r^2}\left(2(M' \cdot \rho) - \frac{3}{r}(M \cdot \rho)\right)\rho .$$

(13.54)

13.6 We conclude from (13.54):

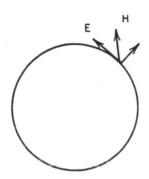

Figure 26

$$E \cdot \rho = 0 ;$$

$$H \cdot \rho = \frac{2}{r^3}(rM' - M) \cdot \rho ,$$

(13.61)

i.e., <u>the electric vector but, in general, not the magnetic vector, is tangent to the sphere.</u>

Let us now assume that $M(\varphi) = 0$ for $\varphi > 0$. This means that the dipole begins to oscillate at the time $t = 0$. It follows that $E \equiv 0$ and $H \equiv 0$ for $\varphi = r - ct > 0$, i.e., for $r > ct$. At the time t, the sphere $r = ct$ thus represents the wave front of the electromagnetic field, i.e., the boundary of the region of penetration. We assume furthermore that

$$M(0) = M'(0) = 0 ,$$

but $M''(0) = m \neq 0$, so that $M(\varphi)$ is a function of the type indicated in Figure 27. On the wave fronts $\varphi = r - ct = 0$; we thus have the boundary values

$$E = \frac{1}{r}(m \times \rho)$$

$$H = \rho \times E = \frac{1}{r}\rho \times (m \times \rho)$$

(13.62)

which represent discontinuities of the electromagnetic field. We immediately verify the relations $E \cdot \rho = H \cdot \rho = 0$, and $E \cdot H = 0$, as our former results required. Hence, only on the wavefront are the vectors E and H of the spherical wave normal to the light ray and perpendicular to each other.

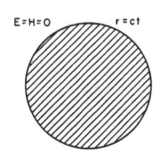

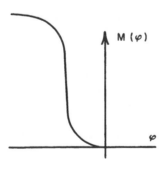

Figure 27

The quantities $|E|^2$ and $|H|^2$ on the wave front decrease in proportion to $1/r^2$; i.e., in proportion to the ratio of corresponding surface elements on the spherical wave fronts.

§14. WAVE FRONTS IN MEDIA OF DISCONTINUOUS OPTICAL PROPERTIES.

We shall not assume in the following that the functions $\epsilon(x,y,z)$ and $\mu(x,y,z)$ are continuous and have continuous derivatives. However, these functions will be sectionally continuous in any finite domain D of the (x,y,z,t) space, i.e., it will be possible to divide D into a finite number of subdomains in which ϵ and μ are continuous, and assume finite boundary values on the bounding surfaces. We furthermore assume that the derivatives of ϵ and μ are also sectionally continuous.

Our first aim is to find the laws according to which wavefronts pass through a refracting surface, i.e., a surface on which ϵ or μ is discontinuous. The result will be Snell's law of refraction and the law of reflection. After that, we can answer the question of how signals, i.e., discontinuities of E and H, are influenced by such surfaces. We will find a system of formulae known as Fresnel's formulae.

14.1 <u>Snell's Law of refraction</u>. Let Σ be a surface in the (x,y,z) space which separates two media D and D' of indices of refraction $n(x,y,z)$ and $n'(x,y,z)$, such that on Σ: $[n] = n' - n \neq 0$. Both n and n' shall have continuous derivatives in their respective domains. A light signal may travel through the (x,y,z) space over a set of wave fronts

$$\varphi(x,y,z,t) = \psi(x,y,z) - ct = 0 . \tag{14.11}$$

In other words, a characteristic hypersurface $\varphi = 0$ in the (x,y,z,t) space is assumed on which E and H are discontinuous. This hypersurface is continuous but does not necessarily have continuous normals, as is indicated in Figure 28. Letting $\varphi = \psi - ct$ we conclude that $\psi(x,y,z)$ must be a continuous solution of the equation

$$\psi_x^2 + \psi_y^2 + \psi_z^2 = n^2 \tag{14.12}$$

even on the surface Σ where n is discontinuous. Its derivatives are sectionally continuous.

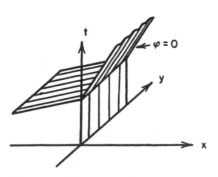

Char. surface $\varphi = 0$ in case n discontinuous on $x = 0$.

Figure 28

We assume that Σ is given in the form

$$x = f(\xi,\eta),\ y = g(\xi,\eta),\ z = h(\xi,\eta) \tag{14.13}$$

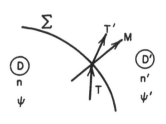

Figure 29

and denote the values of ψ on the two sides of Σ by ψ and ψ'. ψ and ψ' have continuous derivatives in their domains. From this it follows that

$$\psi\left(f(\xi,\eta),\ g(\xi,\eta),\ h(\xi,\eta)\right) =$$
$$\psi'\left(f(\xi,\eta),\ g(\xi,\eta),\ h(\xi,\eta)\right)$$

and that both sides have continuous derivatives. Consequently

$$(\psi_x' - \psi_x)f_\xi + (\psi_y' - \psi_y)g_\xi + (\psi_z' - \psi_z)h_\xi = 0\ ,$$
$$(\psi_x' - \psi_x)f_\eta + (\psi_y' - \psi_y)g_\eta + (\psi_z' - \psi_z)h_\eta = 0\ . \tag{14.14}$$

These equations state that the vector

$$\text{grad }\psi' - \text{grad }\psi = (\psi_x' - \psi_x;\ \psi_y' - \psi_y;\ \psi_z' - \psi_z)$$

is normal to the surface Σ. If M is a unit vector in direction of the surface normal, we can write

$$\text{grad }\psi' - \text{grad }\psi = \Gamma M\ . \tag{14.15}$$

Grad ψ' and grad ψ give the direction of the orthogonal trajectories of the surfaces $\psi' = $ const. and $\psi = $ const., i.e., of the light rays. Let T and T' be unit vectors along the light ray; then grad $\psi = nT$ and grad $\psi' = n'T'$. It follows that

$$n'T' - nT = \Gamma M\ , \tag{14.16}$$

where Γ is a scalar factor.

We conclude that the refracted ray leaves the surface Σ in the plane formed by the incident ray and the surface normal M.

From (14.16) it follows that

$$n'(T' \times M) = n(T \times M)\ . \tag{14.17}$$

The length of the vector on the left side is $n'\sin\vartheta'$; on the right side, $n\sin\vartheta$, where ϑ is the angle of incidence and ϑ' is the angle of refraction. This yields Snell's law of refraction

$$n'\sin\vartheta' = n\sin\vartheta\ . \tag{14.18}$$

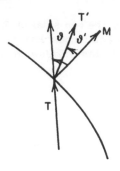

Figure 30

We find the factor Γ in (14.16) by forming the scalar product of M with (14.16):

$$n'(T' \cdot M) - n(T \cdot M) = \Gamma ,$$

or

$$\Gamma = n' \cos \vartheta' - n \cos \vartheta . \qquad (14.19)$$

The relations

$$n'T' - nT = (n' \cos \vartheta' - n \cos \vartheta)M ,$$
$$(14.191)$$
$$n' \sin \vartheta' = n \sin \vartheta$$

allow us to find T' if T and M are given.

14.2 <u>The law of reflection</u>. We know by experience that a light signal when reaching a surface of discontinuity of n is not only transmitted, but also reflected. Mathematically, this possibility is suggested by the quadratic character of the equation

$$\psi_x^2 + \psi_y^2 + \psi_z^2 = n^2 .$$

We have seen in §9 that, on account of this, there exist two solutions ψ which attain given boundary values on a given surface Σ.

That a surface of discontinuity must actually produce a set of reflected wave fronts will be seen in the next section by deriving Fresnel's formulae. Let us here assume the existence of a reflected signal. This means that the characteristic hypersurface $\varphi = 0$ consists of two branches $\varphi = \psi - ct = 0$ and $\varphi^* = \psi^* - ct$ in the neighborhood D of the surface Σ. These two branches are joined together on Σ, as is indicated in Figure 31 for the case when Σ is the plane x = 0. This implies that the two functions ψ and ψ^* must have the same boundary values on Σ, i.e.,

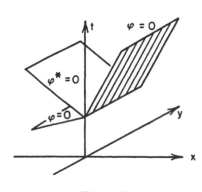

Figure 31

$$\psi\left(f(\xi,\eta), \; g(\xi,\eta), \; h(\xi,\eta)\right) =$$
$$\psi^*\left(f(\xi,\eta), \; g(\xi,\eta), \; h(\xi,\eta)\right) .$$
$$(14.21)$$

As above we conclude that

$$\text{grad } \psi^* - \text{grad } \psi = \Gamma M \qquad (14.22)$$

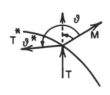

Figure 32

or, by introducing the unit vector T^* on the reflected ray,

$$n(T^* - T) = \Gamma M . \tag{14.23}$$

It follows: The reflected ray leaves the surface Σ in the plane determined by T and M. We find as above that

$$T^* \times M = T \times M , \tag{14.24}$$

and hence

$$\sin \vartheta^* = \sin \vartheta , \tag{14.25}$$

where ϑ^* is the angle of reflection. For Γ we obtain

$$\Gamma = n(\cos \vartheta^* - \cos \vartheta) .$$

The equations

$$T^* - T = (\cos \vartheta^* - \cos \vartheta)M ,$$
$$\sin \vartheta^* = \sin \vartheta \tag{14.26}$$

can be satisfied by two vectors T^* if T is given. From $\sin \vartheta^* = \sin \vartheta$ it follows

$$\vartheta^* = \vartheta , \qquad\qquad \vartheta^* = \pi - \vartheta ,$$
$$\text{or} \tag{14.27}$$
$$\cos \vartheta^* = \cos \vartheta , \qquad \cos \vartheta^* = -\cos \vartheta .$$

The first solutions give $T^* = T$, i.e., the incident ray; the second solutions yield

$$T^* - T = -2 \cos \vartheta \, M . \tag{14.28}$$

This is the reflected ray. T^* and T are symmetrical with respect to the tangent plane of Σ at the point of incidence.

§15. TRANSPORT OF SIGNALS IN MEDIA OF DISCONTINUOUS OPTICAL PROPERTIES. FRESNEL'S FORMULAE.

15.1 In 14.2 we have seen that, on a surface Σ: $\omega(x,y,z) = 0$ on which the functions ϵ and μ are discontinuous, a set of wave fronts $\psi = ct$ must be expected to split up into two sets of wave fronts; the transmitted wave fronts $\psi - ct = 0$ and the reflected wave fronts $\psi^* - ct = 0$. The corresponding characteristic hypersurface then consists of three branches; the incident branch $\varphi = \psi - ct = 0$, the reflected branch $\varphi* = \psi* - ct = 0$, and the transmitted branch $\varphi' = \psi' - ct = 0$. All these branches intersect each other in a common manifold J which lies on the cylindrical hypersurface $\omega(x,y,z) = 0$.

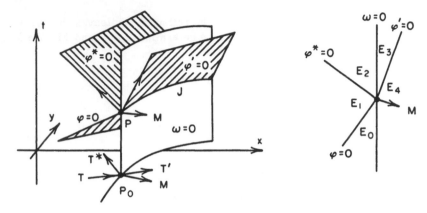

Figure 33

In this section we study an electromagnetic field which is discontinuous on these three branches and, of course, also on the cylindrical hypersurface $\omega = 0$. The four-dimensional neighborhood of the common manifold of intersections, J, is divided into five parts separated by the four hypersurfaces $\varphi = 0$, $\varphi* = 0$, $\omega = 0$, $\varphi' = 0$. Let us assume that the vectors E and H are continuous in these five parts and that they attain finite limits.

$$E_0, E_1, E_2, E_3, E_4,$$

$$H_0, H_1, H_2, H_3, H_4,$$

(15.11)

if a point P of the manifold J is approached.

We denote the surface normal of $\omega = 0$ by M, i.e., a unit vector proportional to the vector $(\omega_x, \omega_y, \omega_z, 0)$ at P. By applying the conditions derived in §6 for E and H on a surface of refraction, we find

$$\epsilon(E_2 \cdot M) = \epsilon'(E_3 \cdot M); \quad \epsilon(E_0 \cdot M) = \epsilon'(E_4 \cdot M);$$

$$E_2 \times M = E_3 \times M; \quad E_0 \times M = E_4 \times M;$$

(15.12)

where ϵ and ϵ' are the boundary values of ϵ on $\omega = 0$.

Similarly,

$$\mu(H_2 \cdot M) = \mu'(H_3 \cdot M); \quad \mu(H_0 \cdot M) = \mu'(H_4 \cdot M);$$

$$H_2 \times M = H_3 \times M; \quad H_0 \times M = H_4 \times M. \tag{15.13}$$

Since the boundary values of the discontinuities on J are given by

$$U = E_1 - E_0 \qquad V = H_1 - H_0$$

$$U* = E_2 - E_1 \qquad V* = H_2 - H_1 \tag{15.14}$$

$$U' = E_3 - E_4, \qquad V' = H_3 - H_4.$$

We find from (15.12) and (15.13) readily that

$$\epsilon(U + U*) \cdot M = \epsilon'U' \cdot M,$$

$$(U + U*) \times M = U' \times M;$$

$$\mu(V + V*) \cdot M = \mu'V' \cdot M, \tag{15.15}$$

$$(V + V*) \times M = V' \times M.$$

15.2 We consider now the light rays which belong to the point P_0 on Σ, i.e., to the projection of P in the (x,y,z) space. The direction of these rays is given by

$$T = \frac{1}{n} \text{ grad } \psi$$

$$T* = \frac{1}{n} \text{ grad } \psi* \tag{15.21}$$

$$T' = \frac{1}{n'} \text{grad } \psi'.$$

These vectors and M are related by the equations

$$n'T' = nT + (n' \cos \vartheta' - n \cos \vartheta)M$$

$$T* = T - 2 \cos \vartheta \ M. \tag{15.22}$$

The four vectors T, $T*$, T', and M lie in one and the same plane of the x,y,z space. This plane is normal to the unit vector

$$S = \frac{T \times M}{\sin \vartheta} = \frac{T* \times M}{\sin \vartheta*} = \frac{T' \times M}{\sin \vartheta'} \tag{15.23}$$

This fact suggests the introduction of the following orthogonal system of unit vectors attached to each of the three rays:

$$T \; , \; N \;\; = \; S \times T \; , \; S$$

$$T^* , \; N^* \; = \; S \times T^* , \; S \qquad\qquad (15.24)$$

$$T' , \; N' \; = \; S \times T' , \; S$$

15.3   We consider the incident ray first. We know that the discontinuities U and V on this ray are related by the equations

$$\text{grad } \psi \times V + \epsilon U \; = \; 0 \; ,$$

$$\text{grad } \psi \times U - \mu V \; = \; 0 \; , \qquad\qquad (15.31)$$

or, on account of (15.21):

$$\sqrt{\mu} \; (T \times V) + \sqrt{\epsilon} \; U \; = \; 0$$

$$\sqrt{\epsilon} \; (T \times U) - \sqrt{\mu} \; V \; = \; 0 \; . \qquad\qquad (15.32)$$

From these equations it follows that U and V can be represented in the form

$$\sqrt{\epsilon} \; U \; = \; \alpha N + \beta S$$

$$\sqrt{\mu} \; V \; = \; -\beta N + \alpha S \; , \qquad\qquad (15.33)$$

as linear combinations of N and S.

In order to apply the conditions (15.15) let us determine the products $U \cdot M$ and $U \times M$. We obtain

$$\epsilon U \cdot M \; = \; \alpha \sqrt{\epsilon} \, M \cdot N \; = \; \alpha \sqrt{\epsilon} \, \sin \vartheta$$

and

$$U \times M = \frac{\alpha}{\sqrt{\epsilon}} \; (N \times M) + \frac{\beta}{\sqrt{\epsilon}} \; (S \times M) \; .$$

However,

$$N \times M \; = \; (S \times T) \times M \; = \; - (M \cdot T)S \; = \; - \cos \vartheta \, S \; .$$

Hence

$$\epsilon U \cdot M \; = \; \alpha \sqrt{\epsilon} \; \sin \vartheta$$

$$U \times M \; = \; - \frac{\alpha}{\sqrt{\epsilon}} \; \cos \vartheta \, S + \frac{\beta}{\sqrt{\epsilon}} \; (S \times M) \; . \qquad\qquad (15.34)$$

15.4 Equations identical with (15.34) can be found for the products $\epsilon U^* \cdot M$, $U^* \times M$, and $\epsilon U' \cdot M$, $U' \times M$. Since S and $S \times M$ are orthogonal, and thus independent, we obtain from (15.15) the equations

$$\sqrt{\epsilon}\,(\alpha \sin \vartheta + \alpha^* \sin \vartheta^*) = \sqrt{\epsilon'}\,\alpha' \sin \vartheta'$$

$$\frac{1}{\sqrt{\epsilon}}\,(\alpha \cos \vartheta + \alpha^* \cos \vartheta^*) = \frac{1}{\sqrt{\epsilon'}}\alpha' \cos \vartheta' \qquad (15.41)$$

$$\frac{1}{\sqrt{\epsilon}}\,(\beta + \beta^*) = \frac{\beta'}{\sqrt{\epsilon'}}\;.$$

On account of $\vartheta^* = \pi - \vartheta$ and $\sqrt{\epsilon\mu}\,\sin\vartheta = \sqrt{\epsilon'\mu'}\,\sin\vartheta'$ these equations become

$$\alpha + \alpha^* = \sqrt{\frac{\mu}{\mu'}}\,\alpha'\;,$$

$$\beta + \beta^* = \sqrt{\frac{\epsilon}{\epsilon'}}\,\beta'\;, \qquad (15.42)$$

$$\alpha - \alpha^* = \sqrt{\frac{\epsilon}{\epsilon'}}\,\frac{\cos\vartheta'}{\cos\vartheta}\,\alpha'\;.$$

Instead of carrying out analogous calculations for the vector V, we may simply replace α by $-\beta$, β by α, and ϵ by μ in (15.42). The first equations (15.42) do not give any new conditions. The last equation, however, yields

$$\beta - \beta^* = \sqrt{\frac{\mu}{\mu'}}\,\frac{\cos\vartheta'}{\cos\vartheta}\,\beta'\;. \qquad (15.43)$$

The four equations (15.42) and (15.43) allow us to express α', β', and α^*, β^* as functions of α and β. The result is

$$\frac{\alpha'}{\alpha} = \frac{2}{\sqrt{\dfrac{\mu}{\mu'}} + \sqrt{\dfrac{\epsilon}{\epsilon'}}\,\dfrac{\cos\vartheta'}{\cos\vartheta}}\;, \qquad \frac{\beta'}{\beta} = \frac{2}{\sqrt{\dfrac{\epsilon}{\epsilon'}} + \sqrt{\dfrac{\mu}{\mu'}}\,\dfrac{\cos\vartheta'}{\cos\vartheta}}\;,$$

$$\qquad (15.44)$$

$$\frac{\alpha^*}{\alpha} = \frac{\sqrt{\dfrac{\mu}{\mu'}} - \sqrt{\dfrac{\epsilon}{\epsilon'}}\,\dfrac{\cos\vartheta'}{\cos\vartheta}}{\sqrt{\dfrac{\mu}{\mu'}} + \sqrt{\dfrac{\epsilon}{\epsilon'}}\,\dfrac{\cos\vartheta'}{\cos\vartheta}}\;, \qquad \frac{\beta^*}{\beta} = \frac{\sqrt{\dfrac{\epsilon}{\epsilon'}} - \sqrt{\dfrac{\mu}{\mu'}}\,\dfrac{\cos\vartheta'}{\cos\vartheta}}{\sqrt{\dfrac{\epsilon}{\epsilon'}} + \sqrt{\dfrac{\mu}{\mu'}}\,\dfrac{\cos\vartheta'}{\cos\vartheta}}\;.$$

In relation to the plane of incidence, it is customary to call the components of the electric discontinuity U with respect to N and S, the parallel component and the normal component of U, and to use the notation:

$$A_p = \frac{\alpha}{\sqrt{\epsilon}} \qquad A_s = \frac{\beta}{\sqrt{\epsilon}}$$

$$R_p = \frac{\alpha^*}{\sqrt{\epsilon}} \qquad R_s = \frac{\beta^*}{\sqrt{\epsilon}} \qquad\qquad (15.45)$$

$$D_p = \frac{\alpha'}{\sqrt{\epsilon'}} \qquad D_s = \frac{\beta'}{\sqrt{\epsilon'}}$$

It follows that, letting $\mu = \mu' = 1$ and $\sqrt{\epsilon} = n$; $\sqrt{\epsilon'} = n'$;

$$R_p = A_p \frac{1 - \dfrac{n}{n'}\dfrac{\cos \vartheta'}{\cos \vartheta}}{1 + \dfrac{n}{n'}\dfrac{\cos \vartheta'}{\cos \vartheta}} \qquad R_s = A_s \frac{\dfrac{n}{n'} - \dfrac{\cos \vartheta'}{\cos \vartheta}}{\dfrac{n}{n'} + \dfrac{\cos \vartheta'}{\cos \vartheta}}$$

$$\qquad\qquad (15.46)$$

$$D_p = A_p \frac{2\dfrac{n}{n'}}{1 + \dfrac{n}{n'}\dfrac{\cos \vartheta'}{\cos \vartheta}} \qquad D_s = A_s \frac{2\dfrac{n}{n'}}{\dfrac{n}{n'} + \dfrac{\cos \vartheta'}{\cos \vartheta}}$$

The above formulae are identical with Fresnel's formulae for the reflection and transmission of plane waves on a plane surface of refraction. We have seen, however, that their significance is more general since they give also the reflection and transmission of any discontinuity of the vectors E and H on such a surface. We have to consider these equations as the supplement to the transport equations (11.38) for U and V in case of a continuous medium.

In the case of normal incidence, we have $\vartheta = \vartheta' = 0$ and hence:

$$R_p = A_p \frac{n' - n}{n' + n}, \qquad R_s = A_s \frac{n - n'}{n' + n},$$

$$\qquad\qquad (15.47)$$

$$D_p = A_p \frac{2n}{n' + n}, \qquad D_s = A_s \frac{2n}{n' + n}.$$

§16. PERIODIC WAVES OF SMALL WAVE LENGTH.

16.1 The simplest type of radiation from a point source in a vacuum can be represented by the electromagnetic field

$$E = \left(\frac{M''}{r} - \frac{M'}{r^2} \right) \times \rho \, ,$$

(16.11)

$$H = \rho \times E + \frac{M}{r^3} + \frac{1}{r^2} \left[2(M' \cdot \rho) - \frac{3}{r}(M \cdot \rho) \right] \rho \, ,$$

where $M = M(\varphi)$, the moment of the dipole, is an arbitrary vector depending on $\varphi = r - ct$, and ρ is the unit vector $\rho = \frac{1}{r}(x,y,z)$. This was the result of §13.

Let us consider a dipole, which is periodic in t, so that $M = M(\varphi)$ is a periodic vector function of $\varphi = r - ct$. We assume that M is of the form

$$M = -\frac{1}{k^2} m e^{ik\varphi} \, , \quad k = \frac{2\pi}{\lambda} \, ,$$

(16.12)

where m is a constant complex vector $m = a + ia^*$. We introduce (16.12) in (16.11) and obtain E and H in the form

$$E = U e^{ik(r - ct)} \, ,$$

(16.13)

$$H = V e^{ik(r - ct)} \, .$$

U and V are the complex vectors

$$U = \left(\frac{1}{r} + \frac{i}{kr^2} \right) (m \times \rho)$$

(16.14)

$$V = \rho \times U - \frac{m}{k^2 r^3} + (m \cdot \rho) \left(\frac{3}{k^2 r^3} - \frac{2i}{kr^2} \right) \rho \, .$$

In the case of small wave lengths λ, the quantity $k = \frac{2\pi}{\lambda}$ becomes very great so that in (16.14) only those terms are significant which are independent of k. In the limit $\lambda \to 0$ we obtain for $r \neq 0$:

$$U = \frac{1}{r}(m \times \rho) \, ,$$

(16.15)

$$V = \rho \times U = \frac{1}{r} \rho \times (m \times \rho) \, .$$

These expressions are formally identical with the equations (13.62) for the vectors E and H on the wave fronts. The mathematical reason for this is that the terms in (16.11) which are given by the derivatives of M of highest order determine the discontinuities of E and H of highest order, and, in case of periodic waves, the terms of lowest power of $\frac{1}{k}$.

We shall demonstrate in the following that this relation is true in general: Periodic electromagnetic fields of small wave lengths obey the same laws as discontinuities, i.e., signals.

16.2 Let us assume that a dipole is oscillating at the point (0,0,0) of a nonhomogeneous medium. The oscillation of the dipole shall be the same as in the preceding case; namely, that given by the momentum

$$M \; = \; - \frac{m}{k^2} \; e^{-ikct} \; = \; - \frac{m}{k^2} \; e^{-i\omega t} \; . \tag{16.21}$$

On the basis of the exact solution in case of a homogeneous medium, it is justifiable to attempt to solve the case of a nonhomogeneous medium with an electromagnetic field

$$E \; = \; U e^{ik(\psi - ct)} \; , $$
$$H \; = \; V e^{ik(\psi - ct)} \; , \tag{16.22}$$

where the surfaces $\psi(x,y,z) - ct = 0$ represent the "spherical" wave fronts of the medium around the point (0,0,0). The vectors (16.22) are of the type which we have investigated in §2, the relation of u,v and U,V being given by

$$u \; = \; U e^{ik\psi} \; , $$
$$v \; = \; V e^{ik\psi} \; . \tag{16.23}$$

If we introduce these expressions in (2.33) we obtain the equations for U and V:

$$\frac{1}{ik} \; \text{curl} \; V \; + \; \left[\text{grad} \; \psi \; \times \; V \; + \; \epsilon U \right] \; = \; 0 \; , $$
$$\frac{1}{ik} \; \text{curl} \; U \; + \; \left[\text{grad} \; \psi \; \times \; U \; - \; \mu V \right] \; = \; 0 \; . \tag{16.24}$$

Let us denote the quantity $\frac{1}{ik}$ by σ; hence

$$\sigma \; \text{curl} \; V \; + \; \left[\text{grad} \; \psi \; \times \; V \; + \; \epsilon U \right] \; = \; 0 \; , $$
$$\sigma \; \text{curl} \; U \; + \; \left[\text{grad} \; \psi \; \times \; U \; - \; \mu V \right] \; = \; 0 \; . \tag{16.25}$$

The solution of (16.25) in the case of a homogeneous medium is given by (16.14), i.e., by vectors of the form

$$U = \sum_{\nu=0}^{1} U_\nu \, \sigma^\gamma \,,$$

$$V = \sum_{\nu=0}^{2} V_\nu \, \sigma^\gamma \,,$$

(16.26)

where U_ν and V_ν are vectors independent of $\sigma = \dfrac{1}{ik}$. This leads us to solve the equations (16.25) by the power series in σ:

$$U = \sum_{\nu=0}^{\infty} U_\nu \, \sigma^\gamma$$

$$V = \sum_{\nu=0}^{\infty} V_\nu \, \sigma^\gamma$$

(16.27)

with vectors U_ν and V_ν as coefficients.

If we introduce these series in (16.25) we find the following conditions for the vectors U_ν, V_ν:

$$\text{grad } \psi \times V_0 + \epsilon U_0 = 0 \,,$$

$$\text{grad } \psi \times U_0 - \mu V_0 = 0 \,,$$

(16.28)

and in case $\nu \geqq 1$:

$$\text{grad } \psi \times V_\nu + \epsilon U_\nu = - \text{ curl } V_{\nu-1} \,,$$

$$\text{grad } \psi \times U_\nu - \mu V_\nu = - \text{ curl } U_{\nu-1} \,.$$

(16.29)

16.3 We are especially interested in the vectors U_0 and V_0 since they determine the electromagnetic field in the form

$$E = U_0 e^{ik(\psi - ct)} \,,$$

$$H = V_0 e^{ik(\psi - ct)} \,,$$

(16.31)

for small wave lengths, i.e., the field which belongs to the realm of Geometrical Optics. We call the function $\psi(x,y,z)$ the <u>phase function</u> of the wave (16.31) so that the wave fronts ψ = const. represent surfaces of equal phase. The vectors U_0 and V_0 may be called the <u>amplitude vectors</u> of the wave.

The equations (16.28) for U_0 and V_0 are formally identical with the conditions for discontinuities on the wave fronts. They can be satisfied by vectors $U_0 \neq 0$ and $V_0 \neq 0$ only if the phase function ψ satisfies the equations $\psi_x^2 + \psi_y^2 + \psi_z^2 = n^2$. The only difference from the equations for discontinuities is that U_0 and V_0 are in general complex vectors. We conclude as before that

$$U_0 \cdot \text{grad } \psi \ = \ V_0 \cdot \text{grad } \psi \ = \ U_0 \cdot V_0 \ = \ 0 \ , \qquad (16.32)$$

i.e., U_0 and V_0 are tangential to the wave fronts. The complex vectors U_0 and $\overline{V}_0$ are orthogonal to each other in accordance with our definition of orthogonal complex vectors in §2.

It follows furthermore that

$$\epsilon U_0 \cdot \overline{U}_0 \ = \ (V_0 \times \text{grad } \psi) \cdot \overline{U}_0$$

$$\mu V_0 \cdot \overline{V}_0 \ = \ (\text{grad } \psi \times U_0) \cdot \overline{V}_0 \ = \ (\overline{V}_0 \times \text{grad } \psi) \cdot U_0 \ .$$

The two right sides of these equations are conjugate complex numbers. They must be equal because they are real as are the left sides. Hence

$$\epsilon U_0 \cdot \overline{U}_0 \ = \ \mu V_0 \cdot \overline{V}_0 \ . \qquad (16.33)$$

The average energy density, according to §2, is given by

$$W \ = \ \frac{1}{16\pi} \, (\epsilon U_0 \cdot \overline{U}_0 + \mu V_0 \cdot \overline{V}_0) \ . \qquad (16.34)$$

We conclude: The average electric energy is equal to the average magnetic energy. Hence

$$W \ = \ \frac{1}{8\pi} \, \epsilon U_0 \cdot \overline{U}_0 \ = \ \frac{1}{8\pi} \, \mu V_0 \cdot \overline{V}_0 \ . \qquad (16.35)$$

The average flux vector S is given by the formula (2.53):

$$S \ = \ \frac{c}{16\pi} \, (U_0 \times \overline{V}_0 + \overline{U}_0 \times V_0) \ .$$

From (16.28) it follows that

$$\epsilon(U_0 \times \overline{V}_0) \ = \ (V_0 \times \text{grad } \psi) \times \overline{V}_0 \ ,$$

$$= \ V_0 \cdot \overline{V}_0 \ \text{grad } \psi \ .$$

Hence

$$U_0 \times \overline{V}_0 + \overline{U}_0 \times V_0 = \frac{2V_0 \cdot \overline{V}_0}{\epsilon} \text{ grad } \psi = \frac{2\mu V_0 \cdot \overline{V}_0}{n^2} \text{ grad } \psi$$

and

$$S = \frac{c}{8\pi n^2} \mu V_0 \cdot \overline{V}_0 \text{ grad } \psi \, ,$$

or

$$S = c \frac{W}{n^2} \text{ grad } \psi \, . \tag{16.36}$$

We conclude that <u>the vector S of average flux is normal to the wave fronts</u>. Its absolute value is related to the average energy W by the equation

$$|S| = \frac{c}{n} W \, . \tag{16.37}$$

16.4 We consider next the equations (16.29) in the case $\nu = 1$; and assume continuity of $\epsilon(x,y,z)$ and $\mu(x,y,z)$. We have

$$\text{grad } \psi \times V_1 + \epsilon U_1 = - \text{ curl } V_0 \, ,$$
$$\text{grad } \psi \times U_1 - \mu V_1 = - \text{ curl } U_0 \, . \tag{16.38}$$

These equations are formally identical with the equations (11.24) for $[E_t]$ and $[H_t]$ in §11. By the same argument we conclude that they are solvable only if the right sides satisfy certain conditions. The former method of deriving these conditions can be repeated literally. We obtain the result: The amplitude vectors U_0 and V_0 satisfy the following differential equations along the light rays:

$$\frac{dU_0}{d\tau} + \frac{1}{2} \Delta_\mu \psi U_0 + \left(U_0 \cdot \frac{\text{grad } n}{n} \right) \text{grad } \psi = 0 \, ,$$
$$\frac{dV_0}{d\tau} + \frac{1}{2} \Delta_\epsilon \psi V_0 + \left(V_0 \cdot \frac{\text{grad } n}{n} \right) \text{grad } \psi = 0 \, . \tag{16.39}$$

Again the only difference is that U_0 and V_0 are complex vectors.

16.4 We solve the equations (16.39) as follows. We introduce

$$U_0(\tau) = |U_0(0)| P e^{-\frac{1}{2} \int_0^\tau \Delta_\mu \psi \, d\tau} \, , \qquad V_0(\tau) = |V(0)| Q e^{-\frac{1}{2} \int_0^\tau \Delta_\epsilon \psi \, d\tau} \, , \tag{16.41}$$

where P and Q satisfy the equations

$$\frac{dP}{d\tau} + \frac{1}{n}\ (P \cdot grad\ n)\ grad\ \psi = 0\ ,$$

$$\frac{dQ}{d\tau} + \frac{1}{n}\ (Q \cdot grad\ n)\ grad\ \psi = 0\ . \tag{16.42}$$

The three vectors P, Q and $T = \frac{1}{n} grad\ \psi$ form a system of unitary vectors along the ray, i.e., complex vectors which satisfy the relations

$$T \cdot \overline{P} = \overline{T} \cdot P = 0\ , \qquad T \cdot \overline{T} = 1\ , \qquad T = \overline{P} \times \overline{Q}\ ,$$

$$P \cdot \overline{Q} = \overline{P} \cdot Q = 0\ , \qquad P \cdot \overline{P} = 1\ , \qquad P = \overline{Q} \times \overline{T}\ , \tag{16.43}$$

$$Q \cdot \overline{T} = \overline{Q} \cdot T = 0\ , \qquad Q \cdot \overline{Q} = 1\ , \qquad Q = \overline{T} \times \overline{P}\ .$$

By introducing on the ray the orthogonal system of unit vectors T,N (Principal normal), and S (Binormal), and the geometrical length of the ray as parameter, the equations (16.41) become

$$P' + \frac{1}{\rho}\ (N \cdot P)T = 0\ ,$$

$$Q' + \frac{1}{\rho}\ (N \cdot P)T = 0\ . \tag{16.44}$$

A complex vector P, normal to T, for which $P \cdot \overline{P} = 1$ can be expressed in the following form as a linear combination of N and S;

$$P = \cos \Lambda\ (N \cos \vartheta + S \sin \vartheta) + i \sin \Lambda\ (N \cos \vartheta* + S \sin \vartheta*)\ , \tag{16.45}$$

where Λ, ϑ and $\vartheta *$ are arbitrary real numbers. The solution of the equations (16.44) belonging to a vector P_0, at a point s = 0 of the ray, then is given by the equation

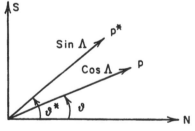

Figure 34

$$\Lambda = \Lambda_0\ ,$$

$$\vartheta = \vartheta_0 - \theta_0\ , \tag{16.46}$$

$$\vartheta* = \vartheta_0* - \theta\ ,$$

where $\theta = \int_0^s \frac{ds}{\tau}$, $\frac{1}{\tau}$ being the torsion of the ray. In other words, if $P = p + ip*$ then both vectors p and p* are rotated relative to N by the same angle $-\theta$ and are unchanged in length.

16.5 The exponential factors in (16.41) can be replaced by the expressions (12.27). We obtain a result, analagous to (12.28): The quantities

$$\frac{K}{n} \epsilon \, |U_0|^2 \quad \text{and} \quad \frac{K}{n} \mu \, |V_0|^2$$

and hence $\dfrac{K}{n}W$, where W is the average energy density, are constant along a light ray in media of continuous index of refraction.

On account of the relations (16.36) and (16.37) this means: The energy flux $|S|$ do through corresponding surface elements of the wave fronts is constant in a medium of continuous index of refraction.

We finally consider the polarization of the light along a light ray. The ellipticity ϵ of the polarization, according to (2.68), is given by the formula

$$\epsilon^2 = \frac{|U_0|^2 - \sqrt{U_0^2 \, \overline{U}_0^2}}{|U_0|^2 + \sqrt{U_0^2 \, \overline{U}_0^2}} \; . \tag{16.47}$$

On account of (16.41) this reduces to

$$\epsilon^2 = \frac{1 - \sqrt{P^2 \, \overline{P}^2}}{1 + \sqrt{P^2 \, \overline{P}^2}} \tag{16.48}$$

However, from (16.42) we readily obtain $\dfrac{d}{d\tau} P^2 = \dfrac{d}{d\tau} \overline{P}^2 = 0$, i.e., P^2 and $\overline{P}^2$ are constants along the ray. Hence: The polarization along the light ray is not changed if n varies continuously.

16.5 Media of discontinuous optical properties. Results, quite different from the above, are obtained if the optical properties are discontinuous. Let us assume that a surface $\Omega(x,y,z) = 0$ separates two media with continuous ϵ, μ and ϵ', μ' in such a way that on $\Omega = 0$ these functions assume different boundary values. We have seen, that in this case, a reflected set of wave fronts has to be introduced so that we have to deal with two sets of wave fronts $\psi = ct$ and $\psi* = ct$ on one side of $\Omega = 0$, and with one set $\psi' = ct$ on the other side. This suggests solving the problem 16.2 by an electromagnetic field of the type

$$E = U \, e^{ik(\psi - ct)} + U* \, e^{ik(\psi* - ct)}$$

$$H = V \, e^{ik(\psi - ct)} + V* \, e^{ik(\psi* - ct)} \tag{16.51}$$

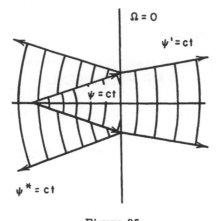

Figure 35

on the one side of $\Omega = 0$, and by

$$E = U' \, e^{ik(\psi' - ct)}$$

$$H = V' \, e^{ik(\psi' - ct)} \qquad (16.52)$$

on the other side. On $\Omega = 0$ we have $\psi = \psi* = \psi'$ and, on account of the conditions (6.1), the relations

$$\epsilon(U + U*) \cdot M = \epsilon'(U' \cdot M)$$

$$(U + U*) \times M = U' \times M , \qquad (16.53)$$

$$\mu(V + V*) \cdot M = \mu'(V' \cdot M)$$

$$(V + V*) \times M = V' \times M . \qquad (16.54)$$

By introducing the power series (16.27) in these conditions, we find that each one of the vector coefficients U_ν, V_ν; $U_\nu*$, $V_\nu*$; U_ν', V_ν' must satisfy the conditions (16.53) and (16.54). In the case of the vectors U_0, V_0; U_0*, V_0*; U_0', V_0' which represent the field for small wave lengths, not only the above conditions are valid but also the relations (16.28) written down for each of the three functions ψ and their associated vectors U_0, V_0. However, this set of conditions is identical with the conditions from which, in §15, we have derived Fresnel's formulae. By introducing normal and parallel components, as defined in (15.45), we thus have the result, for the complex vectors U, U*, U':

$$U_{0p}^* = U_{0p} \, \frac{1 - \dfrac{n}{n'} \dfrac{\cos \vartheta'}{\cos \vartheta}}{1 + \dfrac{n}{n'} \dfrac{\cos \vartheta'}{\cos \vartheta}} \qquad U_{0s}^* = U_{0s} \, \frac{\dfrac{n}{n'} - \dfrac{\cos \vartheta'}{\cos \vartheta}}{\dfrac{n}{n'} + \dfrac{\cos \vartheta'}{\cos \vartheta}}$$

$$(16.55)$$

$$U_{0p}' = U_{0p} \, \frac{2 \dfrac{n}{n'}}{1 + \dfrac{n}{n'} \dfrac{\cos \vartheta'}{\cos \vartheta}} \qquad U_{0s}' = U_{0s} \, \frac{2 \dfrac{n}{n'}}{\dfrac{n}{n'} + \dfrac{\cos \vartheta'}{\cos \vartheta}} .$$

Of course, if we consider a given light ray which passes the surface $\Omega = 0$ we cannot expect that the flux through surface elements of the transmitted wave fronts is the same as the flux through the corresponding surface elements of the incident wave front. Part of the incident flux follows the reflected light ray. One also verifies readily that the polarization of the light on the transmitted or reflected ray is not the same as on the incident ray.

16.6 An optical instrument consists in general of several refracting surfaces which separate media of different refractive indices. Consequently,

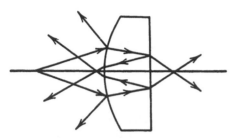

multiple reflections must be expected, such that in every part of the medium infinitely many different sets of wave fronts are to be considered. The electromagnetic fields in each of the media are therefore considerably more complicated than the field given in the formula (16.51); namely an infinite sum

$$E = \Sigma \; U(\gamma) \; e^{ik(\psi_\gamma - ct)} \qquad (16.61)$$

Figure 36

with terms related to the different wave fronts.

However, on each surface of refraction, the boundary values of these different terms can be divided into groups of three corresponding to an incident, transmitted and reflected set of wave fronts, for which $\psi = \psi^* = \psi'$ on the refracting surface. The boundary values of E and H given by infinite series of the type (16.61) then can satisfy the conditions (6.1) of §6 only if the conditions (16.53) and (16.54) are satisfied by each of the above groups individually. This means that Fresnel's formulae can be applied safely at every single step of the multiple reflection in the instrument.

In practice the internal reflections are seldom of interest but are even carefully eliminated by the absorbing walls of the objective. Only the set of wave fronts which consists of transmitted wave fronts alone is of prime interest. With the aid of Fresnel's formulae — or in case of continuous variation of n — with the aid of the transport equations (16.39) we are now in the position to construct the electromagnetic field U_0, V_0 for small wave lengths, if the paths of light rays in the instrument are known. The field, which by this procedure is obtained in the image space, represents the actual field to a close approximation.

Serious deviation from the exact solution, however, must be expected in

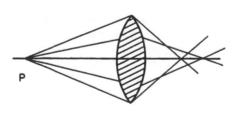

the neighborhood of conjugate points, i.e., at points where the bundle of light rays from a point source P contracts to a narrow region. In a later chapter we shall see how one can use the present first approximation to construct a second approximation which is also valid with great accuracy in the neighborhood of conjugate points.

Figure 37

CHAPTER II.

HAMILTON'S THEORY OF GEOMETRICAL OPTICS

§17. PRINCIPLES OF GEOMETRICAL OPTICS.

The deduction of the principles of Geometrical Optics from the electromagnetic differential equations might tempt us to consider these principles as inherently connected with the special form of wave optics which Maxwell's theory represents. Actually only a few of the general premises of the wave theory are essential for the deduction of the laws of Geometrical Optics. Mathematically, this follows from the fact that the same characteristic equation can be obtained from various forms of wave equations, that, in other words, the same laws for the propagation of discontinuities can be found for quite different forms of wave motion. This applies especially if the field of geometrical optics is limited to the construction of wave fronts and light rays.

In the next chapters we shall be concerned with this special part of optics. The concept of the electromagnetic field or any form of wave motion does not play any part in our investigation; it would, in fact, have been entirely possible to develop this theory from its own premises. However, in order to develop the diffraction theory of optical instruments, we then would be forced either to develop an entirely different theory or to carry out similar considerations to those in Chapter I.

17.1 <u>Huyghens' Principle</u>. Huyghens' construction of wave fronts as envelopes of wavelets, i.e., of spherical waves, can be considered as the principle of Geometrical Optics which is most closely related to the concept of wave motion. In Chapter I this principle has been formulated mathematically as follows: If $\psi(x,y,z) - ct = 0$ represents a set of wave fronts then $\psi(x,y,z)$ must be a solution of the differential equation

$$\psi_x^2 + \psi_y^2 + \psi_z^2 = n^2 \ . \tag{17.11}$$

The direct connection with Huyghens' construction was given by the theorem: If $\psi(x,y,z; a,b)$ is an integral of (17.11) which depends on two arbitrary parameters a,b, then the envelope of the wave fronts $\psi(x,y,z; a,b) - ct$ is also a wave front $\psi(x,y,z) - ct = 0$ and $\psi(x,y,z)$ is a solution of (17.11). Indeed, Huyghens' wavelets $V(x,y,z; x_0,y_0,z_0)$ where (x_0,y_0,z_0) is a point on a surface Γ represent integrals of the above type.

It can be seen quite easily that Huyghens' construction leads to the above mathematical formulation. Let us assume that light is a wave motion of finite velocity. A light signal is given at a point (x_0,y_0,z_0) at the time t_0 and has penetrated at a time $t > t_0$ into a domain of the space which is enclosed by a surface given in the form

$$V(x_0,y_0,z_0; x,y,z) = c(t - t_0) . \tag{17.12}$$

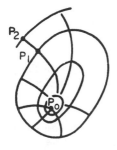

P_2

P_1

P_0

$V = ct_0$

Figure 38

This set of wave fronts has a two-parameter manifold of orthogonal trajectories. We measure the velocity of the disturbance by the velocity of the wave fronts along the orthogonal trajectories. For two points P_1 and P_2 on such a trajectory we have

$$V(x_0,y_0,z_0; x_2,y_2,z_2) - V(x_0,y_0,z_0; x_1,y_1,z_1)$$

$$= c(t_2 - t_1)$$

or in differential form

$$V_x\, dx + V_y\, dy + V_z\, dz = c\, dt ;$$

i.e.,

$$V_x\, \frac{dx}{dt} + V_y\, \frac{dy}{dt} + V_z\, \frac{dz}{dt} = c . \tag{17.13}$$

The velocity of the point P_1 along the trajectory is given by the vector $\left(\dfrac{dx}{dt}, \dfrac{dy}{dt}, \dfrac{dz}{dt}\right)$:

$$\frac{dx}{dt} = v\, \frac{V_x}{\sqrt{V_x^2 + V_y^2 + V_z^2}} \quad ; \quad \frac{dy}{dt} = v\, \frac{V_y}{\sqrt{V_x^2 + V_y^2 + V_z^2}} \quad ;$$

$$\frac{dz}{dt} = v\, \frac{V_z}{\sqrt{V_x^2 + V_y^2 + V_z^2}} \quad ; \tag{17.14}$$

where v is the absolute value of the velocity. By introducing these expressions in (17.13) we find

$$\sqrt{V_x^2 + V_y^2 + V_z^2} = \frac{c}{v} . \tag{17.15}$$

We finally assume that the medium is isotropic: the velocity v at the point (x,y,z) shall be independent of the direction of the particular trajectory, and also independent of (x_0,y_0,z_0), the point where the light signal was released. We thus introduce

$$\frac{c}{v} = n(x,y,z) \tag{17.16}$$

and obtain

$$V_x^2 + V_y^2 + V_z^2 = n^2(x,y,z) .$$ (17.17)

This means that $V(x_0,y_0,z_0; x,y,z)$ is a solution of (17.17) which depends on the arbitrary parameters x_0,y_0,z_0. Huyghens' Principle, that all other wave fronts $\psi(x,y,z) - ct = 0$ can be found as envelopes of wavelets $V(x,y,z; x_0,y_0,z_0)$, then characterizes $\psi(x,y,z)$ also as a solution of this partial differential equation (17.17).

We remark explicitly that, with this interpretation of $v(x,y,z)$ as the velocity at points on the wave fronts, the velocity is <u>smaller</u> in an optical medium of greater n.

17.2 <u>Light rays as paths of corpuscles</u>. We have introduced the light rays of a medium as orthogonal trajectories of the wave fronts and we have found that it is possible to characterize the light rays, independently of the concept of wave fronts, as solutions of a set of ordinary differential equations, namely as those solutions of the system:

$$\ddot{x} = \frac{\partial}{\partial x}\left(\frac{1}{2}n^2\right) ,$$

$$\ddot{y} = \frac{\partial}{\partial y}\left(\frac{1}{2}n^2\right) ,$$ (17.21)

$$\ddot{z} = \frac{\partial}{\partial z}\left(\frac{1}{2}n^2\right) ,$$

which satisfy the condition

$$\dot{x}^2 + \dot{y}^2 + \dot{z}^2 = n^2 .$$ (17.22)

These equations allow us to employ a radically different interpretation of the phenomenon of light. By considering the parameter τ as a time parameter we find that the light rays are nothing but the paths of corpuscles which move in a potential field $\phi = -\frac{1}{2}n^2$ with a velocity $w = n$. This interpretation, however, forces us to admit that the velocity of the corpuscles is <u>greater</u> in the medium of greater n contrary to Huyghens' interpretation.

The first derivation of Snell's Law was given by Descartes on the corpuscular basis by using the principle that the tangential component of the velocity is unchanged when a medium of different refractive index is entered. This means that the difference $\vec{w}' - \vec{w}$ of the two velocity vectors $\vec{w}'$ and $\vec{w}$ on the boundary of the medium has the direction of the normal unit vector

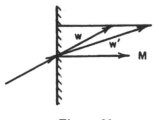

Figure 39

M. Hence $\vec{w}' = \vec{w} + \Gamma M$; or, since $\vec{w}' = n'T'$ and $\vec{w} = nT$ on account of (17.22), where T and T' are unit vectors:

$$n'T' = nT + \Gamma M \ . \tag{17.23}$$

This we recognize as our former formulation (14.16) of Snell's Law.

17.3 This second interpretation of Geometrical Optics is the one which we have to adopt in Electron Optics. If a charged particle moves in an electrostatic field of potential $\phi(x,y,z)$ we have the equations of movement

$$\ddot{x} = - K\phi_x$$

$$\ddot{y} = - K\phi_y \tag{17.31}$$

$$\ddot{z} = - K\phi_z$$

where K is a constant. It follows in the usual manner that

$$\dot{x}^2 + \dot{y}^2 + \dot{z}^2 = 2(C - K\phi) \tag{17.32}$$

with a constant C depending on the original kinetic energy of the electron and the location of its origin. If the origin of the electron lies in a part of the field where the potential ϕ is negligible then C is proportional to the original kinetic energy. For such electrons it follows that the possible paths of electrons with given original kinetic energy are the same as the light rays in a medium of refractive index

$$n = \sqrt{2(C - K\phi)} \ . \tag{17.32}$$

This statement allows us to apply the theory of geometrical optics of continuous media directly to electron optical instruments which employ electrostatic fields but not magnetic fields. The velocity of the electron is greater in a medium of greater n and therefore is not to be identified with the velocity of light on the rays according to Huyghens' definition.

17.4 Let us consider, finally, all solutions of the equations (17.21), and not only those compatible with the condition (17.22). In other words all paths of corpuscles of equal mass in the potential field $\phi = - \frac{1}{2} n^2$. We have for an individual corpuscle

$$\dot{x}^2 + \dot{y}^2 + \dot{z}^2 = n^2 + C \ , \tag{17.41}$$

where C is a constant. Let us now consider all solutions of (17.21) for which (17.41) is satisfied, C being given arbitrarily. These solutions represent a four-parameter manifold of curves. If we denote

$$n^{*2} = n^2 + C ,$$ (17.42)

we can write the equations (17.21) in the form

$$\ddot{x} = \frac{\partial}{\partial x} \left(\frac{1}{2} n^{*2} \right) ,$$

$$\ddot{y} = \frac{\partial}{\partial y} \left(\frac{1}{2} n^{*2} \right) ,$$ (17.43)

$$\ddot{z} = \frac{\partial}{\partial z} \left(\frac{1}{2} n^{*2} \right) ,$$

and (17.41):

$$\dot{x}^2 + \dot{y}^2 + \dot{z}^2 = n^{*2} .$$ (17.44)

It follows that <u>solutions of (17.21) which satisfy the condition</u>

$$\dot{x}^2 + \dot{y}^2 + \dot{z}^2 = n^2 + C$$ (17.45)

<u>where C is a fixed but arbitrary constant determine the light rays in a medium of refractive index</u>

$$n^* = \sqrt{n^2 + C} .$$ (17.46)

17.5 <u>Fermat's Principle</u>. It is quite customary to deduce the laws of geometrical optics from Fermat's Principle of shortest optical path. In fact we have seen that the extremals of the problem of variation

$$V(P_0, P_1) = \int_{P_0}^{P_1} n \, ds = \text{Min.}$$ (17.51)

are the solutions of the differential equations

$$(nx')' = n_x ,$$

$$(ny')' = n_y ,$$ (17.52)

$$(nz')' = n_z ,$$

if we chose s the geometrical length on the extremals as a parameter. These differential equations have the same integral curves as (17.21) and (17.22). Euler's equations (17.52) represent necessary but not sufficient conditions for the solutions of the problem (17.51). It is perfectly possible that curves between two given points P_0 and P_1 exist with an optical length smaller than the optical length of the extremal between P_0 and P_1, i.e., the solution of (17.52) which goes through both points P_0 and P_1. The light ray, determined by P_0 and P_1, then is given by the extremal and not by the curve of shorter or shortest optical path.

This makes it necessary to formulate Fermat's principle more cautiously. The existence of curves of shorter optical path than the length of the light ray can be expected if the two points P_0 and P_1 on the light ray are too far apart. To a given point P_0 on a given light ray there exists, however, a neighboring section of the light ray such that the light ray is the curve of shortest optical path between P_0 and a point P_1 on this section.

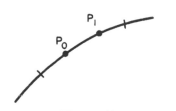

Figure 40

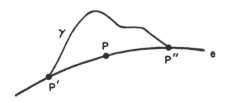

Figure 41

On account of this, Caratheodory gives the following formulation of Fermat's principle: A curve e can coincide with a light ray if and only if each point P of e is an interior point of at least a partial section of e with the following property: Fermat's integral taken along this partial section, and between its end points, P' and P'', has a smaller value than the same integral calculated for a curve ν different from e which has the same end points P' and P'', and lies in a certain neighborhood of e.

17.6 A simple example for the case that the light ray between two points is not the curve of shortest optical path is given by the concave spherical mirror. Let Q be the vertex of the mirror and M its center. We consider two points, P_0 and P_1, symmetrically located to the mirror axis and in the plane through M which is perpendicular to the axis. The light ray between P_0 and P_1 is

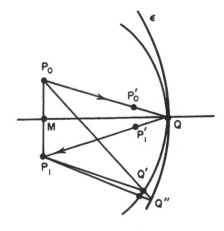

Figure 42

given by the line $P_0 Q P_1$. Its optical path has the length $P_0 Q + Q P_1$. We can show that any other line $P_0 Q' + P_1 Q'$, where Q' lies on the mirror and in the plane $P_0 Q P_1$ has a shorter optical path. Let us construct the ellipse ϵ through Q which has P_0 and P_1 as focal points. Its radius of curvature at Q is certainly greater than MQ, the radius of the mirror. Hence it lies outside the mirror. We now extend the line $P_0 Q'$ until it intersects the ellipse at Q''. Then $P_1 Q'' + Q'' Q' > P_1 Q'$ and hence

$$P_1 Q'' + Q'' P_0 > P_1 Q' + Q' P_0 . \tag{17.61}$$

On the other hand

$$P_1 Q'' + Q'' P_0 = P_1 Q + Q P_0 . \tag{17.62}$$

Hence

$$P_1 Q + Q P_0 > P_1 Q' + Q' P_0 \tag{17.63}$$

which demonstrates our statement.

The line $P_1 Q' + Q' P_0$ certainly cannot be a light ray because the angles of reflection can only be equal if $P_1 Q' = Q' P_0$ which is not the case. The optical path $P_0 Q + Q P_1$ of the actual light ray, however, has an extreme value, a maximum, so that the light does not choose the path of shortest but of longest optical length in this example.

If, on the other hand, two symmetrical points P_0' and P_1' on the light ray are considered which are near enough to the mirror, then the ellipse through Q with P_0' and P_1' as focal points lies <u>inside</u> the mirror and the optical path $P_0'Q + Q P_1'$ becomes a minimum. This illustrates Caratheodory's formulation of Fermat's principle, that the light ray is the curve of shortest optical path between two of its points which are not too far apart.

§18. THE CANONICAL EQUATIONS.

18.1 In most optical instruments there exists an axis with respect to which the instrument is symmetrical. On account of this it is advantageous to orient the coordinate system so that one axis, for example the z-axis, coincides with the axis of symmetry. In the case of rotational symmetry with respect to the z-axis this axis represents a light ray. Only those light rays are of practical significance which lie in a more or less extended cylindrical neighborhood of the z-axis. We can describe these light rays by two single-valued functions

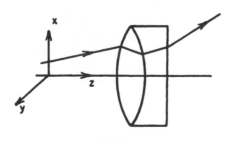

Figure 43

$$x = x(z) , \qquad y = y(z) , \qquad (18.11)$$

thus excluding the case that a refracted light ray reverses direction, i.e., returns towards the side where it originated. We shall in general assume that the light is directed from left to right through the instrument. We do not assume in the following that the optical medium is rotationally symmetrical. Let us, however, limit our investigation to instruments in which the light rays can be represented in the form (18.11) in a sufficiently large neighborhood of the z-axis.

18.2 We derive the following results with the aid of Fermat's integral and interpret the light rays as the extremals of Fermat's problem of variation:

$$V = \int_{z_0}^{z_1} n\sqrt{1 + \dot{x}^2 + \dot{y}^2} \, dz . \qquad (18.21)$$

We use the notation $\dot{x} = \dfrac{dx}{dz}$; $\dot{y} = \dfrac{dy}{dz}$.

With this choice of the parameter Euler's equations of V become,

$$\frac{d}{dz}\left[\frac{n\dot{x}}{\sqrt{1 + \dot{x}^2 + \dot{y}^2}}\right] - n_x \sqrt{1 + \dot{x}^2 + \dot{y}^2} = 0 ,$$

$$\frac{d}{dz}\left[\frac{n\dot{y}}{\sqrt{1 + \dot{x}^2 + \dot{y}^2}}\right] - n_y \sqrt{1 + \dot{x}^2 + \dot{y}^2} = 0 . \qquad (18.22)$$

The quantities

$$\cos a = \frac{\dot{x}}{\sqrt{1 + \dot{x}^2 + \dot{y}^2}} ,$$

$$\cos b = \frac{\dot{y}}{\sqrt{1 + \dot{x}^2 + \dot{y}^2}} , \qquad (18.23)$$

$$\cos c = \frac{1}{\sqrt{1 + \dot{x}^2 + \dot{y}^2}} ,$$

are the direction cosines of the light ray with respect to the x,y,z axes. The form of the equations (18.22) suggests the introduction of the notation

$$p = n \cos a = n \frac{\dot{x}}{\sqrt{1 + \dot{x}^2 + \dot{y}^2}} \quad ,$$

$$\tag{18.24}$$

$$q = n \cos b = n \frac{\dot{y}}{\sqrt{1 + \dot{x}^2 + \dot{y}^2}} \quad .$$

These quantities are called the optical direction cosines.

We can express the derivatives $\dot{x}$ and $\dot{y}$ in terms of p and q. From (18.24) it follows that

$$n^2 - p^2 - q^2 = \frac{n^2}{1 + \dot{x}^2 + \dot{y}^2} \quad , \tag{18.25}$$

and hence

$$\dot{x} = \frac{p}{\sqrt{n^2 - p^2 - q^2}} = \frac{\partial}{\partial p} \sqrt{n^2 - p^2 - q^2} \quad ,$$

$$\tag{18.26}$$

$$\dot{y} = \frac{q}{\sqrt{n^2 - p^2 - q^2}} = \frac{\partial}{\partial q} \sqrt{n^2 - p^2 - q^2} \quad .$$

The Euler equations (18.22) become with the aid of (18.25):

$$\dot{p} = \frac{n n_x}{\sqrt{n^2 - p^2 - q^2}} = \frac{\partial}{\partial x} \sqrt{n^2 - p^2 - q^2} \quad ,$$

$$\tag{18.27}$$

$$\dot{q} = \frac{n n_y}{\sqrt{n^2 - p^2 - q^2}} = \frac{\partial}{\partial y} \sqrt{n^2 - p^2 - q^2} \quad .$$

The equations (18.26) and (18.27) demonstrate that the functions $x(z)$, $y(z)$, $p(z)$, $q(z)$ satisfy a system of <u>canonical equations</u>

$$\dot{x} = H_p \ , \qquad\qquad \dot{p} = - H_x \ ,$$

$$\tag{18.28}$$

$$\dot{y} = H_q \ , \qquad\qquad \dot{q} = - H_y \ ,$$

where the Hamiltonian function

$$H(x,y;p,q) - \sqrt{n^2 - p^2 - q^2} = - n \cos c \tag{18.281}$$

is equal to the negative optical direction cosine, $-n \cos c$ of the ray with respect to the z-axis.

We may finally express the optical length, V, of the ray in terms of x(z), y(z); p(z), q(z). It follows, on account of (18.25) that

$$V = \int_{z_0}^{z_1} \frac{n^2}{\sqrt{n^2 - p^2 - q^2}} \, dz = \int_{z_0}^{z_1} \left(\sqrt{n^2 - p^2 - q^2} + \frac{p^2 + q^2}{\sqrt{n^2 - p^2 - q^2}} \right) dz$$

and hence, with the aid of (18.26):

$$V = \int_{z_0}^{z_1} (\dot{x}p + \dot{y}q - H)dz . \tag{18.29}$$

The canonical equations (18.28) are nothing but Euler's equations of this integral (18.29) if the four functions x(z), y(z); p(z), q(z) are considered as unknown. The problem is to find functions x(z), y(z), p(z), q(z) such that x(z), y(z) represents a curve between P_0 and P_1 for which the integral (18.29) has an extreme value.

18.3 The variation problem (18.29) is called the canonical form of Fermat's problem (18.21). It is possible to generalize the above procedure so that it applies to other types of problems of variation.[†] Let us consider the problem

$$V = \int_{z_0}^{z_1} F(x, y, \dot{x}, \dot{y}; z)dz = \text{Extremum} \tag{18.31}$$

where F is a function of the indicated variables with continuous second derivatives. We mention that the limitation to two unknown functions x(z) and y(z) is not essential. Euler's differential equations are

$$\frac{d}{dz} (F_{\dot{x}}) - F_x = 0 ,$$

$$\frac{d}{dz} (F_{\dot{y}}) - F_y = 0 . \tag{18.32}$$

We introduce

$$p = F_{\dot{x}} (x, y; \dot{x}, \dot{y}; z) ,$$

$$q = F_{\dot{y}} (x, y; \dot{x}, \dot{y}; z) . \tag{18.33}$$

[†]Courant-Hilbert, Methoden der Math. Physik. 2nd Ed. of Vol. I, pp. 199 etc. (Transformation of Friedrichs). Vol. II, page 96.

If we assume that the Jacobian

$$\begin{vmatrix} F_{\dot{x}\dot{x}} & F_{\dot{x}\dot{y}} \\ F_{\dot{x}\dot{y}} & F_{\dot{y}\dot{y}} \end{vmatrix} \neq 0 \qquad (18.331)$$

it is then possible to calculate $\dot{x}$ and $\dot{y}$ from (18.33) as functions of x,y; p,q, and z. We now define the function H(x,y; p,q; z) by the equation

$$F + H = \dot{x}p + \dot{y}q \qquad (18.34)$$

with the understanding that $\dot{x}$ and $\dot{y}$ shall be replaced by their expressions in x,y,p,q, and z determined from (18.33). If we differentiate the identity (18.34) with respect to p, we obtain

$$F_{\dot{x}} \frac{\partial \dot{x}}{\partial p} + F_{\dot{y}} \frac{\partial \dot{y}}{\partial p} + H_p = \dot{x} + p \frac{\partial \dot{x}}{\partial p} + q \frac{\partial \dot{y}}{\partial p} \qquad (18.35)$$

or, by (18.33): $H_p = \dot{x}$. In a similar way we find $\dot{y} = H_q$. This demonstrates the interesting fact that the inverse transformation (18.33) is of the form

$$\dot{x} = H_p(x,y; p,q; z) ,$$
$$\dot{y} = H_q(x,y; p,q; z) , \qquad (18.36)$$

i.e., with regard to H,p,q it is of the same form as the original transformation (18.33) with regard to F, $\dot{x}$, $\dot{y}$.

We can summarize these results in the formulae:

$$p = F_{\dot{x}} , \qquad \dot{x} = H_p ,$$
$$\qquad\qquad\qquad\qquad\qquad\qquad F + H = \dot{x}p + \dot{y}q , \qquad (18.37)$$
$$q = F_{\dot{y}} , \qquad \dot{y} = H_q ,$$

which represent the transformation of the function F(x,y; $\dot{x}$,$\dot{y}$; z) into H(x,y; p,q; z) and its inverse. A transformation of this type is known as a Legendre transformation of the function F($\dot{x}$,$\dot{y}$) and the variables $\dot{x}$,$\dot{y}$ into the function H(p,q) and the variables p,q.

The variables x,y and z in this transformation only play the part of parameters. We show, however, that the equations

$$F_x + H_x = 0 ,$$
$$F_y + H_y = 0 , \qquad (18.371)$$
$$F_z + H_z = 0$$

are identities if either $\dot{x}$ and $\dot{y}$ in F are expressed by x,y; p,q; z or p,q in H by x,y; $\dot{x},\dot{y}$; z with the aid of (18.37).

In fact, from $F + H = \dot{x}p + \dot{y}q$ follows with respect to the variables x,y; p,q; z;

$$F_x + F_{\dot{x}} \frac{\partial \dot{x}}{\partial x} + F_{\dot{y}} \frac{\partial \dot{y}}{\partial x} + H_x = p \frac{\partial \dot{x}}{\partial x} + q \frac{\partial \dot{y}}{\partial x} .$$

Hence, on account of $p = F_{\dot{x}}$ and $q = F_{\dot{y}}$, $F_x + H_x = 0$. Furthermore, if the variables x,y; $\dot{x},\dot{y}$; z are considered

$$F_x + H_x + H_p \frac{\partial p}{\partial x} + H_q \frac{\partial q}{\partial x} = \dot{x} \frac{\partial p}{\partial x} + \dot{y} \frac{\partial q}{\partial x} ,$$

i.e., again $F_x + H_x = 0$. The other identities (18.371) follow in a similar manner.

Euler's equations (18.32) in terms of x,y; p,q; z become, by (18.33) and (18.371);

$$\dot{p} = - H_x ,$$
$$\dot{q} = - H_y .$$

(18.38)

On the other hand, from (18.37), we have

$$\dot{x} = H_p ,$$
$$\dot{y} = H_q ,$$

(18.381)

which shows that x,y; p,q are solutions of the canonical equations with H(x,y; p,q; z) as the Hamiltonian function.

The variation integral (18.31) assumes the form

$$V = \int_{z_0}^{z_1} (\dot{x}p + \dot{y}q - H) \, dz$$

(18.39)

if F is replaced according to (18.37). The canonical equations (18.38) and (18.381) are Euler equations of this integral as we readily verify.

By applying the transformation (18.37) to Fermat's problem, i.e., to the function

$$F = n(x,y,z) \sqrt{1 + \dot{x}^2 + \dot{y}^2}$$

we find $H = -\sqrt{n^2 - p^2 - q^2}$, in agreement with our former result.

§19. HAMILTON'S CHARACTERISTIC FUNCTION $V(x_0,y_0,z_0; x,y,z)$.

19.1 The numerical investigation of a given optical instrument is carried out in general along the following lines. One considers a point

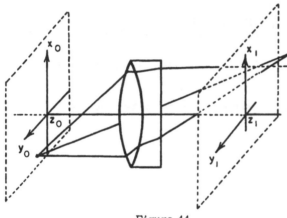

Figure 44

(x_0,y_0) of a certain plane $z = z_0$ which is called the object plane. A suitable number of rays which originate at the point (x_0,y_0) are traced through the instrument with the aim of calculating their intersection (x_1,y_1) with another plane $z = z_1$, the image plane. In a corrected instrument all these rays should intersect the image plane at points (x_1,y_1) which are as near as possible to the ideal image point

$$x_1 = Mx_0 ,$$
$$y_1 = My_0 ,$$

(19.11)

where M is a constant of the instrument depending only upon the choice of the object plane, and not on the data of the incident ray.

The incident ray is completely determined by (x_0,y_0), the intersection with the object plane, and by its optical direction cosines

$$p_0 = n_0 \cos a_0 ,$$
$$q_0 = n_0 \cos b_0 .$$

(19.12)

The intersection (x_1,y_1) of the refracted ray with the image plane $z = z_1$, and its optical direction cosines p_1,q_1 at this plane, are functions of the initial data at $z = z_0$:

$$x_1 = x(z_0,z_1; x_0,y_0; p_0,q_0) ,$$

$$y_1 = y(z_0,z_1; x_0,y_0; p_0,q_0) ,$$

$$p_1 = p(z_0,z_1; x_0,y_0; p_0,q_0) ,$$ (19.13)

$$q_1 = q(z_0,z_1; x_0,y_0; p_0,q_0) .$$

To determine these four functions is a problem of numerical computation in practical optics. The optical image of the plane $z = z_0$ on the plane $z = z_1$ is considered as perfect if the first two functions (19.13) reduce to

$$x_1 = Mx_0 ,$$

$$y_1 = My_0 ,$$ (19.14)

M being a constant for all points x_0,y_0 of the plane $z = z_0$, and all directions p_0,q_0. The main problem of optical design is to find a distribution of optical media, $n = n(x,y,z)$, such that the conditions (19.14) are satisfied, or at least approached to a high extent. The deviations

$$\Delta x_1 = x_1 - Mx_0 ,$$

$$\Delta y_1 = y_1 - My_0 ,$$ (19.15)

are called the aberrations of the optical system.

19.2 The main result of Hamilton's theory is that it is possible to reduce the above problem of finding four functions to the problem of finding one function alone. From this function, then, the four functions (19.13) can be deduced by differentiation and elimination. The derivation of this result can be carried out for canonical equations in general just as readily as for the special optical Hamiltonian $H = -\sqrt{n^2 - p^2 - q^2}$. Therefore we do not assume this special form of H in the following:

Let us assume that

$$x = x(z_0,z; x_0,y_0,p_0,q_0) ,$$

$$y = y(z_0,z; x_0,y_0,p_0,q_0) ,$$

$$p = p(z_0,z; x_0,y_0,p_0,q_0) ,$$ (19.21)

$$q = q(z_0,z; x_0,y_0,p_0,q_0) ,$$

is the solution of the canonical equations

$$\dot{x} = H_p , \qquad\qquad \dot{p} = - H_x ,$$
$$\dot{y} = H_q , \qquad\qquad \dot{q} = - H_y ,$$

(19.22)

which for $z = z_0$ has the initial values x_0, y_0, p_0, q_0. Hence we have the identities

$$x_0 = x(z_0, z_0; x_0, y_0, p_0, q_0) ,$$
$$y_0 = y(z_0, z_0; x_0, y_0, p_0, q_0) ,$$
$$p_0 = p(z_0, z_0; x_0, y_0, p_0, q_0) ,$$
$$q_0 = q(z_0, z_0; x_0, y_0, p_0, q_0) .$$

(19.23)

At the image plane $z = z_1$, the functions (19.21) assume the values

$$x_1 = x(z_0, z_1; x_0, y_0, p_0, q_0) ,$$
$$y_1 = y(z_0, z_1; x_0, y_0, p_0, q_0) ,$$
$$p_1 = p(z_0, z_1; x_0, y_0, p_0, q_0) ,$$
$$q_1 = q(z_0, z_1; x_0, y_0, p_0, q_0) .$$

(19.24)

If the Jacobian of the first two of these equations with respect to (p_0, q_0) is not zero,

$$\frac{\partial(x_1, y_1)}{\partial(p_0, q_0)} \neq 0 ,$$

(19.25)

we can calculate (p_0, q_0) from these equations as functions of $z_0, z_1; x_0, y_0,$ x_1, y_1. By introducing these functions in (19.21), we obtain four functions $x, y; p, q$ of the variables $z_0, z_1; x_0, y_0, x_1, y_1$, and z; namely

$$x = x(z_0, z_1; x_0, y_0, x_1, y_1; z) ,$$
$$y = y(z_0, z_1; x_0, y_0, x_1, y_1; z) ,$$
$$p = p(z_0, z_1; x_0, y_0, x_1, y_1; z) ,$$
$$q = q(z_0, z_1; x_0, y_0, x_1, y_1; z) ,$$

(19.26)

which represent the light ray which goes through the point x_0, y_0, z_0 of the object plane, and the point x_1, y_1, z_1 of the image plane.

These functions now are introduced in the integral for the optical path

$$V = \int_{z_0}^{z_1} (\dot{x}p + \dot{y}q - H) \, dz \, , \qquad (19.27)$$

so that a function

$$V = V(z_0, z_1; x_0, y_0, x_1, y_1) \qquad (19.28)$$

is found which determines the optical distance between the points (x_0, y_0, z_0) and (x_1, y_1, z_1).

19.3 Our aim is to show that the total differential dV of this function has the form

$$dV = p_1 dx_1 + q_1 dy_1 - p_0 dx_0 - q_0 dy_0 - H_1 dz_1 + H_0 dz_0 \, , \qquad (19.31)$$

where the coefficients $p_1, q_1; p_0, q_0$ are functions of $(z_0, z_1; x_0, y_0, x_1, y_1)$. From these functions the desired "image functions" (19.24) can be found by elimination. By H_1 and H_0 we denote the expressions

$$H_1 = H(x_1, y_1, z_1; p_1, q_1) \, ,$$
$$\qquad (19.32)$$
$$H_0 = H(x_0, y_0, z_0; p_0, q_0) \, .$$

From (19.31) it follows that

$$p_0 = -\frac{\partial V}{\partial x_0} \, , \qquad p_1 = \frac{\partial V}{\partial x_1} \, , \qquad H_1 = -\frac{\partial V}{\partial z_1} \, ,$$
$$\qquad (19.33)$$
$$q_0 = -\frac{\partial V}{\partial y_0} \, , \qquad q_1 = \frac{\partial V}{\partial y_1} \, , \qquad H_0 = +\frac{\partial V}{\partial z_0} \, .$$

The first and second column of equations are equivalent to the equations (19.24) and demonstrate the fact that the four functions (19.24) can be found from one function $V(z_0, z_1; x_0, y_0, x_1, y_1)$ by differentiation and elimination.

The last column of (19.33) represents two partial differential equations for V which we obtain by introducing the partial derivatives of V for p_0, q_0, p_1, q_1 in H_0 and H_1. We find

$$\frac{\partial V}{\partial z_0} = H\left(x_0, y_0, z_0; -\frac{\partial V}{\partial x_0} \, , -\frac{\partial V}{\partial y_0}\right) \, ,$$
$$\qquad (19.34)$$
$$\frac{\partial V}{\partial z_1} = -H\left(x_1, y_1, z_1; \frac{\partial V}{\partial x_1} \, , \frac{\partial V}{\partial y_1}\right) \, .$$

19.4 The proof of the above theorem is not difficult. First we consider the integrand

$$\mathscr{L} = \dot{x}p + \dot{y}q - H \tag{19.41}$$

of (19.27). This becomes a function of $(z_0, z_1; x_0, y_0, x_1, y_1)$ and z if the functions (19.26) are introduced. Let σ denote any one of the parameters $(z_0, z_1; x_0, y_0, x_1, y_1)$. Hence

$$\frac{\partial \mathscr{L}}{\partial \sigma} = (\dot{x} - H_p)\frac{\partial p}{\partial \sigma} + (\dot{y} - H_q)\frac{\partial q}{\partial \sigma} + p\frac{\partial \dot{x}}{\partial \sigma} - H_x\frac{\partial x}{\partial \sigma}$$

$$+ q\frac{\partial \dot{y}}{\partial \sigma} - H_y\frac{\partial y}{\partial \sigma} \tag{19.42}$$

or, by the canonical equations (19.22):

$$\frac{\partial \mathscr{L}}{\partial \sigma} = p\frac{\partial \dot{x}}{\partial \sigma} + q\frac{\partial \dot{y}}{\partial \sigma} + \dot{p}\frac{\partial x}{\partial \sigma} + \dot{q}\frac{\partial y}{\partial \sigma} . \tag{19.43}$$

This, however, is equal to

$$\frac{\partial \mathscr{L}}{\partial \sigma} = \frac{d}{dz}\left(p\frac{\partial x}{\partial \sigma} + q\frac{\partial y}{\partial \sigma}\right), \tag{19.44}$$

and it follows that

$$\int_{z_0}^{z_1} \frac{\partial \mathscr{L}}{\partial \sigma}\, dz = \left(p\frac{\partial x}{\partial \sigma} + q\frac{\partial y}{\partial \sigma}\right)_{z=z_1} - \left(p\frac{\partial x}{\partial \sigma} + q\frac{\partial y}{\partial \sigma}\right)_{z=z_0} . \tag{19.45}$$

Let us now assume that σ is one of the parameters x_0, y_0, x_1, y_1. Then

$$\frac{\partial V}{\partial \sigma} = \int_{z_0}^{z_1} \frac{\partial \mathscr{L}}{\partial \sigma}\, dz , \tag{19.46}$$

and we conclude immediately from (19.45), and from the relations

$$x(z_1) = x_1 , \qquad\qquad x(z_0) = x_0 ,$$

$$y(z_1) = y_1 , \qquad\qquad y(z_0) = y_0 ,$$

the four relations

$$\frac{\partial V}{\partial x_1} = p_1 , \qquad \frac{\partial V}{\partial x_0} = -p_0 , \qquad \frac{\partial V}{\partial y_1} = q_1 , \qquad \frac{\partial V}{\partial y_0} = -q_0 . \tag{19.47}$$

If, however, $\sigma = z_1$, we have

$$\frac{\partial V}{\partial z_1} = \mathcal{L}(z_1) + \int_{z_0}^{z_1} \frac{\partial \mathcal{L}}{\partial z_1} \, dz$$

$$= -H_1 + p_1 \left(\dot{x} + \frac{\partial x}{\partial z_1} \right)_{z=z_1} + q_1 \left(\dot{y} + \frac{\partial y}{\partial z_1} \right)_{z=z_1}$$

$$- p_0 \left(\frac{\partial x}{\partial z_1} \right)_{z=z_0} - q_0 \left(\frac{\partial y}{\partial z_1} \right)_{z=z_0} .$$

From the identities

$$x_1 = x(z_0, z_1; x_0, y_0, x_1, y_1, z_1) ,$$

$$x_0 = x(z_0, z_1; x_0, y_0, x_1, y_1, z_0) ,$$

it follows that

$$0 = \left(\frac{\partial x}{\partial z_1} + \dot{x} \right)_{z=z_1} ; \qquad 0 = \left(\frac{\partial x}{\partial z_1} \right)_{z=z_0}$$

and similarly: $\quad 0 = \left(\frac{\partial y}{\partial z_1} + \dot{y} \right)_{z=z_1} ; \qquad 0 = \left(\frac{\partial y}{\partial z_1} \right)_{z=z_0} .$

Hence

$$\frac{\partial V}{\partial z_1} = -H_1 . \tag{19.48}$$

By analagous methods in case $\sigma = z_0$, we obtain from

$$\frac{\partial V}{\partial z_0} = -\mathcal{L}(z_0) + \int_{z_0}^{z_1} \frac{\partial \mathcal{L}}{\partial \sigma} \, dz , \tag{19.49}$$

the relation $\dfrac{\partial V}{\partial z_0} = +H_0$, which, together with (19.48) and (19.47), represents the hypothesis (19.33).

19.5 In the case of the optical Hamiltonian $H = -\sqrt{n^2 - p^2 - q^2}$, the function $V(z_0, z_1; x_0, y_0; x_1, y_1)$ is identical with Huyghens' wavelet function. The surfaces $V = $ const. are the spherical wave fronts around the object point x_0, y_0, z_0. In our present interpretation it measures the optical distance of two points, one of which is lying on the object plane, the other on the image plane. V is completely determined by the instrument and the position z_0, z_1 of the two planes. Hence it is called the <u>Characteristic</u> of the instrument; especially the <u>Point characteristic</u>, since it is a function of two points.

If the point characteristic V for a given pair of planes z_0 and z_1 is known, we can find the image functions (19.24) by elimination from the formulae

$$p_0 = - V_{x_0}(z_0,z_1; x_0,y_0,x_1,y_1) \, ,$$

$$p_1 = V_{x_1}(z_0,z_1; x_0,y_0,x_1,y_1) \, ,$$

$$q_0 = - V_{y_0}(z_0,z_1; x_0,y_0,x_1,y_1) \, ,$$ (19.51)

$$q_1 = V_{y_1}(z_0,z_1; x_0,y_0,x_1,y_1) \, .$$

Finally, from (19.34) it follows that V can be found as the solution of the two partial differential equations

$$V_{z_0} = - \sqrt{n^2 - V_{x_0}^2 - V_{y_0}^2} \, , \quad V_{z_1} = + \sqrt{n^2 - V_{x_1}^2 - V_{y_1}^2} \, . \quad (19.52)$$

§20. HAMILTON'S CHARACTERISTIC FUNCTIONS, W AND T.

20.1 We have obtained the point characteristic $V(z_0,z_1; x_0,y_0,x_1,y_1)$ by introducing the functions (19.26) in the integral (19.27). Let us now consider the image functions (19.24) again, and assume that the Jacobian of the last two equations with respect to p_0 and q_0 is not zero:

$$\frac{\partial(p_1, q_1)}{\partial(p_0, q_0)} \neq 0 \, . \quad (20.11)$$

Then we can calculate p_0 and q_0 from these two equations as functions of $(z_0,z_1; x_0,y_0,p_1,q_1)$ and introduce these functions in (19.21). We obtain four functions

$$x = x(z_0,z_1; x_0,y_0,p_1,q_1) \, ,$$

$$y = y(z_0,z_1; x_0,y_0,p_1,q_1) \, ,$$ (20.12)

$$p = p(z_0,z_1; x_0,y_0,p_1,q_1) \, ,$$

$$q = q(z_0,z_1; x_0,y_0,p_1,q_1) \, .$$

They represent the light ray from a point (x_0, y_0) of the object plane which intersects the image plane with the direction (p_1, q_1). If the functions (20.12) are introduced in the integral (19.27), we obtain V as the function

$$V = V(z_0, z_1; x_0, y_0, p_1, q_1) \tag{20.13}$$

From (19.31) it follows that

$$d(V - x_1 p_1 - y_1 q_1) = - p_0 dx_0 - q_0 dy_0 - x_1 dp_1 - y_1 dq_1$$

$$- H_1 dz_1 + H_0 dz_0 . \tag{20.14}$$

On the left side we replace x_1 and y_1 by

$$x_1 = x(z_0, z_1; x_0, y_0, p_1, q_1; z_1) ,$$

$$y_1 = y(z_0, z_1; x_0, y_0, p_1, q_1; z_1) , \tag{20.15}$$

and obtain a function

$$W = V - x_1 p_1 - y_1 q_1 , \tag{20.16}$$

of the six variables $z_0, z_1; x_0, y_0, p_1, q_1$, which has the differential

$$dW = - p_0 dx_0 - q_0 dy_0 - x_1 dp_1 - y_1 dq_1 - H_1 dz_1 + H_0 dz_0 . \tag{20.17}$$

It follows that

$$p_0 = - \frac{\partial W}{\partial x_0} , \qquad x_1 = - \frac{\partial W}{\partial p_1} , \qquad H_1 = - \frac{\partial W}{\partial z_1} ,$$

$$q_0 = - \frac{\partial W}{\partial y_0} , \qquad y_1 = - \frac{\partial W}{\partial q_0} , \qquad H_0 = + \frac{\partial W}{\partial z_0} . \tag{20.18}$$

By comparing these equations with (19.33) we find that only the second columns are formally different. We recognize immediately that these four equations

$$p_1 = \frac{\partial V}{\partial x_1} , \qquad x_1 = - \frac{\partial W}{\partial p_1} ,$$

$$q_1 = \frac{\partial V}{\partial y_1} , \qquad y_1 = - \frac{\partial W}{\partial q_1} , \tag{20.181}$$

together with (20.16), i.e.,

$$V - W = x_1 p_1 + y_1 q_1 , \tag{20.182}$$

represent a Legendre transformation and its inverse. Thus we may formulate the statement:

The function – W can be obtained from V by transforming V, and the variables x_1, y_1 with a Legendre transformation.

The last column of (20.18) allows us again to characterize W as a solution of two partial differential equations:

$$\frac{\partial W}{\partial z_1} = - H \left(- \frac{\partial W}{\partial p_1}, - \frac{\partial W}{\partial q_1}; p_1, q_1; z_1 \right),$$

$$\frac{\partial W}{\partial z_0} = H \left(x_0, y_0, - \frac{\partial W}{\partial x_0}, - \frac{\partial W}{\partial y_0}, z_0 \right).$$

(20.19)

In the special case of optics we obtain the two equations:

$$\frac{\partial W}{\partial z_1} = + \sqrt{n^2 \left(- \frac{\partial W}{\partial p_1}, - \frac{\partial W}{\partial q_1}, z_1 \right) - p_1^2 - q_1^2}$$

$$\frac{\partial W}{\partial z_0} = - \sqrt{n^2(x_0, y_0, z_0) - \left(\frac{\partial W}{\partial x_0} \right)^2 - \left(\frac{\partial W}{\partial y_0} \right)^2}$$

(20.191)

20.2 The function $W(z_0, z_1; x_0, y_0; p_1, q_1)$ is called Hamilton's mixed characteristic because it depends upon one point of the object plane, and on the direction (p_1, q_1) of the ray at the image plane. It is not difficult to give a geometric interpretation to this function. Let us consider a light ray

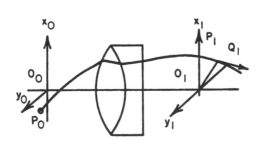

originating at a point P_0: (x_0, y_0) of the object plane. Let P_1: (x_1, y_1) be its intersection point with the image plane and (p_1, q_1) its direction at this point. We construct the tangent of the ray at P_1 and drop the perpendicular O_1Q_1 from the origin O_1 of the (x_1, y_1) plane onto this tangent. The optical length of the straight section P_1Q_1 of the tangent, measured with the index n_1 at P_1, then is given by

Figure 45

$$n_1(P_1Q_1) = - (x_1p_1 + y_1q_1).$$

(20.21)

The optical path between P_0 and P_1, on the other hand, is given by the function V. It follows that

The mixed characteristic $W = V - x_1p_1 - y_1q_1$ represents the optical length of the light ray from its origin (x_0, y_0, z_0) to the foot Q_1 of the perpendicular dropped from O_1 upon the tangent of the light ray at P_1. The optical length of the linear section P_1Q_1 has to be measured with the index n_1 at the point P_1.

20.3 The method described in 20.1 can obviously be varied in many ways and leads to other characteristics. The essential point in all these variations is that among the canonical variables $(x_0, y_0; p_0, q_0)$ and $(x_1, y_1; p_1, q_1)$ any two pairs may be selected, provided that one pair belongs to the object plane, and one pair to the image plane. By excluding the case that one pair consists of variables of different canonical type (for example (x_0, p_0)), we obtain four combinations, i.e., four characteristics:

Point characteristic: $V(z_0, z_1; x_0, y_0, x_1, y_1)$

Mixed characteristics:
$$\begin{cases} W(z_0, z_1; x_0, y_0, p_1, q_1) \\ \\ W^*(z_0, z_1; p_0, q_0, x_1, y_1) \end{cases}$$
                                                                            (20.31)

Angular characteristic: $T(z_0, z_1; p_0, q_0, p_1, q_1)$

All of these different functions have their use in practical optics. We have established the first two V and W. The third function W^* follows by considerations very similar to those in (20.1). The geometrical interpretation of W^* is that of the optical length of the ray from the foot Q_0 of the perpendicular O_0Q_0 dropped from O_0 onto the tangent of the ray at P_0 to the intersection P_1 with the image plane. The section Q_0P_0 of the tangent has to be measured with the index n_0 at P_0, i.e., by $n_0(x_0, y_0)$.

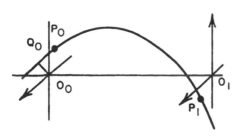

Figure 46

The function W^* is defined by

$$W^* = V + x_0p_0 + y_0q_0 ,\qquad (20.31)$$

and we obtain the relations

$$x_0 = \frac{\partial W^*}{\partial p_0} \qquad p_1 = \frac{\partial W^*}{\partial x_1} \qquad H_1 = -\frac{\partial W^*}{\partial z_1}$$

$$\qquad\qquad\qquad\qquad\qquad\qquad\qquad\qquad (20.32)$$

$$y_0 = \frac{\partial W^*}{\partial q_0} \qquad q_1 = \frac{\partial W^*}{\partial y_1} \qquad H_0 = \frac{\partial W^*}{\partial z_0}$$

It satisfies the partial differential equations

$$\frac{\partial W^*}{\partial z_1} = - H\left(x_1, y_1; \frac{\partial W^*}{\partial x_1}, \frac{\partial W^*}{\partial y_1}, z_1\right)$$

$$\frac{\partial W^*}{\partial z_0} = + H\left(\frac{\partial W^*}{\partial p_0}, \frac{\partial W^*}{\partial q_0}, p_0, q_0, z_0\right)$$

(20.33)

i.e., in the optical case:

$$W_{z_1}^* = \sqrt{n^2(x_1, y_1, z_1) - W_{x_1}^{*2} - W_{y_1}^{*2}}$$

$$W_{z_0}^* = -\sqrt{n^2(W_{p_0}^*, W_{q_0}^*, z_0) - p_0^2 - q_0^2}$$

(20.34)

In order to carry out the necessary eliminations, we have to assume that the Jacobian

$$\frac{\partial(x_1, y_1)}{\partial(x_0, y_0)} \neq 0$$

(20.35)

20.4 **The angular characteristic T.** In order to derive Hamilton's angular characteristic $T(z_0, z_1; p_0, q_0, p_1, q_1)$ we again consider the image functions (19.24) and use the last two equations to express (x_0, y_0) as functions of $(z_0, z_1; p_0, q_0, p_1, q_1)$. In order to be able to do this, we have to assume

$$\frac{\partial(p_1, q_1)}{\partial(x_0, y_0)} \neq 0 \ .$$

(20.41)

By introducing the resulting function x_0, y_0 in (19.21), we obtain

$$x = x(z_0, z_1; p_0, q_0, p_1, q_1; z)$$

$$y = y(z_0, z_1; p_0, q_0, p_1, q_1; z)$$

$$p = p(z_0, z_1; p_0, q_0, p_1, q_1; z)$$

$$q = q(z_0, z_1; p_0, q_0, p_1, q_1; z)$$

(20.42)

and these equations represent a light ray which leaves the object plane with the direction p_0, q_0 and intersects the image plane with the direction p_1, q_1. We introduce these functions in the integral (19.27) and obtain V as the function

$$V = V(z_0, z_1; p_0, q_0, p_1, q_1) \ .$$

(20.421)

From (19.31) it follows that

$$d(V - x_1p_1 - y_1q_1 + x_0p_0 + y_0q_0$$

$$= - x_1dp_1 - y_1dq_1 + x_0dp_0 + y_0dq_0 - H_1dz_1 + H_0dz_0 . \qquad (20.43)$$

By introducing the functions

$$x_0 = x(z_0,z_1; p_0,q_0,p_1,q_1; z_0) ,$$

$$y_0 = y(z_0,z_1; p_0,q_0,p_1,q_1; z_0) ,$$

$$x_1 = x(z_0,z_1; p_0,q_0,p_1,q_1; z_1) , \qquad (20.441)$$

$$y_1 = y(z_0,z_1; p_0,q_0,p_1,q_1; z_1) ,$$

in

$$T = V - x_1p_1 - y_1q_1 + x_0p_0 + y_0q_0 \qquad (20.44)$$

we obtain a function $T = T(z_0,z_1; p_0,q_0,p_1,q_1)$ whose differential, according to (24.43), is given by

$$dT = + x_0dp_0 + y_0dq_0 - x_1dp_1 - y_1dq_1 - H_1dz_1 + H_0dz_0 .$$

$$(20.45)$$

It follows that

$$x_0 = \frac{\partial T}{\partial p_0} , \qquad x_1 = - \frac{\partial T}{\partial p_1} , \qquad H_1 = - \frac{\partial T}{\partial z_1} ,$$

$$(20.451)$$

$$y_0 = \frac{\partial T}{\partial q_0} , \qquad y_1 = - \frac{\partial T}{\partial q_1} , \qquad H_0 = \frac{\partial T}{\partial z_0} .$$

The first two columns (20.451), and (19.33), and (20.44) show that V and $-$ T are related by the Legendre transformation

$$- p_0 = + \frac{\partial V}{\partial x_0} , \qquad p_1 = \frac{\partial V}{\partial x_1} , \qquad x_0 = \frac{\partial T}{\partial p_0} , \qquad x_1 = - \frac{\partial T}{\partial p_1} ,$$

$$(20.46)$$

$$- q_0 = \frac{\partial V}{\partial y_0} , \qquad q_1 = \frac{\partial V}{\partial y_1} , \qquad y_0 = \frac{\partial T}{\partial q_0} , \qquad y_1 = - \frac{\partial T}{\partial q_1} .$$

$$V - T = x_1p_1 + y_1q_1 - x_0p_0 - y_0q_0 ,$$

with respect to the variables (x_0,y_0,x_1,y_1) and $(- p_0, - q_0, p_1, q_1)$, respectively.

The last column (20.451) yields two partial differential equations for T:

$$\frac{\partial T}{\partial z_1} = - H \left(- \frac{\partial T}{\partial p_1}, - \frac{\partial T}{\partial q_1}, p_1, q_1; z_1 \right),$$

$$\frac{\partial T}{\partial z_0} = H \left(\frac{\partial T}{\partial p_0}, \frac{\partial T}{\partial q_0}, p_0, q_0; z_0 \right),$$

(20.47)

which, in the case

$$H(x,y,p,q; z) = - \sqrt{n^2(x,y,z) - p^2 - q^2},$$

becomes

$$\frac{\partial T}{\partial z_1} = \sqrt{n^2(- T_{p_1}, - T_{q_1}, z_1) - p_1^2 - q_1^2},$$

$$\frac{\partial T}{\partial z_0} = - \sqrt{n^2(T_{p_0}, T_{q_0}, z_0) - p_0^2 - q_0^2}.$$

(20.48)

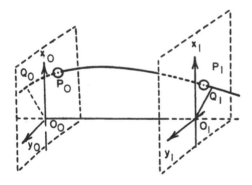

Figure 47

For the geometric interpretation of T, let us construct the tangents of the light ray at P_0 and P_1, and drop the perpendiculars from O_0 and O_1 to these tangents. Let Q_0 and Q_1 be the foot points of these perpendiculars. Then

$$T = V + n_0(Q_0 P_0) + n_1(P_1 Q_1),$$

(20.49)

which is the optical distance from Q_0 to Q_1 if the straight sections on the tangents are measured with the indices n_0 and n_1 at the points P_0, P_1 respectively.

20.5 The special significance of the angular characteristic T rests upon the fact that the dependance on the variables z_0 and z_1 is linear in practical cases. Let us assume that the medium is homogeneous outside a given domain of the space. For example:

$$n = n_0 = \text{constant for } z < l_0,$$

$$n = n_1 = \text{constant for } z > l_1.$$

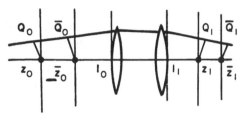

Figure 48

Obviously this is the case in almost all optical instruments. We assume the planes z_0 and z_1 in these homogeneous parts of the space, i.e., $z_0 < l_0$; $z_1 > l_1$.

Let now $T = T(z_0, z_1; p_0, q_0, p_1, q_1)$ be the angular characteristic for the planes z_0 and z_1, i.e., the optical distance between the foot points Q_0 and Q_1.

For another pair of reference planes $\bar{z}_0$ and $\bar{z}_1$ we have an angular characteristic $\bar{T}$, and the relation between both is given by

$$\bar{T} = T - n_0(Q_0\bar{Q}_0) + n_1(Q_1\bar{Q}_1) . \tag{20.51}$$

However, we have

$$Q_0\bar{Q}_0 = (\bar{z}_0 - z_0)\sqrt{1 - \frac{1}{n_0^2}(p_0^2 + q_0^2)} ,$$

$$Q_1\bar{Q}_1 = (\bar{z}_1 - z_1)\sqrt{1 - \frac{1}{n_1^2}(p_1^2 + q_1^2)} . \tag{20.52}$$

Hence it follows that

$$\bar{T} = T - (\bar{z}_0 - z_0)\sqrt{n_0^2 - p_0^2 - q_0^2} + (\bar{z}_1 - z_1)\sqrt{n_1^2 - p_1^2 - q_1^2} \tag{20.53}$$

We can express this result as follows: The function

$$T + z_0\sqrt{n_0^2 - p_0^2 - q_0^2} - z_1\sqrt{n_1^2 - p_1^2 - q_1^2} = T_0 \tag{20.54}$$

is independent of z_0 and z_1 i.e., of the choice of the reference planes. In other words: T is a linear function of z_0 and z_1:

$$T = T_0 - z_0\sqrt{n_0^2 - p_0^2 - q_0^2} + z_1\sqrt{n_1^2 - p_1^2 - q_1^2} \tag{20.55}$$

provided that both "object space" and "image space" are homogeneous.

We can expect that the function T_0 itself will be of significance for finding properties of the optical instrument and its image performance which apply to any choice of object the image plane.

§21. INTEGRAL INVARIANTS.

21.1 Fields of light rays. A set of wave fronts $\psi(x,y,z)$ = a constant defines, by its orthogonal trajectories, a two-parameter manifold of light rays. In general, we call a two-parameter set of light rays a congruence of rays.

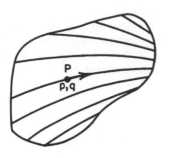

Figure 49

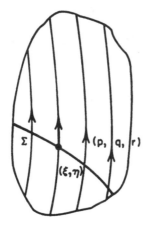

Figure 50

The special congruences which are formed by the orthogonal trajectories of a set of wave fronts are known as <u>normal congruences</u>. If a domain D is covered by the rays of a normal congruence such that, at every point P of D, we have one and only one ray through P then we say that the light rays form a field in D. The quantities p, q, and $\sqrt{n^2 - p^2 - q^2}$ = r determine a vector <u>field in D</u>. (p,q,r) is called the field vector.

Let us derive the condition which determines a normal congruence of light rays. We can characterize a congruence of rays as follows: We consider a surface Σ in space given in parametric form $X = X(\xi,\eta)$; or explicitly:

$$x = f(\xi,\eta) \, ,$$

$$y = g(\xi,\eta) \, , \qquad (21.11)$$

$$z = h(\xi,\eta) \, .$$

Through any point (ξ,η) of this surface we construct a light ray such that its optical direction cosines

$$p, q, \; r = \sqrt{n^2 - p^2 - q^2}$$

are determined by given functions

$$p = p(\xi,\eta), \; q = q(\xi,\eta) \, .$$

These rays can be found in the same manner as in §9 as the solutions, $X(\xi,\eta,\tau)$, of the differential equations

$$\ddot{x} = nn_x$$

$$\ddot{y} = nn_y \quad \text{where } \ddot{x} = \frac{d^2x}{d\tau^2}, \text{ etc.} \qquad (21.12)$$

$$\ddot{z} = nn_z$$

which satisfy the boundary conditions

$$x(\xi,\eta,0) = f(\xi,\eta) , \qquad \dot{x}(\xi,\eta,0) = p(\xi,\eta) ,$$

$$y(\xi,\eta,0) = g(\xi,\eta) , \qquad \dot{y}(\xi,\eta,0) = q(\xi,\eta) , \qquad (21.13)$$

$$z(\xi,\eta,0) = h(\xi,\eta) , \qquad \dot{z}(\xi,\eta,0) = r(\xi,\eta) .$$

We make the assumption that the Jacobian

$$\frac{\partial(x,y,z)}{\partial(\xi,\eta,\tau)} \neq 0$$

in a neighborhood of Σ; all the following considerations refer only to such domains D which lie in this neighborhood.

Let us now assume that the rays $X = X(\xi,\eta,\tau)$ form a field in the neighborhood of Σ. Then there exists a set of surfaces Γ to which the rays are normal. We consider one of these surfaces Γ_0. We can, with the method of §9.4, construct a function $\psi(x,y,z)$ such that $\psi = 0$ on Γ_0 and such that ψ satisfies the differential equation

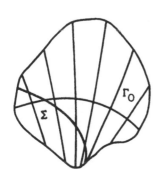

$$\psi_x^2 + \psi_y^2 + \psi_z^2 = n^2 .$$

The surfaces $\psi = $ a constant are normal to the light rays, and we have, at any point (x,y,z), the relations

$$\dot{X} = \text{grad } \psi . \qquad (21.14)$$

Figure 51   On Σ the function ψ assumes the values

$$F(\xi,\eta) = \psi(f,g,h) ,$$

and has the derivatives

$$F_\xi = \psi_x f_\xi + \psi_y g_\xi + \psi_z h_\xi = p f_\xi + q g_\xi + r h_\xi ,$$

$$F_\eta = \psi_x f_\eta + \psi_y g_\eta + \psi_z h_\eta = p f_\eta + q g_\eta + r h_\eta . \qquad (21.15)$$

These last equations demonstrate that

$$dF = pdf + qdg + rdh \qquad (21.16)$$

must be a total differential on Σ .

On the other hand, if on a given surface Σ the differential expression pdf + qdg + rdh is a total differential, $dF = F_\xi \, d\xi + F_\eta \, d\eta$, then the congruence of light rays satisfies the conditions

$$F_\xi = pf_\xi + qg_\xi + rh_\xi \, ,$$

$$F_\eta = pf_\eta + qg_\eta + rh_\eta \, ,$$

(21.17)

on Σ, and by §9.3 we obtain a solution of $\psi_x^2 + \psi_y^2 + \psi_z^2 = n^2$ in the form

$$\psi = F(\xi,\eta) + \int_0^\tau n^2 \, d\tau \, ,$$

such that the surfaces ψ = a constant are intersected at right angles by the rays of the congruence. Hence the congruence is normal.

We can formulate the statement: A congruence of light rays forms a field in D if, and only if, the expression pdf + qdg + rdh is a total differential on an arbitrary surface x = f(ξ,η), y = g(ξ,η), z = h(ξ,η) in D.

From the above considerations it follows that the normality of a congruence of rays is insured if the differential pdf + qdg + rdh is total on one particular surface Σ_0 in D.

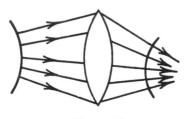

Figure 52

This gives the theorem of Malus:

If a normal congruence of straight lines is submitted to an arbitrary number of refractions then the final congruence is still normal.

Indeed, this follows from the preceding remark if we assume that object and image space are homogeneous, i.e., the light rays are straight lines. We remark that Malus' theorem is true also in case of reflections or combinations of refractions and reflections.

We next consider a closed curve C: $X = X(s) = \big(x(s), y(x), z(s)\big)$ on a surface Σ. Since pdx + qdy + rdz is total, it follows that the line integral

$$\int_C p \, dx + q \, dy + r \, dz = 0 \, .$$

Since Σ is arbitrary, this yields: The line integral

$$\int_C p \, dx + q \, dy + r \, dz$$

(21.18)

is zero for any closed curve C which lies in the field D.

The quantities (p,q,r) at a point (x,y,z) are the optical direction cosines of the field ray which goes through this point. If the closed curve C is given in the form $X = X(s)$, where s is the length on the curve, we can write (21.8) in the form

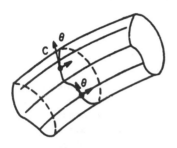

$$\int_C n \cos \theta \; ds = 0 , \qquad (21.181)$$

θ being the angle between the tangent of the curve and the direction of the field ray at this point.

Figure 53

Finally, in (21.18) we may introduce the expression $r = \sqrt{n^2 - p^2 - q^2} = -H$, and consider a curve C which connects two points P_0 and P_1 in the field. It follows that the integral

$$V = \int_{P_0}^{P_1} (p \; dx + q \; dy - H \; dz) \qquad (21.19)$$

is independent of the path between P_0 and P_1, and thus determines a function $V(P_0, P_1)$ of the end points alone. In order to determine this function $V(P_0, P_1)$ let us choose a special path between P_0 and P_1 which consists of a section $P_0 P$ of the field ray through P_0 such that P lies on the surface $\psi = \psi(P_1)$ through P_1. The section PP_1 may be an arbitrary curve on the surface $\psi = \psi(P_1)$. The expression (21.181) shows that the integral on PP_1 is zero and that on $P_0 P$ it determines the optical length $\psi(P) - \psi(P_0)$ of this section. Hence the result

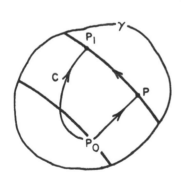

$$\psi(P_1) - \psi(P_0)$$

Figure 54

$$= \int_{P_0}^{P_1} (p \; dx + q \; dy - H \; dz) , \qquad (21.191)$$

is independent of the curve which connects P_0 and P_1.

This integral is nothing but Hilbert's invariant integral determined for our special optical problem of variation. It allows us to find the wave fronts $\psi(x,y,z)$ which belong to a given field of light rays.

Let us for example consider the normal congruence of rays which go through a given point (x_0, y_0, z_0). The problem is to determine the corresponding wave fronts

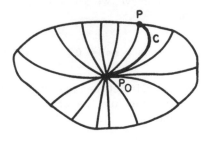

Figure 55

$$V(x_0, y_0, z_0, x, y, z) = ct ,$$

i.e., Huyghens' wavelets.

By our former method in §9.5 we have to determine the light ray through P_0 and a point $P = (x, y, z)$, and then calculate the integral (9.55) along this light ray. By our result (21.191) we can avoid the determination of the light ray, but we are permitted to integrate (21.191) over an arbitrary curve between P_0 and P:

$$V(x_0, y_0, z_0; x, y, z) = \int_{P_0}^{P} (p \ dx + q \ dy - H \ dz) . \qquad (21.192)$$

21.2 In the case of a congruence of rays which is not normal the above results are of course not valid. It is however possible to derive a number of interesting results about integrals on closed curves. We derive these results for solutions of canonical equations in general.

Let us assume that a surface Σ is given in the form

$$z_0 = f(x_0, y_0) . \qquad (21.21)$$

Through the points of this surface we construct a two parameter manifold of curves such that their direction on Σ is given by two arbitrarily chosen functions

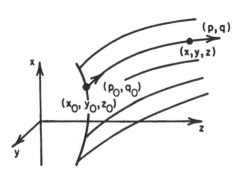

Figure 56

$$p_0 = p_0(x_0, y_0) ,$$
$$q_0 = q_0(x_0, y_0) . \qquad (21.22)$$

The curves and their directions are represented by the solutions $x(z)$, $y(z)$; $p(z)$, $q(z)$ of the canonical equations

$$\dot{x} = H_p , \quad \dot{p} = -H_x ,$$
$$\dot{y} = H_q , \quad \dot{q} = -H_y , \qquad (21.23)$$

which satisfy the above boundary conditions.

We obtain four functions

$$x = x(x_0, y_0; z) \ ,$$

$$y = y(x_0, y_0; z) \ ,$$

$$p = p(x_0, y_0; z) \ , \qquad (21.231)$$

$$q = q(x_0, y_0, z) \ ,$$

which determine a congruence of curves.

We consider a domain D of the space such that through every point P of D goes one and only one ray. We thus assume in D that the Jacobian

$$\frac{\partial(x, y)}{\partial(x_0, y_0)} \neq 0$$

so that we can relate to every point (x,y,z) a corresponding point $\left(x_0, y_0, z_0 = f(x_0, y_0) \right)$ on the surface Σ with the aid of the first two equations (21.231). The optical distance of a point (x,y,z) from its corresponding point (x_0, y_0, z_0) is given by Hamilton's point characteristic $V(x_0, y_0, z_0, x, y, z)$. By introducing in this function

$$x_0 = x_0(x, y, z) \ ,$$

$$y_0 = y_0(x, y, z) \ , \qquad (21.232)$$

$$z_0 = f\left(x_0(x, y, z), y_0(x, y, z) \right) = z_0(x, y, z) \ ,$$

from (21.231) and (21.21), we obtain a function of x,y,z alone, namely,

$$V^*(x, y, z) = V\left(x_0(x, y, z), y_0(x, y, z), z_0(x, y, z); x, y, z \right) \ .$$

The differential of this function can be found with the aid of

$$dV^* = -\, p_0 dx_0 \, - \, q_0 dy_0 \, + \, H_0 dz_0 \, + \, pdx + qdy \, - \, H \, dz \ . \quad (21.24)$$

In this formula we have to interpret p_0, q_0, H_0 as functions of (x,y,z) which can be obtained by introducing (21.232) in (21.22) and in $H_0 = H(x_0, y_0, p_0, q_0, z_0)$. The differentials dx_0, dy_0, dz_0 are the differentials of the functions (21.232) i.e., linear combinations of dx,dy,dz.

We integrate the equation (21.24) over an arbitrary closed curve C in D. It follows that

$$0 = -\, \int_{C_0} (p_0 dx_0 + q_0 dy_0 - H_0 dz_0) + \int_{C} (pdx + qdy - Hdz) \ , \quad (21.25)$$

where C_0 is the curve on Σ which is formed by the corresponding points to the points of C. The equation (21.25) demonstrates that the integral,

$$J = \int_C p \, dx + q \, dy - H \, dz \qquad (21.251)$$

has the same value for all closed curves C which have the same corresponding curve C_0 on Σ. Let us now consider a "tube" of light rays, i.e., the surface which is formed by the rays through a closed curve C_0 on Σ. All closed curves C which go around the cylindrical wall of this tube have C_0 as a corresponding curve. Hence it follows: The integral

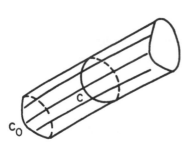

Figure 57

$$J = \int_C (p \, dx + q \, dy - H \, dz)$$

has the same value for all closed curves C around a given tube of rays. It is an invariant of the tube.†

We can easily obtain the results of 21.1 even in a more general form. Indeed, if on Σ the functions p_0, q_0, and $z_0 = f(x_0, y_0)$ are such that the expression $p_0 \, dx_0 + q_0 \, dy_0 - H_0 \, dz_0$ is a total differential, then J has the value zero for all curves C.

Congruences of rays of this type are called transversal; or, the rays form a field. It is possible in this case to construct a set of surfaces $\psi(x,y,z) = $ a constant, such that every surface is intersected in the "transversal" direction p,q. This direction is defined by those p,q on the surface, for which

$$p \, dx + q \, dy - H \, dz = 0 \qquad (21.26)$$

for all increments dx, dy, dz on the surface $\psi = 0$, i.e., for all dx, dy, dz which satisfy the condition

$$\psi_x \, dx + \psi_y \, dy + \psi_z \, dz = 0 . \qquad (21.261)$$

We obtain two equations for p,q:

$$p \, \psi_z + H \, \psi_x = 0 , \qquad q \, \psi_z + H \, \psi_y = 0 . \qquad (21.262)$$

---

† Poincaré's invariant

In case $H = -\sqrt{n^2 - p^2 - q^2}$, we find readily

$$p = n\, \frac{\psi_x}{\sqrt{\psi_x^2 + \psi_y^2 + \psi_z^2}}$$

$$q = n\, \frac{\psi_y}{\sqrt{\psi_x^2 + \psi_y^2 + \psi_z^2}}$$

(21.263)

as a solution of (20.262). This shows that the transversal direction has the direction of the surface normal in the optical case.

We finally mention an interesting geometrical interpretation of the integral invariant (21.251) which was given by Prange. Let us consider the case of optics where, as we have seen above, the integral (21.251) can be written in the form

$$J = \int_C n \cos \theta \, ds \ ,$$

(21.27)

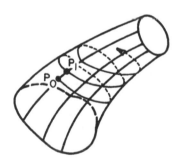

Figure 58

θ being the angle between curve tangent and light ray. We choose a curve C which intersects the rays of the wall of the tube at right angles. We can in general not expect that this curve is closed but will reach the light ray through the starting point P_0 at a point P_1 different from P_0. We close the curve, however, by the section $P_0 P_1$ on the ray, on which $\cos \theta = -1$. It follows that

$$J = -\int_{P_0}^{P_1} n \, ds \ .$$

(21.28)

Since J is invariant, we find that the optical path between P_0 and P_1 is the same wherever the beginning, P_0, of such a curve lies. By continuing the curve C, we obtain a spiral around the tube which divides every light ray in parts of the same optical length. The spiral becomes a closed curve in case of a normal congruence of rays.

We also may consider a spiral which intersects the rays of the tube at an arbitrary but fixed angle θ_0. Let C be the curve which consists of one turn $P_0 P_1$ of the spiral and the section $P_1 P_0$ on the light ray. From (21.27) it follows that

$$\cos \theta \oint_{P_0}^{P_1} n \, ds - \int_{P_0}^{P_1} n \, ds = J \ .$$

(21.29)

Let L be the optical length of one turn of the spiral and ℓ the optical length of the width of the "thread" on the ray through P_0. It follows that

$$L \cos \theta_0 = J + \ell. \tag{21.291}$$

If we consider m turns of the spiral and denote by L the total length of the m turns and by ℓ the length of the thread, we find

$$L \cos \theta_0 = mJ + \ell. \tag{21.292}$$

In case of J = 0, i.e., if the rays of the tube belong to a normal congruence, we obtain the interesting relation

$$L \cos \theta_0 = \ell. \tag{21.293}$$

We observe that the above results refer actually not to congruences of rays, but to any one parameter manifold of rays through a closed curve in space. If the medium is homogeneous (n = 1) then L and ℓ are geometrical lengths and our results refer to tubes whose walls are formed by a set of straight lines through a closed curve.

§22. EXAMPLES.

22.1 <u>Mixed characteristics for stratified media</u>. Let us consider the case of a stratified medium in which n = n(z). The optical designer frequently has to deal with this case. If an optical instrument contains a number of 45° prisms, it is possible to replace it by a fictitious instrument which contains a corresponding number of parallel plates lined up on the same axis. The thickness of the plates must be equal to the path along the axis of the instrument which originally is broken into several reflections. The mixed characteristic $W(z_0,z_1; x_0,y_0,p_1,q_1)$ satisfies the partial differential equations (20.191). The second of these equations becomes, in our case,

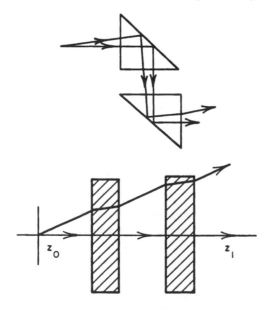

$$\frac{\partial W}{\partial z_0} = -\sqrt{n^2(z_0) - W_{x_0}^2 - W_{y_0}^2}. \tag{22.11}$$

Figure 59

We can determine W as a solution of this equation which satisfies the boundary condition

$$W = - (x_0 p_1 + y_0 q_1) \tag{22.12}$$

at $z_0 = z_1$.

Indeed, in case $z_0 = z_1$ we have $p_0 = p_1$, $q_0 = q_1$, and $x_1 = x_0$, $y_1 = y_0$; hence by (20.18):

$$\frac{\partial W}{\partial x_0} = - p_1 , \qquad \frac{\partial W}{\partial p_1} = - x_0 ,$$

$$\frac{\partial W}{\partial y_0} = - q_1 , \qquad \frac{\partial W}{\partial q_1} = - y_0 ,$$

which yields (22.12).

A solution $W(z_0, z_1; x_0, y_0, p_1, q_1)$ with these boundary values can be found easily. We introduce a function of the type

$$W = a(z_0) - x_0 p_1 - y_0 q_1 \tag{22.13}$$

in (22.11) and find the condition

$$a'(z_0) = - \sqrt{n^2(z_0) - p_1^2 - q_1^2} , \tag{22.14}$$

i.e.,

$$a(z_0) = - \int_{z_1}^{z_0} \sqrt{n^2(z) - p_1^2 - q_1^2} \, dz + a(z_1) . \tag{22.15}$$

The boundary condition (22.12) requires $a(z_1) = 0$; hence

$$W = \int_{z_0}^{z_1} \sqrt{n^2(z) - p_1^2 - q_1^2} \, dz - x_0 p_1 - y_0 q_1 \tag{22.16}$$

is the desired solution.

The image functions of our instrument are determined by the general equations (20.18). The result is:

$$x_1 = x_0 + p_1 \int_{z_0}^{z_1} \frac{dz}{\sqrt{n^2(z) - p_1^2 - q_1^2}} , \quad p_0 = p_1$$

$$y_1 = y_0 + q_1 \int_{z_0}^{z_1} \frac{dz}{\sqrt{n^2(z) - p_1^2 - q_1^2}} , \quad q_0 = q_1$$
$$\tag{22.17}$$

The equation (22.16) is also valid in case n is discontinuous. For example, in the case of a system of parallel plates of thickness ℓ_i and index n_i we find

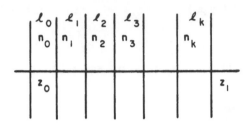

Figure 60

$$W = \sum_{i=0}^{k} \ell_i \sqrt{n_i^2 - p_i^2 - q_i^2}$$

$$- (x_0 p_1 + y_0 q_1) . \quad (22.18)$$

This last result makes it obvious that the optical coordination of rays is completely independent of the order in which the plates are arranged.

We apply our result to the following problem. Let us consider a stratified medium such that $n(z)$ is different from 1 only in the region $L_0 \leq z \leq L_1$. Outside the region we have $n(z) = 1$.

We consider all the light rays which originate at the point $x_0 = y_0 = 0$ of the plane z_0 and determine the corresponding rays after passing the region $L_0 \leq z \leq L_1$. The equations (22.17) demonstrate that the refracted ray is parallel to the incident ray: $p_1 = p_0; q_1 = q_0$, and that

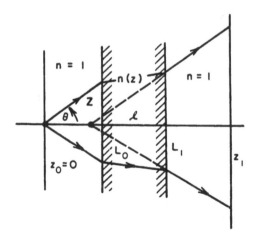

Figure 61

its intersection with the plane $z = z_1$, is given by

$$x_1 = p_0 \left[\frac{L_0 - z_0 + z_1 - L_1}{\sqrt{1 - p_0^2 - q_0^2}} + \int_{L_0}^{L_1} \frac{dz}{\sqrt{n^2 - p_0^2 - q_0^2}} \right] ,$$

$$(22.19)$$

$$y_1 = q_0 \left[\frac{L_0 - z_0 + z_1 - L_1}{\sqrt{1 - p_0^2 - q_0^2}} + \int_{L_0}^{L_1} \frac{dz}{\sqrt{n^2 - p_0^2 - q_0^2}} \right] .$$

The ray in the region $z > L_1$ is clearly a straight line and its equation is:

$$x = p_0 \left[\frac{L_0 - z_0 + z - L_1}{\sqrt{1 - p_0^2 - q_0^2}} + \int_{L_0}^{L_1} \frac{dz}{\sqrt{n^2 - p_0^2 - q_0^2}} \right] ,$$

$$\tag{22.191}$$

$$y = q_0 \left[\frac{L_0 - z_0 + z - L_1}{\sqrt{1 - p_0^2 - q_0^2}} + \int_{L_0}^{L_1} \frac{dz}{\sqrt{n^2 - p_0^2 - q_0^2}} \right] .$$

We wish to investigate whether this two-parameter bundle of straight lines still has a common point of intersection or, if not, how much it departs from such a bundle. From (22.191) it follows that all rays which include the same angle θ with the z-axis intersect each other at a point Z of the z-axis. Introducing

$$\sqrt{1 - p_0^2 - q_0^2} = \cos \theta ,$$

we obtain

$$Z = z_0 + L_1 - L_0 - \cos \theta \int_{L_0}^{L_1} \frac{dz}{\sqrt{n^2 - \sin^2 \theta}} . \tag{22.192}$$

Since Z is a non-constant function of θ, we see that the refracted bundle of rays has no common point of intersection but is of the type illustrated in Figure 62. The intersection point Z_0 in the case where $\theta \to 0$ is given by

Figure 62

$$Z_0 = z_0 + L_1 - L_0 - \int_{L_0}^{L_1} \frac{dz}{n} . \tag{22.193}$$

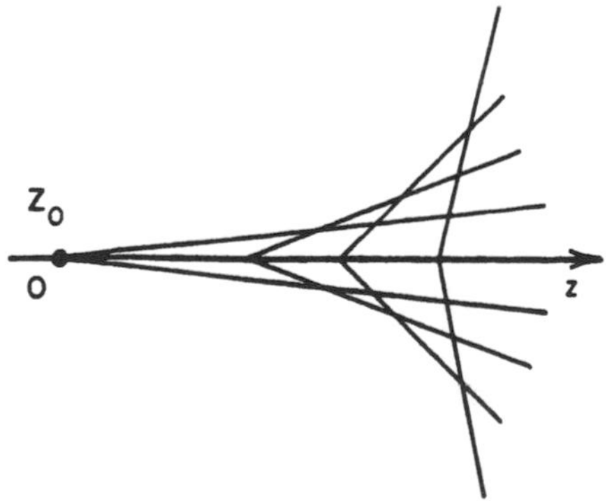

The difference $Z - Z_0$ is called the Spherical Aberration of the bundle. We find

$$\Delta Z = Z - Z_0 = \int_{L_0}^{L_1} \left(\frac{1}{n(z)} - \frac{\cos \theta}{\sqrt{n^2(z) - \sin^2 \theta}} \right) dz . \tag{22.194}$$

This is a monotonic function of θ which increases from 0 to $\int_{L_0}^{L_1} \dfrac{dz}{n}$ if θ goes from 0 to $\pi/2$. We notice that it is independent of the position z_0 of the object point.

In case of a single plane parallel plate of constant n and of thickness $\ell = L_1 - L_0$, we find

$$\Delta Z = Z - Z_0 = \ell \left(\frac{1}{n} - \frac{\cos \theta}{\sqrt{n^2 - \sin^2 \theta}} \right) \qquad (22.195)$$

as the expression for the spherical aberration.

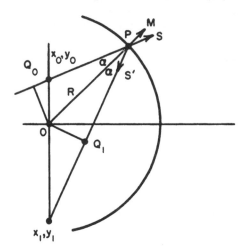

Figure 63

22.2 The angular character-istic in the case of a spherical mirror. We determine next the angular characteristic of a concave spherical mirror. First, let us assume that the two reference planes $z = z_0$, $z = z_1$ coincide at the center 0 of the mirror, i.e., $z_0 = z_1 = 0$. Let T_0 be the angular characteristic for this choice of reference planes. We consider a light ray, from a point x_0, y_0 of this plane, having a direction

$$S_0 = \left(p_0, q_0, \sqrt{1 - p_0^2 - q_0^2} \right) . \qquad (22.21)$$

It reaches the sphere at the point P and is reflected in the direction

$$S_1 = \left(p_1, q_1, -\sqrt{1 - p_1^2 - q_1^2} \right) . \qquad (22.211)$$

We have found in (14.28) that S_1 and S_0 are related by the equation

$$S_1 = S_0 + \Gamma M , \qquad (22.212)$$

where M is the normal unit vector at P, and Γ the quantity

$$\Gamma = - 2 \cos \alpha = - 2(S_0 \cdot M) . \qquad (22.213)$$

The angular characteristic T_0 is given by the optical distance between the base points Q_0 and Q_1 of the perpendiculars dropped from 0 to the ray before and after reflection. We find

$$T_0 = 2(Q_0 P) = 2 R(S_0 \cdot M) = - R \Gamma \qquad (22.22)$$

where R is the radius of the sphere. From (22.212) it follows that

$$S_0 \cdot S_1 = 1 + \Gamma(S_0 \cdot M) = 1 - \frac{1}{2} \Gamma^2 \ ,$$

and hence

$$T_0 = R \sqrt{2} \sqrt{1 - S_0 \cdot S_1} \ . \qquad (22.23)$$

This is the desired expression. With the aid of (22.21) and (22.211) it follows that

$$T_0 = R \sqrt{2} \sqrt{1 + \sqrt{(1 - p_0^2 - q_0^2)(1 - p_1^2 - q_1^2)} - (p_0 p_1 + q_0 q_1)}. \qquad (22.24)$$

We see that T_0 depends only on the three combinations

$$u = p_0^2 + q_0^2 \ ,$$

$$v = p_1^2 + q_1^2 \ , \qquad (22.241)$$

$$w = 2(p_0 p_1 + q_0 q_1) \ ,$$

which, incidentally, is the case in any optical instrument of rotational symmetry. With this notation, we have

$$T_0 = R \sqrt{2} \sqrt{1 - \frac{1}{2} w + \sqrt{(1 - u)(1 - v)}} \ . \qquad (22.25)$$

In order to find the characteristic for any pair of reference planes z_0 and z_1, we apply the same method as in (21.5). From Figure 64 it follows that

$$T = T_0 - (Q_0 \overline{Q}_0) - (Q_1 \overline{Q}_1) \ ,$$

or explicitly,

$$T = T_0 - z_0 \sqrt{1 - u} - z_1 \sqrt{1 - v} \ . \qquad (22.26)$$

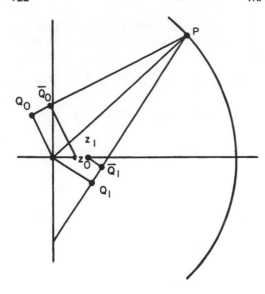

Figure 64

The intersections (x_0, y_0), (x_1, y_1) with the planes z_0, z_1, respectively, are given by the formulae

$$x_0 = 2(T_u p_0 + T_w p_1) ,$$

$$x_1 = -2(T_w p_0 + T_v p_1) ,$$

$$\qquad\qquad\qquad\qquad (22.27)$$

$$y_0 = 2(T_u q_0 + T_w q_1) ,$$

$$y_1 = -2(T_w q_0 + T_v q_1) .$$

On the other hand, these equations represent the coordination of the incident ray (x_0, y_0, p_0, q_0) and the reflected ray (x_1, y_1, p_1, q_1).

We investigate in detail the case of small mirrors where we are allowed to develop T in a power series with respect to u,v,w and neglect all terms of higher than first order.

We find from (22.25) and (22.26):

$$T = B_0 + B_1 u + B_2 w + B_3 v \qquad\qquad (22.28)$$

where

$$B_0 = 2R - z_0 - z_1 ,$$

$$B_1 = \frac{1}{2} z_0 - \frac{1}{4} R ,$$

$$\qquad\qquad\qquad\qquad (22.281)$$

$$B_2 = \qquad - \frac{1}{4} R ,$$

$$B_3 = \frac{1}{2} z_1 - \frac{1}{4} R .$$

In practice, it is customary to measure the position of the reference planes (z_0, z_1) from the vertex of the mirror and to introduce

$$a_0 = z_0 - R ,$$

$$\qquad\qquad\qquad\qquad (22.282)$$

$$a_1 = z_1 - R .$$

The coefficients (22.281) become

$$B_0 = - (a_0 + a_1) ,$$

$$B_1 = \frac{a_0}{2} + \frac{R}{4} ,$$

$$B_2 = - \frac{R}{4} ,$$

$$B_3 = \frac{a_1}{2} + \frac{R}{4} .$$

(22.283)

For this first-order approximation, we obtain from (22.27) the linear image functions

$$x_0 = 2(B_1 p_0 + B_2 p_1) , \qquad x_1 = - 2(B_2 p_0 + B_3 p_1) ,$$

$$y_0 = 2(B_1 q_0 + B_2 q_1) , \qquad y_1 = - 2(B_2 q_0 + B_3 q_1) .$$

(22.29)

We can use these equations to express x_1, y_1, p_1, q_1 as linear functions of (x_0, y_0, p_0, q_0). It follows that

$$x_1 = - \frac{B_3}{B_2} x_0 + \frac{2}{B_2} (B_1 B_3 - B_2^2) p_0 ,$$

$$p_1 = \frac{1}{2 B_2} x_0 - \frac{B_1}{B_2} p_0 ,$$

(22.291)

and similar equations for y_1, q_1 as functions of (y_0, q_0).

These equations show that x_1 and y_1 are independent of the directions (p_0, q_0) of the incident rays from a point (x_0, y_0) if

$$B_1 B_3 - B_2^2 = 0 .$$

(22.292)

In this case all rays from any point (x_0, y_0) of the object plane intersect at the corresponding point (x_1, y_1) of the image plane. Planes of this relationship are called conjugate planes of first-order optics. By introducing the expressions (22.283) in (22.292), we find a condition for conjugate planes:

$$\left(a_0 + \frac{1}{2} R\right) \left(a_1 + \frac{1}{2} R\right) = \frac{R^2}{4}$$

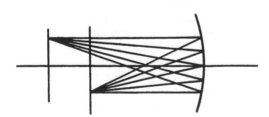

Figure 65

or

$$\frac{1}{a_0} + \frac{1}{a_1} = -\frac{2}{R} .$$

(22.293)

This equation is called the <u>mirror equation</u>.

Finally, we introduce the notation

$$f = -2 B_2 , \qquad \text{(focal length)}$$

(22.294)

$$M = -\frac{B_3}{B_2} = -\frac{B_2}{B_1} , \qquad \text{(magnification)}$$

and obtain (22.291) in the simple form

$$x_1 = M x_0 , \qquad\qquad y_1 = M y_0 ,$$

(22.295)

$$p_1 = -\frac{1}{f} x_0 + \frac{1}{M} p_0 , \qquad q_1 = -\frac{1}{f} y_0 + \frac{1}{M} q_0 ,$$

which determines the image formation for conjugate planes. The location of conjugate planes follows from the equation

$$\frac{1}{a_0} + \frac{1}{a_1} = -\frac{1}{f} .$$

Magnification and focal length are given by

$$M = 1 + \frac{a_1}{f} = \frac{1}{1 + \frac{a_0}{f}} ,$$

(22.296)

$$f = \frac{R}{2} .$$

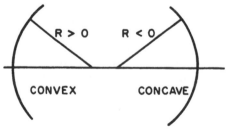

Figure 66

22.3 <u>Angular characteristic for a refracting spherical surface.</u> By a similar method we can find the angular characteristic in the case of a single spherical surface which separates two homogeneous media of index n_0 and n_1. Such a surface can be either convex or concave towards the light. We can treat both cases simultaneously by intrcducing the convention of assigning a positive value to the radius R if the surface is convex and a negative value otherwise. This means, mathematically, that we represent the surface by the equation

$$z = - R \sqrt{1 - \frac{1}{R^2} (x^2 + y^2)} .$$ (22.31)

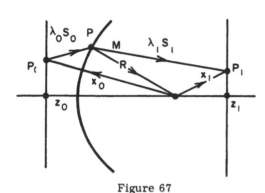

Figure 67

In order to avoid ambiguities, we prefer not to obtain the characteristic T by its geometrical interpretation but by the original definition

$$T = V + x_0 p_0 + y_0 q_0 - x_1 p_1$$

$$- y_1 q_1 .$$

We assume the center of the sphere is at the point $z = 0$ of the axis. The reference planes are arbitrarily chosen at $z = z_0$ and $z = z_1$.

We consider a light ray $P_0 P P_1$ and characterize its directions by the unit vectors

$$S_0 = \frac{1}{n_0} \left(p_0, q_0, \sqrt{n_0^2 - p_0^2 - q_0^2} \right)$$ (22.32)

$$S_1 = \frac{1}{n_1} \left(p_1, q_1, \sqrt{n_1^2 - p_1^2 - q_1^2} \right) .$$

At P the surface normal M points toward the right so that the scalar products $(S_0 \cdot M) = \cos \vartheta_0$ and $(S_1 \cdot M) = \cos \vartheta_1$ are not negative. The vectors S_0 and S_1 are then related by the equation

$$n_1 S_1 = n_0 S_0 + \Gamma M ; \quad \Gamma = n_1 (S_1 \cdot M) - n_0 (S_0 \cdot M) .$$ (22.321)

Furthermore, we introduce the vectors $X_0 = (x_0, y_0, z_0)$ and $X_1 = (x_1, y_1, z_1)$ directed toward the object and the image points respectively. We easily verify the relations

$$X_0 + \lambda_0 S_0 + MR = 0 ,$$

$$X_1 - \lambda_1 S_1 + MR = 0 ,$$ (22.33)

where λ_0 and λ_1 measure the length of the sections $P_0 P$ and $P P_1$ respectively. We remark explicitly that the relations (22.33) are valid for concave and convex surfaces if the above sign convention is applied.

The optical path V between P_0 and P_1 is given by $V = n_0 \lambda_0 + n_1 \lambda_1$. From (22.33) it follows that

$$n_0 \lambda_0 + n_0 (X_0 \cdot S_0) + R n_0 (M \cdot S_0) = 0 ,$$

$$- n_1 \lambda_1 + n_1 (X_1 \cdot S_1) + R n_1 (M \cdot S_1) = 0 , \tag{22.331}$$

and hence

$$V + n_0 (X_0 \cdot S_0) - n_1 (X_1 \cdot S_1) = R \Big[n_1 (M \cdot S_1) - n_0 (M \cdot S_0) \Big] = R \, \Gamma .$$

On account of (22.32) this yields

$$T = R \Gamma - z_0 \sqrt{n_0^2 - p_0^2 - q_0^2} + z_1 \sqrt{n_1^2 - p_1^2 - q_1^2} . \tag{22.34}$$

It remains to express Γ in terms of p_0, q_0; p_1, q_1. From (22.321) it follows that

$$n_1 S_0 \cdot S_1 - n_0 \qquad = \Gamma (S_0 \cdot M) ,$$

$$n_1 \qquad - n_0 S_0 \cdot S_1 = \Gamma (S_1 \cdot M) , \tag{22.35}$$

and hence

$$n_0^2 + n_1^2 - 2 n_0 n_1 (S_0 \cdot S_1) = \Gamma \Big(n_1 (M \cdot S_1) - n_0 (M \cdot S_0) \Big) , \tag{22.351}$$

or

$$n_0^2 + n_1^2 - 2 n_0 n_1 (S_0 \cdot S_1) = \Gamma^2 . \tag{22.352}$$

The last equation shows that Γ^2 never reaches the value zero, its minimum being given if $S_0 \cdot S_1 = 1$, i.e., $\Gamma_{min}^2 = (n_1 - n_0)^2$. We conclude that Γ is either always positive or always negative; from (22.321) it follows, for the case $S_0 \cdot S_1 = 1$,

$$(M \cdot S_0) \, \Gamma = n_1 - n_0 . \tag{22.353}$$

Since $M \cdot S_0 > 0$, this yields

$$\text{sign } \Gamma = \text{sign} (n_1 - n_0) , \tag{22.354}$$

and hence

$$\Gamma = \text{sign} (n_1 - n_0) \sqrt{n_0^2 + n_1^2 - 2 n_0 n_1 \, S_0 \cdot S_1} . \tag{22.355}$$

We again introduce the notation

$$u = p_0^2 + q_0^2 ,$$

$$w = 2(p_0 p_1 + q_0 q_1) , \qquad (22.36)$$

$$v = p_1^2 + q_1^2 ,$$

and obtain the following expression for the angular characteristic T:

$$T = T_0 - z_0 \sqrt{n_0^2 - u} - z_1 \sqrt{n_1^2 - v} , \qquad (22.37)$$

where

$$T_0 = R \ \text{sign} \ (n_1 - n_0) \sqrt{n_0^2 + n_1^2 - w - 2\sqrt{(n_0^2 - u)(n_1^2 - v)}} .$$

The case of surfaces of small diameter can be treated as the case of the small mirror. We develop T with respect to u,w,v and use only the linear terms:

$$T = B_0 + B_1 u + B_2 w + B_3 v . \qquad (22.38)$$

We find:

$$B_0 = n_1 z_1 - n_0 z_0 + (n_1 - n_0)R ,$$

$$B_1 = \frac{1}{2n_0} \left(z_0 + \frac{n_1}{n_1 - n_0} R \right) ,$$

$$B_2 = -\frac{1}{2} \frac{R}{n_1 - n_0} , \qquad (22.381)$$

$$B_3 = \frac{1}{2n_1} \left(- z_1 + \frac{n_0}{n_1 - n_0} R \right) .$$

If the position of the reference planes is measured from the vertex of the surface, i.e., if we introduce

$$z_0 = a_0 - R ,$$

$$z_1 = a_1 - R , \qquad (22.382)$$

we have

$$B_0 = n_1 a_1 - n_0 a_0 , \qquad \qquad B_1 = \frac{1}{2} \left(f + \frac{a_0}{n_0} \right) ,$$

$$\qquad \qquad (22.383)$$

$$B_2 = -\frac{1}{2} f , \qquad \qquad B_3 = \frac{1}{2} \left(f - \frac{a_1}{n_1} \right) ,$$

in which

$$f = \frac{R}{n_1 - n_0} \quad .$$

We now can repeat the same considerations as in 22.2. The equations (22.29) and (22.291) apply directly to our present case. The condition for conjugate planes (22.292) gives the <u>lens equation</u>,

$$\left(f - \frac{a_1}{n_1} \right) \left(f + \frac{a_0}{n_0} \right) = f^2 \quad \text{or} \quad \frac{n_1}{a_1} - \frac{n_0}{a_0} = \frac{1}{f} \quad . \tag{22.39}$$

If conjugate planes are chosen as reference planes, we obtain the first order image functions

$$x_1 = M x_0 \, , \qquad\qquad\qquad y_1 = M y_0 \, ,$$

$$p_1 = -\frac{1}{f} x_0 + \frac{1}{M} p_0 \, , \qquad q_1 = -\frac{1}{f} y_0 + \frac{1}{M} q_0 \, , \tag{22.391}$$

where $f = \dfrac{R \cdot}{n_1 - n_0}$ is the focal length of the refracting surface, and

$$M = \frac{1}{1 + \dfrac{a_0}{f n_0}} = 1 - \frac{a_1}{f n_1} \, , \tag{22.392}$$

the <u>magnification</u> of the conjugate planes.

CHAPTER III

APPLICATION OF THE THEORY TO SPECIAL PROBLEMS

§23. PERFECT CONJUGATE POINTS. CARTESIAN OVALS.

23.1 We call two points P_0 and P_1 perfect conjugate points if all rays through P_0 intersect each other at P_1. When we speak about "all the rays through P_0" we shall mean all the rays of a finite bundle $a \leqq p_0 \leqq b$; $a \leqq q_0 \leqq b$ with $b > a$. The limits a and b are determined by the optical instrument, for example, by the aperture of the lenses.

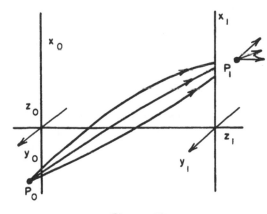

Figure 67

Let us assume that two points (x_0, y_0) and (x_1, y_1) of the object plane $z = z_0$ and the image plane $z = z_1$, respectively, are perfectly conjugate. We consider Hamilton's mixed characteristic $W = W(x_0, y_0; p_1, q_1; z_0, z_1)$ for the point (x_0, y_0). If x_0, y_0, z_0, and z_1 are fixed, then W is a function of p_1 and q_1. The intersection of a ray, originating at (x_0, y_0) with the image plane is given by

$$x_1 = -\frac{\partial W}{\partial p_1}; \qquad y_1 = -\frac{\partial W}{\partial q_1}. \qquad (23.11)$$

Our assumption means that x_1 and y_1 are constant for a finite (p_1, q_1) region. Hence it follows:

$$W = -(x_1 p_1 + y_1 q_1) + C \qquad (23.12)$$

where C is a constant with respect to p_1 and q_1.

The expression $W + x_1 p_1 + y_1 q_1$ is, according to (20.182), nothing but the point characteristic V, i.e., the optical length of the light ray (x_0, y_0, p_1, q_1) from (x_0, y_0, z_0) to the intersection point (x_1, y_1, z_1) with the image plane.

129

From (23.12) follows: If P_0 and P_1 are perfect conjugate points then all the rays through these points have the same optical length, i.e., Fermat's integral

$$V = \int_{z_0}^{z_1} n\sqrt{1 + \dot{x}^2 + \dot{y}^2}\, dz \qquad (23.13)$$

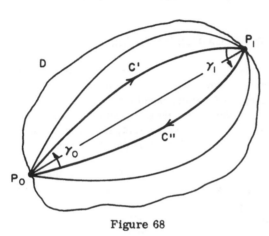

has the same value for all rays of the bundle.

We can also derive this theorem with the aid of the integral invariant

$$J = \int_C n \cos \theta\, ds \qquad (23.14)$$

Figure 68

of §21. Let us assume that the rays of the bundle determine a field of light rays in the interior of a domain D which has P_0 and P_1 on its boundary. This is the case if, through every interior point of D, there goes one and only one ray. The other condition that the congruence of rays is normal is satisfied since the rays are the orthogonal trajectories of the spherical wave fronts around P_0. It follows that $\int_C n \cos \theta\, ds = 0$ for any closed curve in D. If we choose a curve C formed by two rays C' and C'' connected by two arbitrarily short arcs γ_0 and γ_1 in the neighborhood of P_0 and P_1 it follows that

$$\int_{C'} n\, ds - \int_{C''} n\, ds + \int_{\gamma_0 + \gamma_1} n \cos \theta\, ds = 0 \qquad (23.15)$$

and hence, in the limit $\gamma_0 \to 0$; $\gamma_1 \to 0$:

$$\int_{C'} n\, ds = \int_{C''} n\, ds . \qquad (23.16)$$

23.2 **Cartesian ovals.** We can use the above theorem to construct simple optical systems in which two given points P_0 and P_1 are perfect conjugate points. Without loss of generality, we consider the two points $z_0 = 0$ and $z = a > 0$ on the z-axis. The optical instrument consists of two homogeneous media of refractive index n_0 and n_1 which are separated by a surface $\omega(x,y,z) = 0$. The problem is to find ω such that all rays from (0,0,0) in the medium of index n_0, after refraction, converge towards the point (0,0,a) in the medium of index n_1. The optical length of a ray which intersects the surface at the point x,y,z is given by

$$V = n_0\sqrt{x^2 + y^2 + z^2} + n_1\sqrt{x^2 + y^2 + (z - a)^2} \ . \qquad (23.21)$$

Since this length must be constant, we find the equation for $\omega = 0$:

$$n_0\sqrt{x^2 + y^2 + z^2} + n_1\sqrt{x^2 + y^2 + (z - a)^2} = C \ . \qquad (23.22)$$

We can choose the vertex of the surface at an arbitrary point A of the axis between 0 and a. If A is chosen, we have explicitly

$$n_0\sqrt{x^2 + y^2 + z^2} + n_1\sqrt{x^2 + y^2 + (z - a)^2} = n_0A + n_1(a - A). \quad (23.23)$$

This is the equation of a surface of revolution which is obtained by rotating the curve

$$n_0\sqrt{x^2 + z^2} + n_1\sqrt{x^2 + (z - a)^2} = n_0A + n_1(a - A) \qquad (23.24)$$

around the z-axis. We can eliminate the radicals and obtain an algebraic curve of the fourth order which is known as the Cartesian Oval. We remark however that only those branches of the algebraic curve can be used for our purpose which satisfy the condition (23.24) with a positive sign for the radicals.

We finally show that the necessary condition (23.22) is also sufficient, i.e., that on the surface (23.23) the rays are refracted according to Snell's law. This is an immediate consequence of the following interpretation of the problem. The wave fronts in the first medium are given by the function

$$\psi = n_0\sqrt{x^2 + y^2 + z^2} \qquad (23.25)$$

and in the second medium by a function of the type

$$\psi' = C - n_1\sqrt{x^2 + y^2 + (z - a)^2} \qquad (23.26)$$

which represents spherical waves converging towards the point (0,0,a). If, on the surface $\omega = 0$, $\psi = \psi'$, then ψ and ψ' represent a solution of $\psi_x^2 + \psi_y^2 + \psi_z^2 = n^2$ which is continuous in the neighborhood of $\omega = 0$. We have seen that Snell's law is a consequence of this continuity of ψ. However, $\psi = \psi'$ on $\omega = 0$ yields the equation (23.22).

Cartesian ovals can easily be obtained graphically by the intersection points of circles around P_0 and P_1 with radii r_0 and $r_1 = \frac{c}{n_1} - r_0\frac{n_0}{n_1}$, respectively. The case $n_0 = 1$, $n_1 = \frac{3}{2}$, $a = 1$; $A = \frac{1}{2}$ is shown in Figure 69.

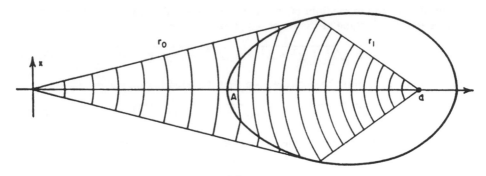

Figure 69

23.3 <u>Object at infinity</u>. We consider next the problem of constructing a refracting surface such that the rays of a parallel bundle intersect each other at a given point P_1 after refraction. The wave fronts in the first medium of index n_0 are given by the function

$$\psi = n_0 z \tag{23.31}$$

and in the second medium by the converging spherical waves

$$\psi' = C - n_1 \sqrt{x^2 + y^2 + (z - a)^2} . \tag{23.32}$$

The condition of the continuity of the solution ψ of $\psi_x{}^2 + \psi_y{}^2 + \psi_z{}^2 = n^2$ leads directly to the equation of the surface: $\psi = \psi'$ or

$$n_0 z + n_1 \sqrt{x^2 + y^2 + (z - a)^2} = C . \tag{23.33}$$

By taking the vertex of this surface of revolution at $z = 0$, we obtain

$$n_0 z + n_1 \sqrt{x^2 + y^2 + (z - a)^2} = n_1 a . \tag{23.34}$$

This is a surface generated by rotating the curve

$$n_0 z + n_1 \sqrt{x^2 + (z - a)^2} = n_1 a \tag{23.35}$$

around the z-axis. One sees immediately that (23.35) represents a conic section. By eliminating the radical, we find

$$\frac{\left(z - a \dfrac{n_1}{n_1 + n_0}\right)^2}{a^2 \left(\dfrac{n_1}{n_1 + n_0}\right)^2} + \frac{x^2}{a^2 \dfrac{n_1 - n_0}{n_1 + n_0}} = 1 \tag{23.36}$$

which is an ellipse if $n_1 > n_0$ and a hyperbola if $n_1 < n_0$.

The axes A and B of the <u>ellipse</u>, in case $n_1 > n_0$, are given by

$$A^2 = a^2 \left(\frac{n_1}{n_1 + n_0}\right)^2 , \quad B^2 = a^2 \frac{n_1 - n_0}{n_1 + n_0} \qquad (23.37)$$

and the center M and the eccentricity $e = \sqrt{A^2 - B^2}$ by

$$M = a \frac{n_1}{n_1 + n_0} , \qquad e^2 = a^2 \left(\frac{n_0}{n_1 + n_0}\right)^2 . \qquad (23.371)$$

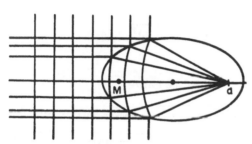

Figure 70

It follows that M + e = a_1. Hence the point a coincides with the second focal point of the ellipse.

Similar formulae can be found for <u>the hyperbola in case</u> $n_1 < n_0$. We have

$$A^2 = a^2 \left(\frac{n_1}{n_1 + n_0}\right)^2 ;$$

$$B^2 = a^2 \frac{n_0 - n_1}{n_1 + n_0} ;$$

$$(23.38)$$

$$M = a \frac{n_1}{n_1 + n_0} ;$$

$$e^2 = A^2 + B^2$$

$$= a^2 \left(\frac{n_0}{n_1 + n_0}\right)^2 .$$

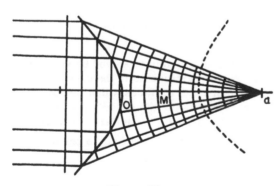

Figure 71

Hence M + e = a, i.e., the point a coincides with the focal point of the hyperbola, which is enclosed by the other branch.

23.4 <u>Virtual conjugate points.</u> If a bundle of rays, originating at a point P_0, is transformed by refraction into a bundle such that the backwards extensions of the rays intersect at one and the same point P_1, then we call P_1 a virtual conjugate point to P_0. In order to determine a refracting surface $\omega = 0$ to which two given points P_0 and P_1 belong as virtual conjugate points we consider the diverging spherical waves

$$\psi = n_0 \sqrt{x^2 + y^2 + z^2}$$

$$\psi' = n_1 \sqrt{x^2 + y^2 + (z - a)^2} + C \tag{23.41}$$

in the two media separated by the surface $\omega = 0$. The condition of continuity on $\omega = 0$ yields immediately the equation

$$n_0 \sqrt{x^2 + y^2 + z^2} - n_1 \sqrt{x^2 + y^2 + (z - a)^2} = C . \tag{23.42}$$

By choosing the vertex of the surface at an arbitrary point $z = A > a$ of the axis, we obtain

$$n_0 \sqrt{x^2 + y^2 + z^2} - n_1 \sqrt{x^2 + y^2 + (z - a)^2} = n_0 A - n_1(A - a) ,$$

$$\tag{23.43}$$

i.e., a surface of revolution generated by the curve

$$n_0 \sqrt{x^2 + z^2} - n_1 \sqrt{x^2 + (z - a)^2} = n_0 A - n_1(A - a) . \tag{23.44}$$

Elimination of the radicals leads again to an algebraic curve of the fourth order. We notice that the same algebraic equation is obtained from

$$n_0 \sqrt{x^2 + z^2} - n_1 \sqrt{x^2 + (z - a)^2} = C$$

and

$$n_0 \sqrt{x^2 + z^2} + n_1 \sqrt{x^2 + (z - a)^2} = C . \tag{23.45}$$

This demonstrates that our present problem is solved by surface sections which belong to the same algebraic surfaces as in the case of the problem (23.2). It is however another branch of these surfaces which we have to use now. An example is illustrated in Figure 72 for the case: $n_0 = 1$, $n_1 = 3/2$; $a = 1/2$; $A = 1$. The construction is carried out similarly as in 23.2 with the aid of two circles of radius r_0 and $r_1 = \frac{2}{3} r_0 - \frac{1}{6}$ about the points $z = 0$ and $z = \frac{1}{2}$, respectively.

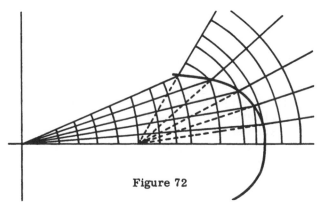

Figure 72

23.5 If the object is at infinity, i.e., if parallel wave fronts have to be transformed into diverging spherical wave fronts, we obtain by our principle of continuity the equation

$$n_0 z - n_1 \sqrt{x^2 + y^2 + (z + a)^2} = -n_1 a . \qquad (23.51)$$

We take the point P_1 at $z = -a$ with $a > 0$. The curve

$$n_0 z - n_1 \sqrt{x^2 + (z + a)^2} = -n_1 a \qquad (23.52)$$

which generates the surface of revolution (23.51) is an <u>ellipse</u> in the case $n_1 > n_0$ and a <u>hyperbola</u> if $n_1 < n_0$. The equation of the conic is given by

$$\frac{\left(z + a \dfrac{n_1}{n_1 + n_0}\right)^2}{a^2 \left(\dfrac{n_1}{n_1 + n_0}\right)^2} + \frac{x^2}{a^2 \dfrac{n_1 - n_0}{n_1 + n_0}} = 1 , \qquad (23.53)$$

which differs from (23.36) only in the quantity a which is replaced by - a.

Consequently, the axes A, B and the eccentricity e are determined by the formulae (23.37) and (23.38). The center M is found by replacing a by - a in these formulae, i.e.,

$$M = -a \frac{n_1}{n_1 + n_0} . \qquad (23.54)$$

The point $z = -a$ is one of the focal points as is illustrated in Figures 73 and 74.

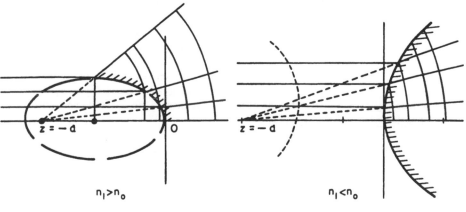

$n_1 > n_0$ $n_1 < n_0$

Figure 73   Figure 74

23.6 <u>The aplanatic points of a sphere</u>. If the vertex A in (23.44) is chosen so that the right side is zero, i.e., if

$$A = \frac{n_1}{n_1 - n_0} a \qquad (23.61)$$

then the curve (23.44) is a circle and therefore the surface (23.43) is a sphere. We find from (23.43) and (23.61):

$$x^2 + y^2 + \left(z - \frac{n_1}{n_1 + n_0} A\right)^2 = \left(\frac{n_0}{n_1 + n_0}\right)^2 A^2 . \qquad (23.62)$$

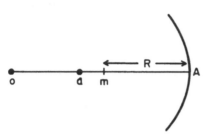

The sphere has its center at the point

$$m = \frac{n_1}{n_1 + n_0} A < A , \qquad (23.63)$$

its radius is

Figure 75

$$R = \frac{n_0}{n_1 + n_0} A < A . \qquad (23.64)$$

This shows that the spherical surface is concave.

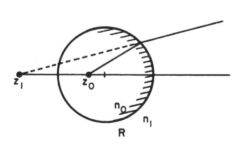

Figure 76

Let us now consider a given concave spherical surface of radius R, which separates two optical media of index n_0 and n_1, respectively. The center of the sphere is taken at $z = 0$. From the above it follows that there exist two points $z = z_0$ and $z = z_1$ on the z-axis such that a spherical wave from z_0 is transformed into a spherical wave diverging apparently from z_1. In other words z_0 and z_1 are virtual conjugate points. One calls these points the <u>aplanatic points</u> of the sphere.

In our above notation we have $z_0 = -m$ and $z_1 = a - m$. Hence it follows from (23.61), (23.63), and (23.64) that

$$z_0 = -\frac{n_1}{n_0} R ,$$

$$\qquad (23.65)$$

$$z_1 = -\frac{n_0}{n_1} R .$$

The point z_0 is called the aplanatic object point and z_1 the aplanatic image point. The location of these points for the two cases $n_1 > n_0$ and $n_1 < n_0$ is shown in Figures 77 and 78.

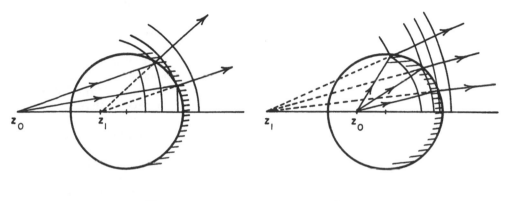

Figure 77   Figure 78

By reversing the direction of the light in the above figures, we see that convex spherical surfaces can be used to transform converging spherical waves into perfect spherical waves of the same type but with a different point of convergence.

On account of the symmetry of the sphere we have not only two aplanatic points but infinitely many, located on two concentric spheres of radius $\frac{n_1}{n_0}$ R and $\frac{n_0}{n_1}$ R, respectively. We can use this result for a simple graphical method of constructing the refracted rays on a sphere if the incident rays are given. Let us consider a convex spherical surface S and let $n_1 > n_0$. We construct the two aplanatic spheres of radius $\frac{n_1}{n_0}$ R and $\frac{n_0}{n_1}$ R. We extend the incident ray until it intersects the aplanatic object sphere at A_0. We know that all incident rays aiming towards A_0 must be refracted towards the conjugate aplanatic point A_1. We find this point by connecting A_0 with M. This line intersects the aplanatic image sphere of radius $\frac{n_0}{n_1}$ R at the point A_1. Hence PA_1 is the refracted ray.

The aplanatic points of a sphere are used in the construction of microscope objectives of high aperture. The object is submerged in oil which has nearly the same refractive index as the front lens of the objective. Hence we may consider the object as being located in a medium of an index of refraction greater than one. The front surface of the objective can be chosen arbitrarily but is in general made plane for practical reasons. The back surface of the front lens is chosen in such a manner that the object point P_0 coincides with

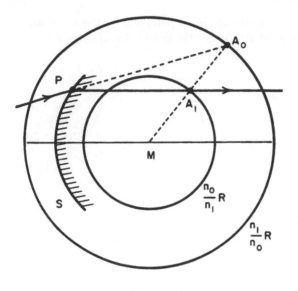

Figure 79

the aplanatic object point
of this sphere. The
result is that a wide
bundle of rays with a
numerical aperture
$\sin \theta_0$ almost equal to 1
leaves the front lens
without any spherical
aberration but with a
smaller aperture
$\sin \theta_1 = 1/n_1$. The next
surface is chosen con-
centric to the point P_1
from which this bundle
seems to diverge such
that the bundle enters
the glass again without
spherical aberration.
We then insert again a
spherical surface with
P_1 as the aplanatic object
point. Hence the bundle
leaves this surface with decreased aperture $\sin \theta = \dfrac{1}{n_1 \, n_2}$. By this construction one is in the position to decrease the steepness of the bundle without
introducing spherical aberration until finally one or two cemented lenses
transform the resulting bundle into a converging bundle. This succession of
aplanatic lenses explains the characteristic construction of a microscope
objective as shown in Figure 80. In practice this construction is not carried
out rigorously. However the departures from the above principle are never
considerable.

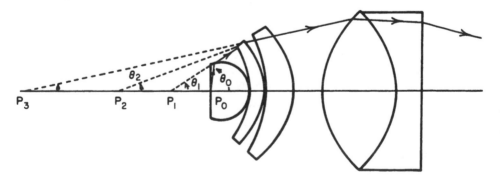

Figure 80

§24. FINAL CORRECTION OF OPTICAL INSTRUMENTS BY ASPHERIC SURFACES.

The usefulness of single aspheric surfaces of the type discussed in §23 is very limited in practice. It is true that we are in the position to eliminate one aberration with these surfaces, but the effect on other aberrations is not favorable. In other words, if we study the refraction of spherical wave fronts which originate at a point (x_0, y_0) of the object plane different from $x_0 = y_0 = 0$, we would find great departures from perfectly converging spherical waves after refraction, even if the surface produces a perfectly sharp image of the point $x_0 = y_0 = 0$.

However, an aspheric surface can be used successfully in combination with spherical surfaces in order to eliminate the last trace of a remaining spherical aberration, for example. Let us assume that an optical system of spherical surfaces has been found which is nearly correct in the sense that a considerable part of the object plane is sharply imaged if the system is used with small apertures. The use of large apertures is prohibited by a rapidly increasing spherical aberration. If it is then possible to eliminate this aberration by replacing the last spherical surface of the system by an aspheric surface which departs only slightly from the original spherical surface, we can expect that no detrimental effect will be introduced with regard to the aberrations of oblique bundles of rays. This procedure can be applied successfully, for example, in certain simple types of photographic objectives known as <u>Cooke Triplets</u>. These objectives consist of two positive lenses with a negative lens between them. It is easily possible to find combinations of this type which would produce excellent images if the objective is used at a "speed" not greater than F: 4.5[†]. The design of such objectives for greater speeds however is difficult because of the spherical aberration which rapidly assumes large values. On the other hand, it is of course desirable for the photographer to have an objective of as great a diameter as possible, in order to decrease the time of exposure. Hence we are led to the attempt to prevent the rapid increase of spherical aberration by employing an aspheric surface as the boundary of the last lens.

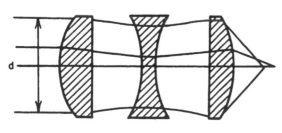

Figure 81

[†]The "speed" or F-number of a photo-objective is defined by the ratio f:d of focal length to diameter of the front lens.

24.2 In principle the solution of the above problem is simple. Let us consider the spherical wave fronts originating at the point P_0 on the axis. We remove the last surface of the instrument and determine the wave fronts of the bundle through P_0 in the still unlimited glass medium of the last lens. Let n be the index of refraction of this medium. These wave fronts are given by the function

$$V = V(P_0; x,y,z) = ct , \qquad (24.21)$$

where V is nothing but Huyghens' wavelet function or Hamilton's point char-

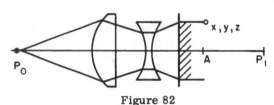

Figure 82

acteristic for the points P_0 and a point (x,y,z) in the glass. Let us assume that this function is known. The problem is to determine a surface $\omega(x,y,z) = 0$ such that, by refraction on this surface, the wave fronts $V = V(x,y,z)$ are transformed into wave fronts

$$\psi = C - \sqrt{x^2 + y^2 + (z - z_1)^2} \qquad (24.22)$$

which converge towards the point $P_1 = (0,0,z_1)$. The condition of continuity of V and ψ on $\omega = 0$ yields the equation of the aspheric surface directly:

$$V(x,y,z) + \sqrt{x^2 + y^2 + (z - z_1)^2} = C . \qquad (24.23)$$

The constant C is determined when the vertex A of the aspheric surface has been chosen. We have

$$C = V(0,0,A) + z_1 - A , \qquad (24.24)$$

and hence the equation

$$V(x,y,z) + \sqrt{x^2 + y^2 + (z - z_1)^2} = V(0,0,A) + z_1 - A . \qquad (24.25)$$

It is quite clear that $V(x,y,z)$ depends only on $\rho = \sqrt{x^2 + y^2}$ and z if the optical instrument is symmetrical with respect to the axis. In this case (24.25) represents a surface of revolution generated by the curve

$$V(x,0,z) + \sqrt{x^2 + (z - z_1)^2} = V(0,0,A) + z_1 - A. \qquad (24.26)$$

24.3 In general it is not easy to find explicit expressions for the function $V(x,y,z)$. Therefore, another procedure of finding the surface $\omega = 0$, in which Hamilton's mixed characteristic $W(z_0, z; x_0, y_0, p,q)$ is used, is preferable. The surface, $\omega = 0$, is then obtained in parametric form. We have the relation

$$V = W + xp + yq \tag{24.31}$$

where W is a function of (z,p,q) since (x_0,y_0,z_0) are considered as given.

From

$$\frac{\partial W}{\partial z} = \sqrt{n^2 - p^2 - q^2} \, , \tag{24.32}$$

it follows that

$$W = W_0 + z\sqrt{n^2 - p^2 - q^2} \, , \tag{24.33}$$

in which $W_0 = W_0(p,q)$ is independent of z. Furthermore, we have

$$x = -\frac{\partial W}{\partial p} = -\frac{\partial W_0}{\partial p} + z\frac{p}{\sqrt{n^2 - p^2 - q^2}} \, ,$$

$$y = -\frac{\partial W}{\partial q} = -\frac{\partial W_0}{\partial q} + z\frac{q}{\sqrt{n^2 - p^2 - q^2}} \, . \tag{24.34}$$

Hence we obtain

$$V = W_0 - p\frac{\partial W_0}{\partial p} - q\frac{\partial W_0}{\partial q} + z\frac{n^2}{\sqrt{n^2 - p^2 - q^2}} \, , \tag{24.35}$$

as a function of p, q and z.

We finally introduce (24.35) in (24.23) and get

$$W_0 - p\frac{\partial W_0}{\partial p} - q\frac{\partial W_0}{\partial q} + z\frac{n^2}{\sqrt{n^2 - p^2 - q^2}}$$

$$+ \sqrt{x^2 + y^2 + (z - z_1)^2} = C \, . \tag{24.36}$$

The equations (24.34) and (24.36) represent two linear and one quadratic equation for x,y,z as functions of p and q. The solution gives the surface $\omega = 0$ in parametric representation.

Let us now assume that our instrument is symmetrical with respect to the z-axis. The surface $\omega = 0$ is a surface of revolution, and it is sufficient to determine its cross section with the xz-plane. We thus assume $y = q = 0$ in the above formulae and obtain

$$W_0(p) - p\,W_0'(p) + z\frac{n^2}{\sqrt{n^2 - p^2}} + \sqrt{x^2 + (z - z_1)^2} = C,$$

(24.37)

$$x = -W_0'(p) + z\frac{p}{\sqrt{n^2 - p^2}} \; ;$$

i.e., one quadratic and one linear equation for x and z as a function of p. The functions $x(p)$ and $z(p)$ give the curve in parametric form.

It is not difficult to determine the function $W_0(p)$ in any actual case. Let us choose the vertex of the aspheric surface as the point $z = 0$. By

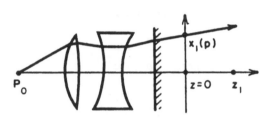

tracing rays from P_0 through the system we are able to calculate the intersection x_1 with the plane $z = 0$ and the direction p of a ray at $z = 0$. By correlating the results x_1 and p for a number of rays, we obtain a function $x_1 = x_1(p)$ which can be fitted, for example, to a polynomial

Figure 83

$$x_1(p) = p(A_0 + A_1 p^2 + A_2 p^4 + \ldots) .$$

(24.38)

If we let $W_0(0) = 0$, it follows from $x_1 = -W_0'(p)$ that

$$W_0(p) = -\int_0^p x_1(p)\,dp$$

(24.381)

$$= p^2\left(\frac{A_0}{2} + \frac{A_1}{4}p^2 + \frac{A_2}{6}p^4 + \ldots\right) .$$

The equations (24.37) become

$$-\int_0^p x_1(p)\,dp + px_1(p) + z\frac{n^2}{\sqrt{n^2 - p^2}} + \sqrt{x^2 + (z - z_1)^2} = C$$

(24.39)

$$x = x_1(p) + z\frac{p}{\sqrt{n^2 - p^2}} .$$

Let us develop $x = x(p)$ and $y = y(p)$ in power series

$$x = p(B_0 + B_1 p^2 + \ldots),$$

(24.391)

$$y = C_0 + C_1 p^2 + \ldots .$$

The coefficients of the power series can be readily determined from (24.39).

24.4 Instead of the characteristic W we can also use the angular characteristic T. This is especially advantageous if the point P_0 is at infinity as in a photographic objective. We have

$$V = T + xp + yq - x_0 p_0 - y_0 q_0 \ ,$$

and hence, since $p_0 = q_0 = 0,$

$$V = T + xp + yq \ . \tag{24.41}$$

We can repeat the same considerations for T as for the case of W. The result is formally the same as before. The cross section of the aspheric surface with the xz-plane is given by the equations

$$T_0(p) - pT_0'(p) + z \frac{n^2}{\sqrt{n^2 - p^2}} + \sqrt{x^2 + (z - z_1)^2} = C \ ,$$
$$\tag{24.42}$$
$$x = - T_0'(p) + z \frac{p}{\sqrt{n^2 - p^2}} \ .$$

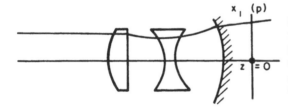

Figure 84

In order to find $T_0(p)$ a number of parallel incident rays have to be traced through the system. Let x_1 and p be the intersection and direction of a ray at the plane z = 0. We obtain

$$T_0 = - \int_0^p x_1(p) \, dp \ , \tag{24.43}$$

and hence the same formulae as in (24.39).

24.5 We shall apply the preceding methods to a number of simple examples. A lens of index of refraction n has to be found which consists of a plane surface and an aspheric surface. It is required that a spherical wave from a point P_0 be transformed into a plane wave after refraction. Let a be the distance of P_0 from the plane surface and t the thickness of the lens on the axis.

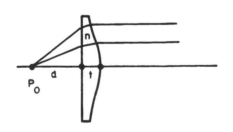

Figure 85

The characteristic $W(p,z)$ in the medium n is given by

$$W = a\sqrt{1 - p^2} + z\sqrt{n^2 - p^2} \qquad (24.51)$$

as we have seen in §22.1. The point characteristic V follows from

$$V = W - p\frac{\partial W}{\partial p} = \frac{a}{\sqrt{1 - p^2}} + z\frac{n^2}{\sqrt{n^2 - p^2}} \qquad (24.52)$$

as a function of p and z. Since $V = $ a constant represents the wave fronts in the medium n, and $\psi = z + C$ represents the required wave fronts in the final medium, we can apply the condition of continuity and find

$$V - z = C \qquad (24.53)$$

as the equation for the aspheric surface. With the aid of (24.52) this yields

$$\frac{a}{\sqrt{1 - p^2}} - z\left(1 - \frac{n^2}{\sqrt{n^2 - p^2}}\right) = C . \qquad (24.54)$$

From $x = -\dfrac{\partial W}{\partial p}$ it follows that

$$x = \frac{ap}{\sqrt{1 - p^2}} + \frac{p}{\sqrt{n^2 - p^2}} z . \qquad (24.55)$$

From (24.54) and (24.55) we find $x = x(p)$ and $z = z(p)$, i.e., the cross section of the aspheric surface in parametric form. The result is

$$x = p\left[\frac{a}{\sqrt{1 - p^2}} + \frac{C - \dfrac{a}{\sqrt{1 - p^2}}}{n^2 - \sqrt{n^2 - p^2}}\right] \qquad (24.56)$$

$$z = \frac{C - \dfrac{a}{\sqrt{1 - p^2}}}{\dfrac{n^2}{\sqrt{n^2 - p^2}} - 1} .$$

The constant C is given by the condition that $z = a + t$ for $p = 0$. We obtain

$$C = n a + (n - 1) t . \qquad (24.57)$$

24.6 As the next example we consider a lens which consists of one spherical surface and one aspheric surface. It is required to find a lens which transforms a plane wave into a converging spherical wave. Let us assume that the center of the sphere is at $z = 0$. The angular characteristic $T(z,p)$ in the case where $p_0 = 0$ is given by

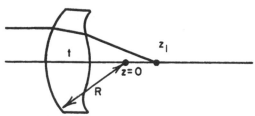

Figure 86

$$T = R\sqrt{1 + n^2 - 2\sqrt{n^2 - p^2}}$$
$$+ z\sqrt{n^2 - p^2} \qquad (24.61)$$

which follows from (22.3). If z_1 is the desired focal point, we obtain directly by (24.42) the equation of the aspheric surface:

$$T_0(p) - pT_0'(p) + z\frac{n^2}{\sqrt{n^2 - p^2}} + \sqrt{x^2 + (z - z_1)^2} = C ,$$

$$(24.62)$$

$$x = - T_0'(p) + z\frac{p}{\sqrt{n^2 - p^2}} ,$$

where $T_0(p) = R\sqrt{1 + n^2 - 2\sqrt{n^2 - p^2}}$.   (24.621)

These equations can be simplified by introducing

$$p = n \sin \alpha \qquad (24.63)$$

and letting $R = 1$. With this we can write (24.62) in the form

$$n(x \sin \alpha + z \cos \alpha) + \sqrt{x^2 + (z - z_1)^2} = C - \sqrt{1 + n^2 - 2n \cos \alpha} ,$$

$$(24.64)$$

$$n(x \cos \alpha - z \sin \alpha) = \frac{n \sin \alpha}{\sqrt{1 - n^2 - 2n \cos \alpha}} .$$

The constant C is given by the position $z = t - 1$ of the vertex of the aspheric surface. Letting $\alpha = x = 0$; $z = t - 1$ in the first equation of (24.64), we find

$$C = (n - 1)t + z_1 . \qquad (24.65)$$

The formulae (24.64) allow us to find, without difficulty, the coefficients of the development of the functions $x = x(\alpha)$ and $z = z(\alpha)$ in powers of α.

24.7 <u>Schmidt Camera</u>. A combination of an aspheric surface and a
spherical mirror characterizes the construction of the astronomical camera
which is known as the Schmidt camera. This camera, first constructed by B.
Schmidt in 1930, has become justly famous in recent years. This fact can be
readily understood if one considers earlier astronomical cameras of high
speed which could be used for angular fields of only a few degrees. With
Schmidt's construction however, we can photograph an angular field of more
than 15°.

The fundamental principle of this camera can be derived as follows.
Let us consider a spherical mirror of radius $R = 1$ and of large angular

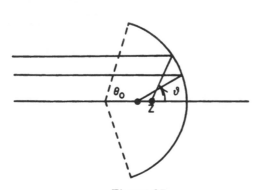

opening θ_0. A bundle of rays
parallel to the z-axis is reflected at the mirror. Consider a ray of this bundle
incident at the angle ϑ with
the z-axis. After reflection
it intersects the z-axis at the
point

$$Z = \frac{1}{2 \cos \vartheta/2} . \tag{24.71}$$

Figure 87   The difference

$$\Delta Z = Z(\vartheta) - Z(0) = \left(\frac{1}{2} \frac{1}{\cos \vartheta/2} - 1 \right) \tag{24.711}$$

measures the spherical aberration of the mirror. This expression increases
rapidly with ϑ and excludes the use of spherical mirrors for critical photographic purposes. However, such a mirror has the advantage that an oblique
bundle of parallel rays is reflected as a bundle which is congruent to the reflected bundle of rays parallel to the z-axis. This follows from the fact that
the z-axis is in no way distinguished from another diameter of the sphere of
the mirror. We can express this fact by stating that the image formation of
the spherical mirror is uniformly bad over a considerable angular field. If
a parabolic mirror is used instead of the spherical mirror, we obtain a flaw-
less image for bundles parallel to the axis, but oblique bundles are worse than
before so that a parabolic mirror can only be used for a very narrow angular
field.

We next consider a spherical mirror in combination with a thin plane
parallel plate placed at the center of the sphere perpendicular to the z-axis.
Obviously the bundle parallel to the axis is undisturbed. Oblique parallel rays
are only shifted sideways by a small amount but arrive at the mirror still

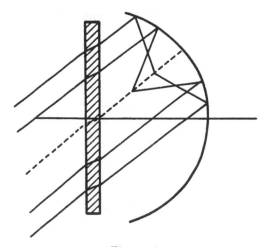

Figure 88

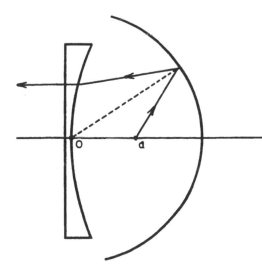

Figure 89

parallel and consequently possess a reflected bundle which is congruent to the reflected bundle of the rays parallel to the z-axis.

Let us now assume that it is possible to eliminate the spherical aberration of the rays parallel to the z-axis by replacing one of the plane surfaces of the plate by an aspheric correction surface which departs only slightly from a plane surface. Then we can expect that the spherical symmetry of the instrument is not seriously impaired and that the correction plate has a similar beneficial effect on the oblique bundles whose inclination is not large. This expectation is indeed verified by experiment, as well as by theoretical investigation.[†]

24.8 **The correction plate of the Schmidt camera.** Let $V(0,0,a; x,y,z) = V(x,y,z)$ be the point characteristic of a spherical mirror of radius R, i.e., the optical distance of a point (x,y,z) from a given point $(0,0,a)$ on the z-axis. Our aim is to find a surface $\omega = 0$ such that the wave fronts $V(x,y,z) = $ a constant are transformed, by refraction on $\omega = 0$, into plane wave fronts $\psi = C - nz$ in the glass part of the correction plate. If the other surface of the correction plate is made plane, then the plane waves leave the

[†]Caratheodory, Elementare Theorie des Spiegelteleskops von B. Schmidt, Hamb. Math. Einzelschrifte, 1940

Synge, Theory of the Schmidt Telescope, Journal Optical Society of America, Vol. 33.3 pp. 129-136.

plate undisturbed. The condition $V = \psi$ on $\omega = 0$ gives the equation of the aspheric surface, namely,

$$V(x,y,z) + nz = C. \qquad (24.81)$$

The surface is a surface of revolution. Therefore it is sufficient to determine its cross section with the xz-plane. If the vertex of the surface is assumed to be at $z = 0$, we obtain $C = 2 - a$ and hence:

$$V(x,0,z) + nz = 2 - a \qquad (24.811)$$

is the equation of the generating curve.

We can find a comparably simple equation for the aspheric surface by expressing $V(x,0,z)$ as a function of z and the distance ρ from the point $(0,a)$ to the point (x_0,y_0). (x_0,y_0) is the point where the ray passing through (x,z) and $(0,a)$ is reflected on the mirror.

Let us assume $R = 1$. From Figure 90 it follows that

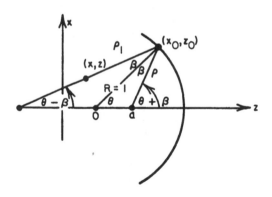

Figure 90

$$\rho^2 = 1 + a^2 - 2a \cos\theta ;$$

$$\text{i.e., } \cos\theta = \frac{1 + a^2 - \rho^2}{2a} .$$

$$(24.82)$$

Therefore the quantity $\cos\theta$ is a rational function of ρ.

We next express the quantities $\sin\beta$ and $\cos\beta$ as functions of ρ. β is the angle of reflection at (x_0,z_0). We have the relations:

$$\rho \sin\beta = a \sin\theta ,$$

$$\rho \cos\beta + a \cos\theta = 1 . \qquad (24.83)$$

It follows that

$$\sin\beta = \frac{a}{\rho} \sin\theta ,$$

$$(24.831)$$

$$\cos\beta = \frac{1}{\rho}(1 - a \cos\theta) .$$

With the aid of these formulae we can express the distance ρ_1 from (x_0, z_0) to (x, z) as a function of z and ρ. We have

$$\rho_1 = \frac{z_0 - z}{\cos(\beta - \theta)} = \frac{\cos\theta - z}{\cos(\beta - \theta)} \, ,$$

since $\cos(\theta - \beta) = \cos(\beta - \theta)$ and hence from (24.831),

$$\rho_1 = \rho \, \frac{\cos\theta - z}{\cos\theta - a\cos 2\theta} \, . \tag{24.84}$$

Since both $\cos\theta$ and $\cos 2\theta = 2\cos^2\theta - 1$ are rational functions of ρ, we observe that ρ_1 itself is a rational function of ρ; thus it follows: The function $V(x, 0, z) = \rho + \rho_1$ is a linear function of z and a rational function of ρ. We have

$$V(z, \rho) = \rho \left[1 + \frac{\cos\theta - z}{\cos\theta - a\cos 2\theta} \right] . \tag{24.85}$$

Finally we introduce (24.85) in (24.811) and obtain z as a rational function of ρ. The result is

$$nz = 2 - a - \rho - \rho \, F(\rho) \, , \tag{24.86}$$

where $F(\rho)$ is defined by

$$F(\rho) = \frac{n\cos\theta + \rho + a - 2}{n(\cos\theta - a\cos 2\theta) - \rho} \, . \tag{24.861}$$

By introducing the relationship (24.82), we obtain explicitly:

$$F(\rho) = \frac{n\left(1 - (\rho^2 - a^2)\right) + 2a(\rho + a - 2)}{n\left((\rho^2 - a^2) - (\rho^2 - a^2)^2\right) + 2a^2 n - 2a\rho} \, . \tag{24.862}$$

This function, incidentally, is nothing but the ratio ρ_1/ρ for the points (x, z) of the aspheric surface. Indeed, this follows from

$$nz + \rho + \rho_1 = 2 - a$$

and from (24.86).

In order to find x as a function of ρ, we write

$$x = \sin\theta - \rho_1 \sin(\theta - \beta) \, . \tag{24.87}$$

With the aid of (24.831) this becomes

$$x = \sin \theta \left[1 + \frac{\rho_1}{\rho} (1 - 2a \cos \theta) \right]^{\dagger} \qquad (24.8"1)$$

$$= \sin \theta \left[1 + (\rho^2 - a^2) F(\rho) \right] . \qquad (24.8"2)$$

We summarize our result: The aspheric surface of the Schmidt camera is given by the equations:

$$z = \frac{1}{n} \left[2 - a - \rho - \rho F(\rho) \right] ,$$
$$x = G(\rho) \left[1 + (\rho^2 - a^2) F(\rho) \right] , \qquad (24.88)$$

where $F(\rho)$ is the rational function (24.862) and $G(\rho) = \sin \theta$ is the function

$$G(\rho) = \frac{1}{2a} \sqrt{(\rho^2 - (1 - a)^2)((1 + a)^2 - \rho^2)} . \qquad (24.831)$$

It can be shown that this curve is a section of an algebraic curve of the tenth order.

We can use the formulae (24.88) to compute points of the curve accurately. It is however preferable to develop the functions $z(\rho)$ and $x(\rho)$ in a power series with respect to the variable

$$t = \rho - (1 - a) , \qquad (24.832)$$

which varies from 0 to 2a if ρ varies from its smallest value $1 - a$ to its largest value $1 + a$; i.e., if the point (x_0, y_0) travels over the circle of the mirror. One can simplify the practical computation even more by determining a sufficiently large number of the coefficients of the development of z in powers of x with the aid of (24.88). In the case where $a = \frac{1}{2}$ this development[††] is given by

$$(n - 1) z = \frac{1}{4} x^4 + \frac{3}{8} x^6 + \frac{45}{64} x^8 + \dots \qquad (24.833)$$

i.e., by a fourth order parabola in the neighborhood of the z-axis obtained by neglecting higher powers of x.

[†] $\dfrac{\rho_1}{\rho} = F(\rho)$.

[††] Caratheodory, loc. cit.

We can easily determine the extreme values of the function $z = z(x)$. The ray which goes through points (x,z) on the curve where $\frac{dz}{dx} = 0$ must be

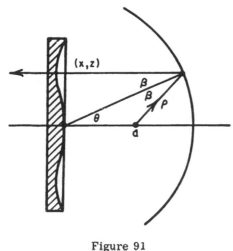

Figure 91

parallel to the z-axis, i.e., $\beta = \theta$. This implies that $\rho = a$. From (24.862) and (24.881) it follows that

$$F(a) = \frac{n + 2a(2a - 2)}{2a^2(n - 1)} ,$$

and   (24.89)

$$G(a) = \sqrt{1 - \frac{1}{4a^2}} .$$

Hence by (24.88):

$$x = \sqrt{1 - \frac{1}{4a^2}} ,$$

and   (24.891)

$$(n - 1)z = 2 - 2a - \frac{1}{2a} .$$

An extreme value is obtained only in the case where $a > \frac{1}{2}$. The corresponding value of z is negative, i.e., the curve has the form indicated in Figure 91. In the case where $a \lessgtr \frac{1}{2}$ the curve has the form shown in Figure 89.

§25. THE ANGULAR CHARACTERISTIC FOR A SINGLE REFRACTING SURFACE.

25.1 In the case of a single reflecting or refracting surface it is possible to formulate a simple method for determining the angular characteristic $T(z_0,z_1; p_0,q_0,p_1,q_1)$. Let us assume that the surface is given in the form

$$z = f(x,y) .$$

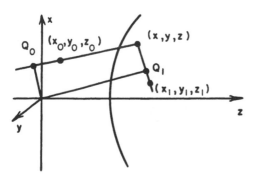

Figure 92

Let the homogeneous media separated by this surface have the indices n_0 and n_1. We consider two points (x_0,y_0,z_0) in the medium of index n_0 and (x_1,y_1,z_1) in the medium of index n_1. If (x,y,z) is the point between (x_0,y_0,z_0) and

(x_1, y_1, z_1) where the light ray intersects the surface $z = f(x, y)$, then the optical path $V(x_0, y_0, z_0, x_1, y_1, z_1)$ is given by the equation

$$V = n_0 \sqrt{(x - x_0)^2 + (y - y_0)^2 + (z - z_0)^2}$$
$$+ n_1 \sqrt{(x - x_1)^2 + (y - y_1)^2 + (z - z_1)^2} . \qquad (25.11)$$

The point (x, y, z) can be found from the condition that V as a function of (x, y, z) has an extremum. By differentiation, we obtain

$$\frac{n_0(x - x_0) + n_0(z - z_0) f_x}{\sqrt{(x - x_0)^2 + (y - y_0)^2 + (z - z_0)^2}}$$

$$+ \frac{n_1(x - x_1) + n_1(z - z_1) f_x}{\sqrt{(x - x_1)^2 + (y - y_1)^2 + (z - z_1)^2}} = 0 ,$$

$$\frac{n_0(y - y_0) + n_0(z - z_0) f_y}{\sqrt{(x - x_0)^2 + (y - y_0)^2 + (z - z_0)^2}}$$

$$+ \frac{n_1(y - y_1) + n_1(z - z_1) f_y}{\sqrt{(x - x_1)^2 + (y - y_1)^2 + (z - z_1)^2}} = 0 .$$

$$(25.12)$$

We now introduce the optical direction cosines (p_0, q_0, r_0) and (p_1, q_1, r_1) of the ray. This yields

$$p_0 - p_1 + (r_0 - r_1) f_x = 0 ,$$
$$q_0 - q_1 + (r_0 - r_1) f_y = 0 . \qquad (25.13)$$

25.2 The angular characteristic $T(z_0, z_1; p_0, q_0, p_1, q_1)$ is given by

$$T = V + x_0 p_0 + y_0 q_0 - x_1 p_1 - y_1 q_1 . \qquad (25.21)$$

We know that T is a linear function of z_0 and z_1 which has the form

$$T = T_0(p_0, q_0, p_1, q_1) - z_0 r_0 + z_1 r_1 , \qquad (25.22)$$

where incidentally T_0 is the optical path between the base points Q_0 and Q_1 of the perpendiculars dropped onto the ray from the point $x = y = z = 0$. By introducing (25.22) in (25.21), we obtain the equation

$$T_0 = V(x_0, y_0, z_0; x_1, y_1, z_1) + x_0 p_0 + y_0 q_0 + z_0 r_0$$

$$- x_1 p_1 - y_1 q_1 - z_1 r_1 . \qquad (25.23)$$

The left side is entirely independent of the position of the points (x_0, y_0, z_0) and (x_1, y_1, z_1) on the ray. Hence we can let both points coincide with the point (x, y, z) on the refracting surface. This yields

$$T_0 = x(p_0 - p_1) + y(q_0 - q_1) + z(r_0 - r_1) , \qquad (25.24)$$

since $V(x, y, z; x, y, z) = 0$ on the surface. We understand that x, y and $z = f(x, y)$ are expressed as functions of (p_0, q_0, r_0) and (p_1, q_1, r_1) with the aid of (25.13).

We can formulate this result in a different way. Let us introduce in (25.24)

$$p_0 - p_1 = - (r_0 - r_1) f_x ,$$

$$q_0 - q_1 = - (r_0 - r_1) f_y , \qquad (25.25)$$

$$z = f (x, y) .$$

We obtain

$$T_0 = (r_1 - r_0)(x f_x + y f_y - f) . \qquad (25.26)$$

The equations

$$\xi = f_x ,$$

$$\eta = f_y , \qquad (25.27)$$

$$\Omega (\xi, \eta) = x f_x + y f_y - f .$$

represent a Legendre transformation of the function $f(x, y)$. Hence we can interpret the conditions (25.25) and (25.26) as follows: <u>The angular character-istic $T_0(p_0, q_0, p_1, q_1)$ of a surface $z = f(x, y)$ is given by the expression</u>

$$T_0 = (r_1 - r_0)\Omega \left(- \frac{p_1 - p_0}{r_1 - r_0} , - \frac{q_1 - q_0}{r_1 - r_0} \right) , \qquad (25.28)$$

<u>where $\Omega = \Omega(\xi, \eta)$ is obtained from $f(x, y)$ by the Legendre transformation</u> (25.27). The quantities r_0 and r_1 are functions of (p_0, q_0) and (p_1, q_1), respectively, namely,

$$r_0 = \sqrt{n_0^2 - p_0^2 - q_0^2} ,$$

$$r_1 = \sqrt{n_1^2 - p_1^2 - q_1^2} . \qquad (25.281)$$

The angular characteristic $T(z_0,z_1,p_0,q_0,p_1,q_1)$ for reference planes $z = z_0$ and $z = z_1$ then follows by (25.22); i.e.,

$$T = (r_1 - r_0)\Omega\left(-\frac{p_1 - p_0}{r_1 - r_0}, -\frac{q_1 - q_0}{r_1 - r_0}\right) - z_0 r_0 + z_1 r_1 . \qquad (25.29)$$

The formulae (25.28) and (25.29) also hold for a reflecting surface $z = f(x,y)$. However, in this case we have to use the relations

$$r_0 = \sqrt{1 - p_0^2 - q_0^2} ,$$

$$r_1 = -\sqrt{1 - p_1^2 - q_1^2} , \qquad (25.291)$$

assuming $n_0 = 1$.

25.3 <u>Surfaces of revolution.</u> By applying Legendre's transformation to a surface of revolution $z = f(\rho)$ where $\rho = \sqrt{x^2 + y^2}$, we obtain a function $\Omega(\lambda)$ where $\lambda = \sqrt{\xi^2 + \eta^2}$. The transformation is given by the equations

$$\lambda = f'(\rho)$$

$$\Omega(\lambda) = \rho f'(\rho) - f(\rho) . \qquad (25.31)$$

The angular characteristic of the surface thus has the form

$$T = (r_1 - r_0)\,\Omega\left(\sqrt{\frac{(p_1 - p_0)^2 + (q_1 - q_0)^2}{(r_1 - r_0)^2}}\right) - z_0 r_0 + z_1 r_1 . \quad (25.32)$$

This is a function of the three combinations

$$u = p_0^2 + q_0^2 ,$$

$$w = 2(p_0 p_1 + q_0 q_1) , \qquad (25.33)$$

$$v = p_1^2 + q_1^2 ,$$

namely,

$$T = (r_1 - r_0)\Omega\left(\sqrt{\frac{u + v - w}{(r_1 - r_0)^2}}\right) - z_0 r_0 + z_1 r_1 . \qquad (25.34)$$

For refracting surfaces: $r_0 = \sqrt{n_0^2 - u}, \ r_1 = \sqrt{n_1^2 - v}$.

For reflecting surfaces: $r_0 = \sqrt{1 - u}, \ \ r_1 = \sqrt{1 - v},$

where n is taken to be equal to unity.

25.4 We consider again the example of a spherical surface. We represent the sphere by the equation

$$z = f(\rho) = a - R\sqrt{1 - \frac{\rho^2}{R^2}} \qquad (25.41)$$

with the convention of §22.3 that the surface is convex if $R > 0$ and concave if $R < 0$. Legendre's transformation leads to

$$\Omega(\lambda) = -a + R\sqrt{1 + \lambda^2} . \qquad (25.42)$$

Hence, we obtain the angular characteristic

$$T = R(r_1 - r_0)\sqrt{1 + \frac{(p_1 - p_0)^2 + (q_1 - q_0)^2}{(r_1 - r_0)^2}} + (z_1 - a)r_1 - (z_0 - a)r_0$$

or

$$T = R \, \mathrm{sign} \, (r_1 - r_0)\sqrt{(p_1 - p_0)^2 + (q_1 - q_0)^2 + (r_1 - r_0)^2}$$
$$+ (z_1 - a)r_1 - (z_0 - a)r_0 . \qquad (25.43)$$

We verify readily that sign $(r_1 - r_0)$ = sign $(n_1 - n_0)$ in the case of refraction, and sign $(r_1 - r_0) = -1$ in the case of reflection. Hence it follows that

For refraction:

$$T = R \, \mathrm{sign} \, (n_1 - n_0)\sqrt{(p_1 - p_0)^2 + (q_1 - q_0)^2 + (r_1 - r_0)^2}$$
$$+ (z_1 - a)r_1 - (z_0 - a)r_0 ,$$
$$r_1 = \sqrt{n_1^2 - p_1^2 - q_1^2}; \qquad r_0 = \sqrt{n_0^2 - p_0^2 - q_0^2}; \qquad (25.44)$$

For reflection:

$$T = -R\sqrt{(p_1 - p_0)^2 + (q_1 - q_0)^2 + (r_1 - r_0)^2} + (z_1 - a)r_1$$
$$- (z_0 - a)r_0 ,$$
$$r_1 = -\sqrt{1 - p_1^2 - q_1^2}; \qquad r_0 = \sqrt{1 - p_0^2 - q_0^2} , \qquad (25.45)$$

where $n_0 = 1$. Both formulae are identical with the expressions derived in §22 by other methods. We remark that the result (25.45) applies to convex and concave mirrors if the above convention with respect to R is applied.

25.5 Let us next apply our method to the elliptic or hyperbolic paraboloid

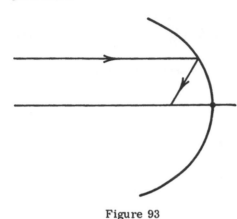

$$z = \frac{1}{2}(Ax^2 + By^2) . \quad (25.51)$$

Legendre's transformation yields

$$\Omega(\xi,\eta) = \frac{1}{2}\left(\frac{\xi^2}{A} + \frac{\eta^2}{B}\right) . \quad (25.52)$$

It follows that

$$T = \frac{1}{2(r_1 - r_0)}\left[\frac{(p_1 - p_0)^2}{A} + \frac{(q_1 - q_0)^2}{B}\right]$$

Figure 93

$$+ z_1 r_1 - z_0 r_0 . \quad (25.53)$$

In the special case $B = A = -a$, where $a > 0$, we obtain a concave paraboloid of revolution. The reflection on such a surface is determined by the characteristic

$$T = \frac{1}{2a}\frac{(p_1 - p_0)^2 + (q_1 - q_0)^2}{\sqrt{1 - u} \div \sqrt{1 - v}} - z_1\sqrt{1 - v} - z_0\sqrt{1 - u}, \quad (25.54)$$

or

$$T = \frac{1}{2a}\frac{u + v - w}{\sqrt{1 - u} + \sqrt{1 - v}} - z_1\sqrt{1 - v} - z_0\sqrt{1 - u} . \quad (25.55)$$

§26. THE ANGULAR CHARACTERISTIC FOR SYSTEMS OF REFRACTING SURFACES.

26.1 Let us consider an optical instrument which consists of a number of surfaces $z = f_1(x,y)$ with homogeneous media between them. We assume that the reference planes, z_0 and z_1, coincide at $z = 0$. Let $T_1(p_0,q_0,p_1,q_1)$ be the angular characteristic of the first surface, $T_2(p_1,q_1,p_2,q_2)$ that of the second surface, and in general $T_i(p_{i-1}, q_{i-1}, p_i, q_i)$ that of the i th surface. T_1 is the optical length of a ray between the base points Q_0 and Q_1, T_2 the path between Q_1 and Q_2, etc. It follows that the angular characteristic of the combined surfaces is given by the sum

$$T = \sum_{i=1}^{k} T_i(p_{i-1}, q_{i-1}, p_i, q_i) . \quad (26.11)$$

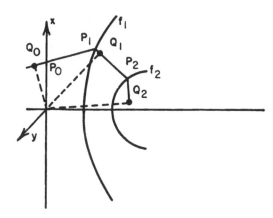

Figure 94

The final function T must be a function of the initial and final variables p_0, q_0 and p_k, q_k. Hence it is necessary to eliminate the intermediate variables $p_\gamma, q_\gamma, \gamma = 1, \ldots, k - 1$ on the right side of (26.11). We find the equations which are necessary for this elimination as follows: Let (x_1, y_1) be the point where the straight line $P_1 P_2$ intersects the plane $z = 0$, (x_2, y_2) the intersection of the line $P_2 P_3$ and in general (x_γ, y_γ) the intersection of $P_\gamma P_{\gamma+1}$. We have the relations

$$x_1 = - \frac{\partial T_1}{\partial p_1} = \frac{\partial T_2}{\partial p_1} ,$$

$$y_1 = - \frac{\partial T_1}{\partial q_1} = \frac{\partial T_2}{\partial q_1} , \tag{26.12}$$

and in general

$$x_\gamma = - \frac{\partial T_\gamma}{\partial p_\gamma} = \frac{\partial T_{\gamma+1}}{\partial p_\gamma} ,$$

$$\gamma = 1, 2, \ldots, k-1 \tag{26.13}$$

$$y_\gamma = - \frac{\partial T_\gamma}{\partial q_\gamma} = \frac{\partial T_{\gamma+1}}{\partial q_\gamma} .$$

It follows that

$$\frac{\partial (T_\gamma + T_{\gamma+1})}{\partial p_\gamma} = 0 ,$$

$$\gamma = 1, 2, \ldots, k-1 \tag{26.14}$$

$$\frac{\partial (T_\gamma + T_{\gamma+1})}{\partial q_\gamma} = 0 .$$

In principle these $2(k - 1)$ conditions allow us to express the $2(k - 1)$ intermediate variables p_γ, q_γ, as functions of the initial and final variables, p_0, q_0, p_k, q_k. The function $T(p_0, q_0, p_k, q_k)$ is then obtained by introducing these functions in (26.11).

We can interpret the conditions (26.14) in a different manner. The variables p_γ and q_γ are contained only in the two characteristics T_γ and $T_{\gamma+1}$ and not in the other functions T_i where i is different from γ and $\gamma+1$. Hence we can write (26.14) as follows:

$$\frac{\partial}{\partial p_\gamma} \sum_{i=1}^{k} T_i = \frac{\partial}{\partial q_\gamma} \sum_{i=1}^{k} T_i = 0 . \qquad (26.15)$$

These equations express the fact that the function $T = \sum_{i=1}^{k} T_i$ has a stationary value with respect to the $2(k-1)$ variables p_γ, q_γ; $\gamma = 1, \ldots, k-1$. Thus the problem of finding the angular characteristic T is equivalent to the problem of finding a stationary value of the sum (26.14) in the domain of the variables p_γ, q_γ. This value is a function of the quantities p_0, q_0, p_k, q_k.

The individual functions T_i are given by the expression

$$T_i = (r_i - r_{i-1})\Omega_i\left(-\frac{p_i - p_{i-1}}{r_i - r_{i-1}}, -\frac{q_i - q_{i-1}}{r_i - r_{i-1}}\right), \qquad (26.16)$$

where $\Omega_i(\xi,\eta)$ is obtained from $f(x,y)$ by Legendre's transformation. Thus we are led to the problem of finding a value of the sum

$$T = \sum_{i=1}^{k} (r_i - r_{i-1})\Omega_i\left[-\frac{p_i - p_{i-1}}{r_i - r_{i-1}}, -\frac{q_i - q_{i-1}}{r_i - r_{i-1}}\right] \qquad (26.17)$$

stationary with respect to variables $p_\gamma, q_\gamma, r_\gamma, \gamma = 1, \ldots, k-1$, which satisfy the conditions

$$p_\gamma^2 + q_\gamma^2 + r_\gamma^2 = n_\gamma^2, \qquad \gamma = 1, \ldots, k \qquad (26.18)$$

and with given boundary values $p_0, q_0, r_0 = \sqrt{n_0^2 - p_0^2 - q_0^2}$ and $p_k, q_k, r_k = \sqrt{n_k^2 - p_k^2 - q_k^2}$.

26.2 In the case of a medium with continuous index $n = n(x,y,z)$ we can consider the surfaces $n(x,y,z) = $ a constant as a continuous set of refracting surfaces. Let us assume that these surfaces are given in the form $z = f(x,y;s)$ in which s is a parameter. The sum (26.17) becomes the integral

$$T = \int_{s_0}^{s_1} \dot{r} \, \Omega\left(-\frac{\dot{p}}{\dot{r}}, -\frac{\dot{q}}{\dot{r}}, s\right) ds , \qquad (26.21)$$

where $\dot{p} = \dfrac{dp}{ds}$, $\dot{q} = \dfrac{dq}{ds}$, $\dot{r} = \dfrac{dr}{ds}$. $\Omega(\xi, \eta, s)$ is obtained from $f(x,y,s)$ by the Legendre transformation

$$\xi = f_x ,$$

$$\eta = f_y , \qquad\qquad (26.22)$$

$$\Omega = x\xi + y\eta - f .$$

The problem is to determine functions $p(s)$, $q(s)$, $r(s)$ which satisfy the relation $p^2 + q^2 + r^2 = n^2(s)$ and the boundary conditions

$$p(s_0) = p_0 , \qquad\qquad p(s_1) = p_1 ,$$

$$q(s_0) = q_0 , \qquad\qquad q(s_1) = q_1 ,$$

$$\qquad\qquad\qquad\qquad\qquad\qquad\qquad\qquad (26.23)$$

$$r(s_0) = r_0 = \sqrt{n_0^2 - p_0^2 - q_0^2} ,$$

$$r(s_1) = r_1 = \sqrt{n_1^2 - p_1^2 - q_1^2} ,$$

such that the integral (26.21) assumes a stationary value. This stationary value $T(p_0, q_0, p_1, q_1)$ is then the angular characteristic of the medium.

26.3 It is possible to characterize the angular characteristic by another problem of variation. We consider first a finite number of refracting surfaces $z = f_i(x,y)$. Let us denote by

$$x_i, y_i, z_i = f_i(x_i, y_i)$$

the coordinates of the point P_i (refer to Figure 94) where a ray intersects the surface $z = f_i(x,y)$. The optical length of the ray between the base points Q_0 and Q_k of the perpendiculars dropped from the point $(0,0,0)$ onto the initial and final ray is given by the expression

$$T = \sum_{i=1}^{k-1} n_i \sqrt{(x_{i+1} - x_i)^2 + (y_{i+1} - y_i)^2 + (z_{i+1} - z_i)^2}$$

$$+ x_1 p_0 + y_1 q_0 + z_1 r_0 - (x_k p_k + y_k q_k + z_k r_k) . \qquad (26.31)$$

We eliminate the variables x_i, y_i, z_i; $(i = 1, \ldots, k)$ by means of the condition that the sum (26.31) shall have a stationary value with respect to the variables x_i, y_i, z_i under the condition

$$z_i = f_i(x_i, y_i) . \qquad\qquad (26.32)$$

This new problem of variation leads to the same function $T(p_0, q_0; p_k, q_k)$ as the problem (26.17). Both problems are related to each other by the transformation of Friedrichs.[†] We demonstrate this fact by the analogy of the problem (26.31) in the case of a continuous medium. Let

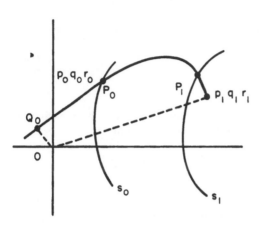

Figure 95

$$z = f(x,y,s) \quad s_0 \leqq s \leqq s_1$$

$$(26.33)$$

be a continuous set of refracting surfaces, i.e., surfaces $n(x,y,z) = $ a constant. We denote by $n(s)$ the value of $n(x,y,z)$ on the surface $z = f(x,y,s)$. A ray intersects the boundary surfaces s_0 and s_1 at P_0 and P_1 with the directions p_0, q_0, r_0 and p_1, q_1, r_1, respectively. The optical path between the base points Q_0 and Q_1 of the perpendiculars from 0 is given by

$$T = \int_{s_0}^{s_1} n\sqrt{\dot{x}^2 + \dot{y}^2 + \dot{z}^2} \; ds - x(s_1) p_1 - y(s_1) q_1 - z(s_1) r_1$$

$$+ x(s_0) p_0 + y(s_0) q_0 + z(s_0) r_0 \; . \qquad (26.34)$$

We now consider the problem of variation of determining a stationary value of (26.34) with regard to the functions $x(s)$, $y(s)$, $z(s)$ which satisfy the condition

$$z(s) = f\left(x(s), y(s), s \right) . \qquad (26.35)$$

26.4 Our aim is to show that the problem can be transformed into the problem (26.21) by the transformation of Friedrichs. For this purpose we introduce four Lagrangian multipliers

$$p(s), q(s), r(s), \text{ and } \lambda(s) , \qquad (26.41)$$

---

† Courant-Hilbert, Methoden der Math. Physik, Vol. 1, 2nd Ed. pp. 199-209.

and write (26.34) in the form

$$T = \int_{s_0}^{s_1} \left[n\sqrt{u^2 + v^2 + w^2} + p(\dot{x} - u) + q(\dot{y} - v) + r(\dot{z} - w) \right.$$

$$\left. + \lambda(f - z) \right] dz + x(s_0)p_0 + y(s_0)q_0 + z(s_0)r_0 - x(s_1)p_1$$

$$- y(s_1)q_1 - z(s_1)r_1 \tag{26.42}$$

in which $x(s)$, $y(s)$, $z(s)$, $u(s)$, $v(s)$, $w(s)$, and $\lambda(s)$, $p(s)$, $q(s)$, $r(s)$ are now treated as variables in the variation problem.

By carrying out the variation with regard to these functions, we obtain the conditions

$$n\frac{u}{\sqrt{u^2 + v^2 + w^2}} - p = 0, \qquad \lambda f_x - \dot{p} = 0,$$

$$n\frac{v}{\sqrt{u^2 + v^2 + w^2}} - q = 0, \qquad \lambda f_y - \dot{q} = 0,$$

$$n\frac{w}{\sqrt{u^2 + v^2 + w^2}} - r = 0, \qquad -\lambda - \dot{r} = 0,$$

$$\tag{26.43}$$

$$p(s_0) = p_0, \qquad p(s_1) = p_1,$$

$$q(s_0) = q_0, \qquad q(s_1) = q_1,$$

$$r(s_0) = r_0, \qquad r(s_1) = r_1,$$

and

$$\dot{x} - u = 0,$$

$$\dot{y} - v = 0, \qquad z - f(x,y,z) = 0. \tag{26.44}$$

$$\dot{z} - w = 0,$$

We can impose any of these conditions upon the problem (26.42) without influencing the solution since the solution of (26.42) must necessarily satisfy all conditions (26.43) and (26.44). If we impose the conditions (26.44), we obtain our original problem (26.34). The equations (26.43) then become the Euler equations of the problem (26.34). If, however, the conditions (26.43) are imposed upon (26.42), we obtain a new problem of variation which has the same extremum V as the problem (26.34). The Euler equations of this problem are precisely the equations (26.44), i.e., the conditions which were formerly imposed. This transformation of the original problem is the transformation of Friedrichs.

We consider first the integral

$$\int_{s_0}^{s_1} \left[n\sqrt{u^2 + v^2 + w^2} - pu - qv - rw \right] ds .$$
(26.45)

By introducing p,q,r from (26.43), we obtain the value zero for this integral. The integral

$$\int_{s_0}^{s_1} (p\dot{x} + q\dot{y} + r\dot{z}) \, ds$$
(26.46)

can be transformed into

$$- \int_{s_0}^{s_1} (x\dot{p} + y\dot{q} + z\dot{r}) ds + \left[xp + yq + rz \right]_{s_0}^{s_1} .$$
(26.461)

Hence it follows, with the aid of the last three rows of (26.43), that the expression (26.42) assumes the form:

$$T = - \int_{s_0}^{s_1} \left[x\dot{p} + y\dot{q} + z\dot{r} - \lambda(f - z) \right] ds .$$
(26.47)

We finally introduce

$$\lambda = \dot{r}, \quad \dot{p} = -\dot{r} f_x, \quad \dot{q} = -\dot{r} f_y,$$

and obtain

$$T = \int_{s_0}^{s_1} \dot{r} (x f_x + y f_y - f) \, ds ,$$
(26.471)

or with the aid of the function $\Omega(\xi, \eta, s)$ defined by (26.22):

$$T = \int_{s_0}^{s_1} \dot{r} \, \Omega \left(-\frac{\dot{p}}{\dot{r}}, -\frac{\dot{q}}{\dot{r}} \right) ds .$$
(26.48)

The functions p,q,r which are admissible in (26.48) must satisfy the relation

$$p^2 + q^2 + r^2 = n^2$$
(26.481)

and the boundary conditions

$$p(s_0) = p_0 , \qquad p(s_1) = p_1 ,$$

$$q(s_0) = q_0 , \qquad q(s_1) = q_1 , \qquad (26.482)$$

$$r(s_0) = r_0 , \qquad r(s_1) = r_1 ,$$

as follows from (26.43). This however is our former variation problem (26.21) for the function T.

It follows from the general theory of the above transformation that the stationary value of (26.48) is a <u>maximum</u> if the stationary value of (26.34) is a <u>minimum</u> and vice versa.

26.5 <u>Systems of spherical surfaces</u>. In the case of spherical surfaces

$$z = a_1 - R_1 \sqrt{1 - \frac{x^2 + y^2}{R_1^2}} \qquad (26.51)$$

we have obtained, for $\Omega_1(\xi, \eta)$, the expression

$$\Omega_1(\xi, \eta) = - a_1 + R_1 \sqrt{1 + \xi^2 + \eta^2} \qquad (26.52)$$

and thus

$$T_1 = - a_1(r_1 - r_{1-1})$$

$$+ R_1 \, \text{sign} \, (r_1 - r_{1-1}) \sqrt{(p_1 - p_{1-1})^2 + (q_1 - q_{1-1})^2 + (r_1 - r_{1-1})^2}.$$

By introducing the quantity

$$K_1 = R_1 \, \text{sign} \, (r_1 - r_{1-1}) = R_1 \, \text{sign} \, (n_1 - n_{1-1}) \qquad (26.53)$$

we are led to the problem of variation of finding a stationary value of the sum

$$T = \sum_{1=1}^{k} \left[K_1 \sqrt{(p_1 - p_{1-1})^2 + (q_1 - q_{1-1})^2 + (r_1 - r_{1-1})^2} \right.$$

$$\left. - a_1 (r_1 - r_{1-1}) \right] \qquad (26.54)$$

under the conditions that

$$p_1^2 + q_1^2 + r_1^2 = n_1^2 , \qquad (26.541)$$

and that $p_0, q_0, r_0; p_k, q_k, r_k$ have given values.

For a continuous set of spherical surfaces

$$z = a(s) - R(s)\sqrt{1 - \frac{x^2 + y^2}{R^2(s)}} \qquad (26.55)$$

the problem is to find a stationary value of the integral

$$T = \int_{s_0}^{s_1} \left[K(s)\sqrt{\dot{p}^2 + \dot{q}^2 + \dot{r}^2} - a\dot{r} \right] ds , \qquad (26.56)$$

where

$$p^2 + q^2 + r^2 = n^2(s) , \qquad (26.561)$$

and where $p(s)$, $q(s)$, $r(s)$ have given boundary values

$$p(s_0) = p_0 , \qquad\qquad p(s_1) = p_1 ,$$

$$q(s_0) = q_0 , \qquad\qquad q(s_1) = q_1 , \qquad (26.562)$$

$$r(s_0) = r_0 , \qquad\qquad r(s_1) = r_1 .$$

The function $K(s)$ is defined to be

$$K(s) = R(s) \text{ sign } n'(s) . \qquad (26.57)$$

§27. MEDIA OF RADIAL SYMMETRY.

27.1 Optical media in which the index of refraction is a function of the radius alone are of considerable theoretical interest. It is possible to integrate the differential equations of the light rays by quadratures and to determine optical systems of this type which represent, in a certain sense, perfect optical instruments.

The light rays in a medium of index $n = n(r)$ are plane curves; this can be shown by the same method as for a particle moving in the field of a central force. Without loss of generality we thus can limit the investigation to rays which lie in the xy-plane, i.e., to the problem of integrating the equation

$$\psi_x^2 + \psi_y^2 = n^2(r); \quad r = \sqrt{x^2 + y^2} . \qquad (27.11)$$

Let us assume that $n(r)$ is a continuous function. If $\psi = \psi(x,y;K)$ is a set of solutions of (27.11) which depends on the arbitrary parameter K then all the light rays of the xy-plane are given by Jacobi's theorem in the form

$$\frac{\partial \psi}{\partial K} = \alpha \qquad (27.12)$$

where α is an arbitrary constant. We can easily find such a set of solutions by introducing in polar coordinates in (27.11)

$$x = r \cos \theta$$

$$y = r \sin \theta \; .$$

(27.13)

It follows that:

$$\psi_r^2 + \frac{1}{r^2} \psi_\theta^2 = n^2(r)$$

(27.14)

which possesses the solution

$$\psi = K\theta \pm \int_{r_0}^r \sqrt{n^2 - \frac{K^2}{r^2}} \, dr$$

(27.15)

with an arbitrary constant K.

By applying (27.12) we obtain the result: The light rays in the xy-plane are given by the equation:

$$\theta - \theta_0 = \pm K \int_{r_0}^r \frac{dr}{r \sqrt{n^2 r^2 - K^2}} \; .$$

(27.16)

The integration constants θ_0, r_0 determine the origin of the ray, the constant K its direction at this point. Indeed from (27.16) it follows that

$$\theta' = \frac{d(\theta - \theta_0)}{dr} = \pm \frac{K}{r \sqrt{n^2 r^2 - K^2}}$$

and hence

$$K = \pm n \frac{r^2 \theta'}{\sqrt{1 + r^2 \theta'^2}} = \pm nr \sin \varphi$$

(27.18)

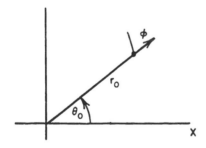

Figure 96

where φ is the angle which the ray makes with the radius vector. The equation (27.18) states that the expression nr sin φ is constant along one and the same ray. We notice that r sin φ is the length OQ of the perpendicular dropped from 0 onto the tangent of the ray at P.

Let us now determine the equation of a ray which originates at a point P_0 of the negative side of the x-axis and includes an angle α_0 with the x-axis.

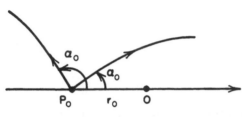

Figure 97

Figure 98

Since $\theta_0 = \pi$ and $K = n_0 r_0 \sin \alpha_0$ we find from (27.16)

$$\theta - \pi \qquad\qquad (27.19)$$

$$= \pm n_0 r_0 \sin \alpha_0 \int_{r_0}^{r} \frac{dr}{r\sqrt{n^2 r^2 - n_0^2 r_0^2 \sin^2 \alpha_0}}$$

If $\alpha_0 < \frac{\pi}{2}$ it is clear that both θ and r decrease at the beginning; hence the ray is represented by

$$\theta - \pi = + K \int_{r_0}^{r} \frac{dr}{r\sqrt{n^2 r^2 - K^2}} .$$
$$(27.191)$$

However, if $\alpha_0 > \frac{\pi}{2}$ then θ decreases and r increases. Hence

$$\theta - \pi = - K \int_{r_0}^{r} \frac{dr}{r\sqrt{n^2 r^2 - K^2}} . \qquad (27.192)$$

In both cases

$$K = n_0 r_0 \sin \alpha_0 \qquad\qquad (27.193)$$

i.e.,

$$0 \leq K \leq n_0 r_0 . \qquad\qquad (27.194)$$

27.2 We have to expect in general that r reaches a maximum or minimum along a given ray. From

$$\left(\frac{dr}{d\theta}\right)^2 = \frac{r^2}{K^2} (n^2 r^2 - K^2) \qquad\qquad (27.21)$$

if n(r) is continuous it follows that $\frac{dr}{d\theta}$ is a continuous function. Hence $\frac{dr}{d\theta} = 0$ determines a point where $r(\theta)$ reaches an extreme value. We conclude that the extreme values of r must be solutions of the equation

$$n^2 r^2 = K^2 . \qquad\qquad (27.22)$$

This result allows us to determine the general form of the light rays in our medium by the structure of the function

$$\rho(r) = nr .\qquad(27.23)$$

We illustrate this for two types of this function.

1. $\rho = nr$ is an increasing monotonic function.

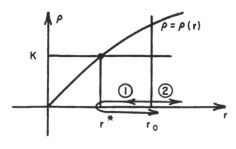

Figure 99

Let $r = r*$ be the solution of the equation $\rho^2 = K^2$. In the case of $\alpha_0 < \frac{\pi}{2}$ the integral in (27.191) has to be taken for values $r < r_0$ until the minimum value $r*$ is reached at a point $P*$. The angle $\theta*$ belonging to $r*$ is given by the integral

$$\theta* = \pi + K \int_{r_0}^{r*} \frac{dr}{r\sqrt{\rho^2 - K^2}} .\qquad(27.24)$$

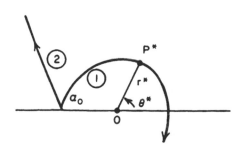

Figure 100

On the section of the ray beyond $P*$ the radius vector r increases again. Consequently this part of the ray is given by the integral

$$\theta = \theta* - K \int_{r*}^{r} \frac{dr}{r\sqrt{\rho^2 - K^2}} .\qquad(27.25)$$

At no point of the ray do we have $\rho^2(r) = K^2$ again, i.e., $r(\theta)$ increases monotonically. The two branches of the ray are symmetrical to the line $OP*$.

In case $\alpha_0 > \frac{\pi}{2}$ the integral (27.192) has to be taken for values $r > r_0$. Consequently no solution $r*$ of $\rho^2(r) = K^2$ is found on the path of integration, i.e., $r(\theta)$ increases monotonically. The two cases are illustrated in Fig. 100.

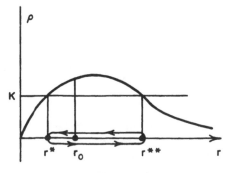

Figure 101

2.   The function $\rho = nr$ has a maximum and converges to zero if $r \rightarrow \infty$.

The equation $\rho^2 = K^2$ has two solutions r^* and r^{**}. The quantity K must be chosen so small that $r^* \leqq r_0 \leqq r^{**}$ which is the case if K satisfies the inequality (27.194). Let us consider the case $\alpha_0 < \frac{\pi}{2}$. On the initial part of the ray, we have

$$\theta - \pi = K \int_{r_0}^{r} \frac{dr}{r\sqrt{\rho^2 - K^2}} \tag{27.26}$$

with decreasing values of r. The radius $r(\theta)$ reaches its minimum $r = r^*$ at the angle given by

$$\theta^* = \pi + K \int_{r_0}^{r^*} \frac{dr}{r\sqrt{\rho^2 - K^2}} \tag{27.261}$$

and then increases according to the equation:

$$\theta = \theta^* - K \int_{r^*}^{r} \frac{dr}{r\sqrt{\rho^2 - K^2}} . \tag{27.27}$$

This equation represents the ray until r reaches its maximum r^{**} at the angle

$$\theta^{**} = \theta^* - K \int_{r^*}^{r^{**}} \frac{dr}{r\sqrt{\rho^2 - K^2}} . \tag{27.271}$$

After this $r(\theta)$ decreases again and the ray is given by

$$\theta = \theta^{**} + K \int_{r^{**}}^{r} \frac{dr}{r\sqrt{\rho^2 - K^2}} . \tag{27.28}$$

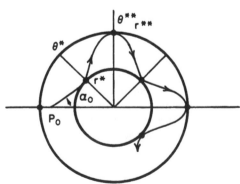

Figure 102

The curve $r = r(\theta)$ is symmetrical to both angular directions θ^* and θ^{**}. This obviously means that $r(\theta)$ is a periodic function of θ which has the period

$$P = 2K \int_{r^*}^{r^{**}} \frac{dr}{r\sqrt{\rho^2(r) - K^2}} . \tag{27.29}$$

The ray is a curve which oscillates back and forth between two circles of radius r^* and r^{**} and which are touched by the ray at equal angular intervals. This curve is closed if $\frac{P}{2\pi}$ is a rational number. In the case where $\alpha_0 > \frac{\pi}{2}$ we obtain in principle the same result. Starting from P_0 the ray now reaches first the maximum r^{**} instead of the minimum r^* as above.

27.3 Let us consider the special case

$$n^2 = C + \frac{1}{r} . \tag{27.31}$$

The light rays in this medium are identical with the paths of particles which move in a Coulomb field of potential $\phi = -1/2r$, and with the energy $C/2$. From (27.16) we obtain

$$\theta - \pi = K \int_{r_0}^{r} \frac{dr}{r\sqrt{Cr^2 + r - K^2}} \tag{27.32}$$

as the equation of the light rays. To evaluate the integral, we introduce

$$z = \frac{\dfrac{K}{r} - \dfrac{1}{2K}}{\sqrt{C + \dfrac{1}{4K^2}}} \tag{27.321}$$

as the variable of integration.

We have

$$dz = - \frac{K}{r^2 \sqrt{C + \dfrac{1}{4K^2}}} dr$$

and   (27.322)

$$1 - z^2 = \frac{C - \dfrac{K^2}{r^2} + \dfrac{1}{r}}{C + \dfrac{1}{4K^2}} .$$

From (27.322) it follows that

$$\theta - \pi = - \int_{z_0}^{z} \frac{dz}{\sqrt{1 - z^2}} = \text{arc sin } z_0 - \text{arc sin } z .$$

We denote the constant arc sin z_0 by $\beta - \pi/2$ and obtain

$$z = \cos(\theta - \beta) \tag{27.33}$$

i.e.

$$r = \frac{2K^2}{1 + \sqrt{1 + 4CK^2}\,\cos(\theta - \beta)} \tag{27.34}$$

as the equation of the light rays. The curves represented by (27.34) are conic sections. In order to determine the type of these conics, let us assume $\beta = 0$. All other light rays can be obtained from this one parameter set of curves by rotating the whole set about the origin.

We introduce $x = r\cos\theta$, $y = r\sin\theta$, and find from (27.34)

$$\sqrt{x^2 + y^2} + \sqrt{1 + 4CK^2}\,x = 2K^2 ,$$

or $\tag{27.35}$

$$4C^2\left(x - \frac{1}{2C}\sqrt{1 + 4CK^2}\right)^2 - \frac{C}{K^2}\,y^2 = 1 .$$

For a given C all these conics have the same principal axes $A = \dfrac{1}{2C}$. The eccentricity e is given by

$$e = \sqrt{\frac{1}{4C^2} + \frac{K^2}{C}} = \frac{1}{2C}\sqrt{1 + 4K^2C} .$$

Hence it follows that the point $x = y = 0$ is a common focal point of all the conics.

In the limit when $C \to 0$ equation (27.35) becomes

$$y^2 = 4K^2(K^2 - x) \tag{27.36}$$

which represents parabolae with the focal point at $x = y = 0$.

We summarize our result as follows: The light rays in a medium of refractive index

$$n^2 = C + \frac{1}{r}$$

are given by the following curves:

C > 0: Hyperbolae with the same principal axes A $= \frac{1}{2}$ C and the point x = y = 0 as common focal point.

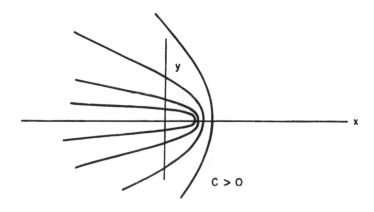

Figure 103

C = 0: Parabolae with x = y = 0 as common focal point.

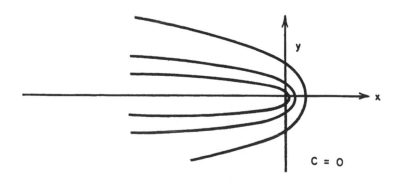

Figure 104

$C < 0$: Ellipses with the same principal axes $A = \dfrac{1}{2|C|}$ and $x = y = 0$ as common focal point

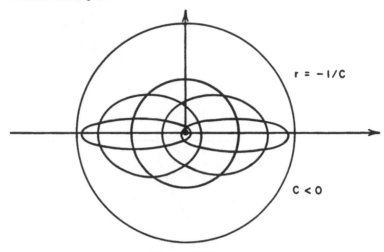

Figure 105

The three different cases are shown in Figs. 103, 104, and 105 with the understanding that all light rays will be obtained by rotating the set of curves about the point $x = y = 0$. In case $C < 0$ no light ray can penetrate into the region $r > -\dfrac{1}{2C}$.

§28. MAXWELL'S FISHEYE.

28.1 We shall consider in this section an optical medium which is characterized by the index of refraction

$$n = a/(b^2 + r^2), \qquad (28.11)$$

where a and b are constants. This optical system is known as Maxwell's fisheye. Without loss of generality we assume $a = b = 1$ and thus have the problem of determining the light rays for the case

$$n = 1/(1 + r^2). \qquad (28.12)$$

There exists a certain relationship between this medium and the example discussed in 27.3 where $n^2 = C + 1/r$. The equation of the wave fronts in the latter case is given by

$$\psi_x^2 + \psi_y^2 + \psi_z^2 = C + \frac{1}{\sqrt{x^2 + y^2 + z^2}}. \qquad (28.13)$$

Let us transform this equation by Legendre's transformation:

$$\xi = \psi_x \qquad\qquad x = \omega_\xi$$

$$\eta = \psi_y \qquad\qquad y = \omega_\eta \qquad\qquad \psi + \omega = x\xi + y\eta + z\zeta \qquad (28.14)$$

$$\zeta = \psi_z \qquad\qquad z = \omega_\zeta$$

whence

$$\xi^2 + \eta^2 + \zeta^2 = C + \frac{1}{\sqrt{\omega_\xi^2 + \omega_\eta^2 + \omega_\zeta^2}} \qquad (28.15)$$

or

$$\omega_\xi^2 + \omega_\eta^2 + \omega_\zeta^2 = \left(\frac{1}{-C + \xi^2 + \eta^2 + \zeta^2} \right)^2 . \qquad (28.16)$$

In the case where $C = -1$ this equation is identical with the equation for the wave fronts in a medium of index of refraction, $n = \dfrac{1}{1 + \xi^2 + \eta^2 + \zeta^2}$, i.e., in Maxwell's fish eye. By submitting the wave fronts of the "potential field" $n^2 = \dfrac{1}{r} - 1$ to a Legendre transformation, we obtain the wave fronts of Maxwell's fish eye.

28.2 By applying the formula (27.191), we obtain the light rays in our medium,

$$\theta - \pi = \int_{r_0}^{r} \frac{K(1 + r^2)dr}{r\sqrt{r^2 - K^2(1 + r^2)^2}} . \qquad (28.21)$$

It is readily verified that the integrand is the derivative of the function $\text{arc sin} \left(\dfrac{r^2 - 1}{r} \dfrac{K}{1 - 4K^2} \right)$. Hence,

$$\frac{r^2 - 1}{r} = \frac{\sqrt{1 - 4K^2}}{K} \sin (\theta - \pi + C) \qquad (28.22)$$

where C is a constant. Let us denote this constant by $\beta - \dfrac{3\pi}{2}$. This yields

$$r^2 - r \frac{\sqrt{1 - 4K^2}}{K} \cos (\theta - \beta) - 1 = 0 \qquad (28.23)$$

as the equation of the light rays in our medium.

In order to determine the type of curves given by (28.23) let us assume $\beta = 0$ and introduce Cartesian coordinates

$$x = r \cos \theta \; ,$$

$$y = r \sin \theta \; .$$

From (28.23) we obtain

$$\left(x - \frac{\sqrt{1 - 4K^2}}{2K} \right)^2 + y^2 = \frac{1}{4K^2} \; . \tag{28.24}$$

Letting $R = \frac{1}{2K}$, we obtain

$$\left(x - \sqrt{R^2 - 1} \right)^2 + y^2 = R^2 \; . \tag{28.25}$$

This is the set of circles which go through the points $y = \pm 1$ of the y-axis. The entire manifold of light rays in the xy-plane thus is found to be given by the set of all circles which go through two points on opposite ends of a diameter of the unit circle.

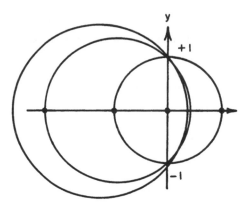

28.3 Let us now consider the bundle of rays through a given point x_0 of the x-axis. By introducing Cartesian coordinates in the general equation (28.23), we find a generalization of (28.25), namely,

$$\left(x - \sqrt{R^2 - 1} \cos \beta \right)^2$$

$$+ \left(y - \sqrt{R^2 - 1} \sin \beta \right)^2 = R^2 \; .$$

Figure 106   (28.26)

The points of intersection of these circles with the x-axis satisfy the equation

$$\left(x - \sqrt{R^2 - 1} \cos \beta \right)^2 + (R^2 - 1) \sin^2 \beta = R^2$$

or

$$x^2 - 2x \sqrt{R^2 - 1} \cos \beta - 1 = 0 \; . \tag{28.27}$$

For the product of the two solutions x_0 and x_1 of this quadratic equation, we obtain

$$x_0 x_1 = -1 \tag{28.271}$$

i.e., a relation independent of the particular parameters R and β of the ray. This demonstrates that all rays through the point x_0 intersect each other at

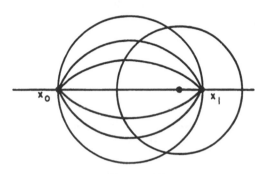

the point $x_1 = -1/x_0$. The same consideration can be carried out for points of any straight line through the origin other than the x-axis. Thus we have the result: All the rays through a point x_0, y_0 intersect each other at the point

Figure 107

$$x_1 = -\frac{x_0}{r_0^2}$$

$$(28.28)$$

$$y_1 = -\frac{y_0}{r_0^2}$$

i.e. every point x_0, y_0 possesses a perfect conjugate point x_1, y_1 which lies on the same radial line as x_0, y_0, but on the opposite side of the origin. The distances of the two points from 0 are related by the equation

$$r_0 r_1 = 1 .$$

$$(28.281)$$

To any sphere of radius r_0 there belongs a conjugate sphere of radius $r_1 = 1/r_0$ which is a perfect and undistorted optical image of the sphere of radius r_0. This image, of course, is inverted; its magnification is $M = -r_1/r_0$.

28.3 The surprising properties of an optical medium of refractive index $n = \frac{1}{1 + r^2}$ find their explanation by the fact that the line element

$$ds^2 = 4 \frac{dx^2 + dy^2}{(1 + r^2)^2}$$

$$(28.31)$$

can be interpreted as the line element of the sphere. We can map the points of a unit sphere by a stereographic projection on the points of the xy-plane. Let us determine the formulae which represent this projection. From Figure 108 it follows that

$$\frac{X}{1 - Z} = x , \qquad \frac{Y}{1 - Z} = y .$$

$$(28.32)$$

Hence

$$X^2 + Y^2 = 1 - Z^2 = r^2(1 - Z)^2$$

$$(28.321)$$

in which $r^2 = x^2 + y^2$.

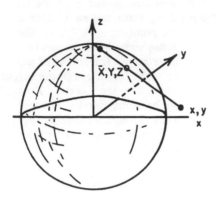

Figure 108

This yields

$$r^2 = \frac{1 + Z}{1 - Z}$$

and   (28.322)

$$Z = \frac{r^2 - 1}{r^2 + 1} .$$

By introducing this expression for Z in (28.32), we obtain

$$X = \frac{2x}{1 + r^2}, \qquad Y = \frac{2y}{1 + r^2},$$

$$Z = \frac{r^2 - 1}{r^2 + 1} . \qquad (28.33)$$

We can consider these formulae as a parametric representation of the unit sphere with the coordinates x,y as parameters. The line element of the sphere in these parameters is given by

$$ds^2 = dX^2 + dY^2 + dZ^2 = 4 \left(d \frac{x}{1 + r^2} \right)^2 + 4 \left(d \frac{y}{1 + r^2} \right)^2 + \left(d \frac{r^2 - 1}{r^2 + 1} \right)^2$$

which gives

$$ds^2 = \frac{4}{(1 + r^2)^2} (dx^2 + dy^2) . \qquad (28.34)$$

We recognize that ds^2 has the form of an optical line element

$$ds^2 = n^2 (dx^2 + dy^2)$$

where $n = \frac{2}{1 + r^2}$ is identical with the function $n(r)$ which characterizes Maxwell's fisheye with the exception of an insignificant factor.

The light rays in this medium therefore must be those curves which are stereographic images of the geodetic lines of the unit sphere, that is, of great circles. We know that the great circles passing through a point (X_0, Y_0, Z_0) of the sphere intersect each other again at the point

$$X_1 = - X_0, \qquad Y_1 = - Y_0, \qquad Z_1 = - Z_0 .$$

The stereographic images of this set of circles are the light rays through the point

$$x_0 = \frac{X_0}{1 - Z_0}, \qquad y_0 = \frac{Y_0}{1 - Z_0} \qquad\qquad (28.35)$$

of the xy-plane. It follows that these rays intersect each other again at the point

$$x_1 = - \frac{X_0}{Z_0 + 1}, \qquad y_1 = - \frac{Y_0}{Z_0 + 1}. \qquad\qquad (28.36)$$

Hence

$$\frac{x_1}{x_0} = \frac{y_1}{y_0} = - \frac{1 - Z_0}{1 + Z_0}$$

or, from (28.322):

$$\frac{x_1}{x_0} = \frac{y_1}{y_0} = - \frac{1}{r_0^2}. \qquad\qquad (28.37)$$

This, however, is the result expressed in (28.28).

From the fact that the stereographic image of an arbitrary circle on the sphere is a circle in the xy-plane it follows that in particular the light rays in Maxwell's fisheye are circles. Since a great circle intersects the equator at two opposite points, we find that all light rays must intersect the unit circle (the image of the equator) in points on opposite ends of a diameter.

Let us now consider a curve C on the sphere. It is clear that the conjugate curve C', i.e., the "antipode" of C has the same length as C. Let c and c' be the stereographic images of C and C' in the xy-plane. This means that c' is the optical image of c, i.e., the curve which is formed by the points conjugate to the points of c. From

$$\int_C ds = \int_{C'} ds'$$

on the sphere we obtain

$$\int_c n\, ds = \int_{c'} n'\, ds',$$

i.e., <u>conjugate curves have the same optical length</u>.

We have demonstrated this result only for curves in the xy-plane. It is, however, easy to prove it more generally for any curve c in the x,y,z space.

28.4 <u>Perfect optical instruments in the xy-plane.</u> The stereographic projection of the sphere is a <u>conformal projection</u>. This follows from the result (28.34) that the line element $ds^2 = dX^2 + dY^2 + dZ^2$ of the sphere has the form $\lambda(x,y)(dx^2 + dy^2)$ which is characteristic of conformal mapping in general. Let us now map the x,y-plane conformally upon itself, i.e., a transformation

$$x = u(\xi, \eta) ,$$

$$y = v(\xi, \eta)$$

$$(28.41)$$

where u and v are the real and imaginary parts of an analytic function

$$z = f(\xi + i\eta) = u(\xi, \eta) + iv(\xi, \eta) .$$

It follows from the Cauchy-Riemann equations that

$$|dz|^2 = dx^2 + dy^2 = |f'|^2 (d\xi^2 + d\eta^2)$$

$$(28.42)$$

where $|f'|^2 = u_\xi^2 + v_\xi^2 = u_\eta^2 + v_\eta^2$.

The line element (28.34) in these new coordinates assumes the form,

$$ds^2 = \frac{4}{(1 + r^2)^2} |f'|^2 (d\xi^2 + d\eta^2) ,$$

$$(28.43)$$

of an optical line element in the $\xi\eta$-plane with the index of refraction

$$n(\xi, \eta) = \frac{2}{(1 + r^2)} |f'| = \frac{2}{1 + |f|^2} |f'| .$$

$$(28.44)$$

The light rays of this medium are the curves into which the great circles of the sphere are transformed by applying stereographic projection and the conformal transformation (28.41) in succession.

Obviously it is still true that all the rays through a point (ξ_0, η_0) intersect each other at a conjugate point (ξ_1, η_1). Hence it follows that <u>every medium in the xy-plane which has an index of refraction</u>

$$n = \frac{2|f'(z)|}{1 + |f^2(z)|} ; \quad z = x + iy$$

$$(28.45)$$

<u>where f(z) is analytic, represents a perfect optical instrument in the following sense:</u> Every point x_0, y_0 of the xy-plane has a perfect conjugate point x_1, y_1 in the plane.

The light rays in all these media can be obtained by conformal mapping of the sphere onto the plane, namely as curves which correspond to the great circles of the sphere. With the aid of (28.26) one can easily prove that these rays are given by the equation

$$|f|^2 + (af + \bar{a}f) - 1 = 0 \tag{28.46}$$

where $a = \alpha + i\beta$ is an arbitrary complex number.

Let us consider the function

$$f(z) = z^\gamma, \quad \gamma \geqq 1 \tag{28.47}$$

as an example. We obtain a medium of radial symmetry, namely

$$n(r) = \frac{2\gamma r^{\gamma-1}}{1 + r^{2\gamma}}, \quad r = \sqrt{x^2 + y^2}. \tag{28.471}$$

The light rays in this medium are the curves

$$r^{2\gamma} + r^\gamma A \cos(\theta - \beta) - 1 = 0 \tag{28.472}$$

where A and β are arbitrary real constants, as follows from (28.46).

Since n(r) is a function of $\sqrt{x^2 + y^2}$ in the xy-plane we conclude that a medium of refractive index

$$n = \frac{r^{\gamma-1}}{1 + r^{2\gamma}}$$

in the xyz-space with $r = \sqrt{x^2 + y^2 + z^2}$ is perfect in the sense of our above definition not only for points of the xy-plane but for any points of the x,y,z space. Obviously, our result is a direct generalization of Maxwell's fisheye which is obtained by taking $\gamma = 1$.

Another example is given by the function

$$f(z) = e^{iz}. \tag{28.48}$$

This gives the refractive index

$$n = \frac{1}{\cosh y}. \tag{28.481}$$

The rays satisfy the equation

$$\sinh y = A \sin(x + \alpha) \tag{28.482}$$

with arbitrary constants A and α.

A straight line $x = x_0$ has a perfect image on the line $x = x_0 + \pi$; a point of the "object line" $x = x_0$ has a perfect image at the point $y_1 = -y_0$ of the "image line", $x = x_0 + \pi$. This example applies only to the xy-plane. Contrary to the former example it is not possible to find a medium in the x,y,z-space which corresponds to this example and forms a perfect optical instrument in the x,y,z-space.

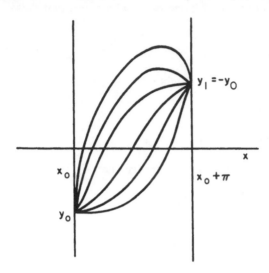

Figure 109

28.5 Light rays which remain in one and the same plane play an important part in the investigation of optical instruments of revolution. Let us assume that a medium is symmetrical with respect to the x-axis so that $n = n\left(x, \sqrt{y^2 + z^2}\right)$. Every ray which intersects the x-axis is plane, i.e., lies in a meridional plane through the axis of the instrument. We may assume that this plane is the xy-plane. The rays in this plane, which is often called the primary plane of the instrument, are given by the geodetic lines of the line element

$$ds^2 = n^2(x,y)(dx^2 + dy^2) .$$

(28.51)

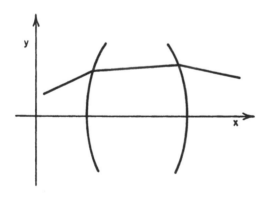

Figure 110

Let $x = x(u,v)$, $y = y(u,v)$, $z = z(u,v)$ be the parametric representation of a surface in the x,y,z space. Its line element ds^2 has the form

$$ds^2 = E\, du^2 + 2F\, du\, dv + G dv^2$$

(28.52)

where E,F,G are functions of u and v.

One can show that it is possible by a transformation of the parameters

$$u = u(x,y)$$

$$v = v(x,y)$$

to transform the line element (28.51) into the form of an optical line element

$$ds^2 = n^2(x,y)(dx^2 + dy^2) \ . \tag{28.53}$$

The corresponding parametric representation of the surface

$$X = X(x,y); \quad Y = Y(x,y); \quad Z = Z(x,y) \tag{28.54}$$

can be considered as a conformal mapping of the surface onto the xy-plane. Thus the light rays of a medium of refractive index $n(x,y)$ in the xy-plane are the images of the geodetic lines of a surface projected conformally onto the plane.

As in the case of the sphere there exist many conformal projections (28.54) for a given surface. From one projection (28.54) we can find others by a conformal transformation of the plane onto itself. If $x = u(\xi,\eta)$, $y = v(\xi,\eta)$ are the real and imaginary parts of an analytic function

$$f(\zeta) = u(\xi,\eta) + iv(\xi,\eta); \quad \zeta = \xi + i\eta$$

we obtain a new conformal projection by introducing u and v in (28.54). The line element (28.53) becomes

$$ds^2 = n^2(u,v) \ |f'|^2 \ (d\xi^2 + d\eta^2) \tag{28.55}$$

which corresponds to an optical medium of refractive index

$$n^*(\xi,\eta) = n\left(u(\xi,\eta) , \ v(\xi,\eta)\right) |f'| \ . \tag{28.56}$$

If the light rays of (28.53) are given in the form

$$g(x,y,a,b) = 0 \tag{28.57}$$

with arbitrary parameters a,b then we obtain the light rays of the medium in the form

$$g\left(u(\xi,\eta), \ v(\xi,\eta), \ a, \ b\right) = 0 \ . \tag{28.571}$$

The results derived in (28.4) are a special application of this theorem.

Let us now consider the case of a homogeneous medium where
$n(x,y) = 1$. The rays of this medium have the form

$$x \cos \vartheta + y \sin \vartheta = C$$

where C and ϑ are arbitrary constants. From (28.56) and (28.571) it follows
that the light rays of a medium of refractive index

$$n = |f'(\xi + i\eta)| \qquad (28.58)$$

are given by the curves

$$u \cos \vartheta + v \sin \vartheta = C \qquad (28.581)$$

in which u and $v(\xi,\eta)$ are real and imaginary part of the arbitrary analytic
function $f = u + iv$.

The problem of optical design is to find a medium such that the light
rays through an arbitrary point of a finite section of the object plane intersect
each other at a conjugate point of the image plane. This necessarily implies
the same condition for the plane rays of the xy-plane and hence for the
geodetic lines of the surface (28.54). We therefore recognize the close
relationship of the problem of optical design to a problem in differential
geometry which could be formulated as follows: To determine surfaces such
that the geodetic lines through at least a one-parameter set of points on the
surface intersect each other in a set of perfect conjugate points. The sphere
is the simplest example of such a surface.

Any surface of this type determines an instrument of revolution such
that the rays of the primary plane produce a sharp image of a certain curve
in this plane. Among these instruments we have to determine those in which
the skew rays, i.e., the rays which do not intersect the axis focus at the same
points as the primary rays. Only if this additional condition is satisfied can
the instrument be considered as perfect.

§29. OTHER OPTICAL MEDIA WHICH IMAGE A SPHERE ONTO A SPHERE.

29.1 We have found in Maxwell's fisheye an optical instrument which
forms a perfect image of an arbitrary sphere about the origin. Both the
object and the image sphere are located in a region where the index of refraction varies. This leads us to the question of whether it is possible to find
a medium which is homogeneous inside this sphere such that two given spheres
in the homogeneous part are perfect conjugate spheres. We shall see that the
answer is in the affirmative and that the problem can be solved in many ways.
We remark that our condition refers only to two given spheres and requires
nothing of other spheres.

Let us assume that the boundary of the non-homogeneous medium is a sphere of radius 1. Let r_0 be the radius of the object sphere and r_1 the radius of the image sphere. The index of refraction shall be a function of $\gamma, n = n(r)$, $r < 1$; $n = 1$, $r \geq 1$. We assume $n(1) = 1$ so that $n(r)$ is a continuous function. On account of the radial symmetry of the medium it is sufficient to require that all rays passing through a point $x_0 = -r_0$; $y_0 = 0$ which enter the glass sphere pass through the point $x_1 = r_1, y_1 = 0$.

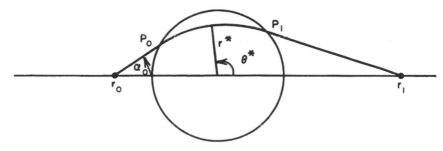

Figure 111

Since $r = 1$ at the points P_0 and P_1 where a ray enters and leaves the unit sphere, we conclude that $r(\theta)$ must reach a minimum r^* at a certain angle θ^*. Let us simplify our problem by excluding functions $n(r)$ which introduce more than one extreme value r^* along the rays. In other words, let us assume that the function $\rho = rn(r)$ increases monotonically so that the equation

$$\rho^2 = n^2 r^2 = K^2 = r_0^2 \sin^2 \alpha_0 \qquad (29.11)$$

has only one positive solution $r^* < 1$. The equation of the light ray for $\theta < \theta^*$ is given by equation (27.25) namely:

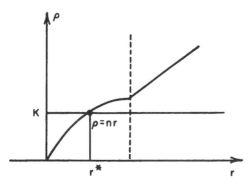

Figure 112

$$\theta = \theta^* - K \int_{r^*}^{r} \frac{dr}{r\sqrt{\rho^2 - K^2}} \qquad (29.12)$$

in which

$$\theta^* = \pi + K \int_{r_0}^{r^*} \frac{dr}{r\sqrt{\rho^2 - K^2}} \,. \qquad (29.13)$$

It follows that the ray intersects the positive side of the x-axis at a point $\theta = 0$, $r = r_1$ where r_1 satisfies the equation

$$- \pi = K \int_{r_0}^{r^*} \frac{dr}{r\sqrt{\rho^2 - K^2}} - K \int_{r^*}^{r_1} \frac{dr}{r\sqrt{\rho^2 - K^2}} . \quad (29.14)$$

If $\rho = nr$ is given then this equation determines the intersection r_1 with the axis as a function of $K = r_0 \sin \alpha_0$, i.e., of the direction of the incident ray. If, however, r_1 is a given constant then (29.14) represents an integral equation for the function $\rho = \rho(r)$ for $r < 1$. For $r \geq 1$ we have $\rho = r$. Since

$$K \int \frac{dr}{r\sqrt{r^2 - K^2}} = - \arcsin \frac{K}{r}$$

from (29.14) we obtain the condition

$$K \int_{r^*}^{1} \frac{dr}{r\sqrt{\rho^2(r) - K^2}} = f(K) \quad (29.15)$$

where $f(K)$ is the function

$$f(K) = \frac{1}{2} \left(\pi + \arcsin \frac{K}{r_1} + \arcsin \frac{K}{r_0} - 2 \arcsin K \right). \quad (29.16)$$

29.2 In order to transform (29.15) into an integral equation of a known type, let us first introduce the variable

$$\tau = \log r . \quad (29.21)$$

Then it follows that

$$K \int_{-\infty}^{0} \frac{d\tau}{\sqrt{\rho^2(\tau) - K^2}} = f(K) . \quad (29.22)$$

We now define the measure function $\Omega(\rho)$ as the measure of all τ-regions in the interval $-\infty < \tau \leq 0$ where $\rho(\tau) > \rho$. Since we have assumed that $\rho = \rho(r)$ and hence $\rho = \rho(\tau)$ is a monotonic function, we see that in our case $\Omega(\rho)$ is nothing but

$$\Omega(\rho) = - \tau(\rho) = - \log r(\rho) , \quad (29.23)$$

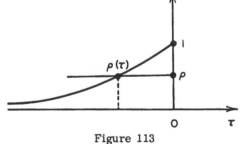

Figure 113

i.e., the abscissa $- \tau(\rho)$ where the curve $\rho(\tau)$ reaches the distance ρ from the τ-axis. Evidently, we have

$\Omega(\rho) = 0$ for $\rho \geqq 1$. The difference

$$\Omega(\rho_1) - \Omega(\rho_2)$$

measures the τ interval where $\rho_1 < \rho(\tau) \leqq \rho_2$.

With the aid of this function $\Omega(\rho)$ we can write (29.22) in the form

$$- K \int_K^1 \frac{d\Omega(\rho)}{\sqrt{\rho^2 - K^2}} = f(K) \ . \tag{29.24}$$

This is an integral equation of Abel's type.

29.3 For the solution of this equation we shall prove the following theorem.

If the function $f(K)$ is defined by the integral

$$f(K) = - K \int_K^1 \frac{d\Omega(\rho)}{\sqrt{\rho^2 - K^2}} \tag{29.31}$$

in the interval $0 \leqq K \leqq \lambda$ then $\Omega(\rho)$ is determined by the integral

$$\Omega(\rho) - \Omega(\lambda) = \frac{2}{\pi} \int_\rho^\lambda \frac{f(K)}{\sqrt{K^2 - \rho^2}} \, dK \ . \tag{29.32}$$

We prove this inversion theorem as follows: We multiply (29.31) by $\dfrac{2}{\sqrt{K^2 - \rho^2}}$ and integrate with respect to K from ρ to λ. It follows that

$$2 \int_\rho^\lambda \frac{f(K)}{\sqrt{K^2 - \rho^2}} \, dK = - \int_\rho^\lambda \frac{2K}{\sqrt{K^2 - \rho^2}} \int_K^\lambda \frac{d\Omega(s)}{\sqrt{s^2 - K^2}} \, dK \ . \tag{29.33}$$

By interchanging the order of integration we obtain

$$2 \int_\rho^\lambda \frac{f(K)}{\sqrt{K^2 - \rho^2}} \, dK = - \int_\rho^\lambda d\Omega(s) \int_\rho^s \frac{2K \, dK}{\sqrt{K^2 - \rho^2} \sqrt{s^2 - K^2}} \ . \tag{29.34}$$

We transform the inner integral by introducing the integration variable z given by

$$K^2 = (s^2 - \rho^2) z + \rho^2 \tag{29.35}$$

and the consequent relations

$$2KdK = (s^2 - \rho^2)dz ,$$

$$K^2 - \rho^2 = (s^2 - \rho^2)z ,$$ (29.36)

$$s^2 - K^2 = (s^2 - \rho^2)(1 - z) .$$

We obtain

$$2 \int_\rho^\lambda \frac{f(K)}{\sqrt{K^2 - \rho^2}} dK = \int_\rho^\lambda d\Omega(s) \int_0^1 \frac{dz}{\sqrt{z(1 - z)}} = \pi \int_\rho^\lambda d\Omega(s) \quad (29.37)$$

whence we obtain (29.32).

29.4 We carry out the integration (29.32) first for the function

$$\varphi(K) = \frac{\pi}{2} - \text{arcsin } K \qquad (29.41)$$

which is a part of the function (29.16). We can write $\varphi(K)$ in the form

$$\varphi(K) = + K \int_K^1 \frac{d\rho}{\rho\sqrt{K^2 - \rho^2}} = + K \int_K^1 \frac{d \log \rho}{\sqrt{K^2 - \rho^2}} . \qquad (29.42)$$

By applying the inversion theorem to the special case of (29.31), in which $\Omega = \log \rho$, we obtain

$$- \log \rho = \frac{2}{\pi} \int_\rho^1 \frac{\varphi(K)}{\sqrt{K^2 - \rho^2}} dK . \qquad (29.43)$$

With the aid of this result and by applying our theorem to the function (29.16) we obtain:

$$\Omega(\rho) - \Omega(1) = - \log \rho + \frac{1}{\pi} \int_\rho^1 \left(\text{arcsin } \frac{K}{r_1} + \text{arcsin } \frac{K}{r_0} \right) \frac{dK}{\sqrt{K^2 - \rho^2}} . \qquad (29.44)$$

Let us introduce the function

$$\omega(\rho,a) = \frac{1}{\pi} \int_\rho^1 \frac{\text{arcsin } \dfrac{t}{a}}{\sqrt{t^2 - \rho^2}} dt . \qquad (29.45)$$

Since $\Omega(1) = 0$ and $\Omega(\rho) = -\log r(\rho)$, from (29.44) we obtain:

$$\log \frac{\rho}{r} = \omega(\rho, r_0) + \omega(\rho, r_1) \ . \tag{29.46}$$

By introducing $\rho = nr$ this yields the relation

$$\log n = \omega(\rho, r_0) + \omega(\rho, r_1) \ . \tag{29.47}$$

This equation together with $\rho = nr$ determines the function $n = n(r)$ in parametric form:

$$r = \rho e^{-\omega(\rho, r_0) - \omega(\rho, r_1)} \ ,$$

$$n = e^{\omega(\rho, r_0) + \omega(\rho, r_1)} \ . \tag{29.48}$$

29.5 Let us consider the case when $r_0 = \infty$ and $r_1 = 1$ as an example. We have $\omega(\rho, \infty) = 0$ and

$$\omega(\rho, 1) = \frac{1}{\pi} \int_\rho^1 \frac{\arcsin t}{\sqrt{t^2 - \rho^2}} \, dt \ . \tag{29.51}$$

We can evaluate this integral explicitly with the aid of the relation (29.43). We obtain the relation

$$2\omega(\rho, 1) = \log \rho + \int_\rho^1 \frac{dK}{\sqrt{K^2 - \rho^2}} = \log\left(1 + \sqrt{1 - \rho^2}\right),$$

i.e.,

$$\omega(\rho, 1) = \frac{1}{2} \log\left(1 + \sqrt{1 - \rho^2}\right) \ . \tag{29.52}$$

The two equations (29.48) give for $n(r)$ the explicit expression

$$n(r) = \sqrt{2 - r^2} \ . \tag{29.53}$$

This is a function which decreases gradually from the central value $n = \sqrt{2}$ to $n = 1$ at $r = 1$.

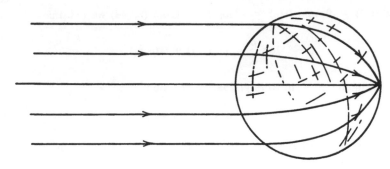

Figure 114

§30. OPTICAL INSTRUMENTS OF REVOLUTION.

Almost every optical instrument is symmetrical with respect to an axis. Let us assume this axis to be the z-axis of our coordinate system. The index of refraction is a function

$$n = n(\rho, z)$$

of z and $\rho = \sqrt{x^2 + y^2}$.

30.1 Let us first consider Fermat's integral

$$V = \int_{z_0}^{z_1} n(\rho, z) \sqrt{1 + \dot{x}^2 + \dot{y}^2} \, dz \tag{30.11}$$

and introduce polar coordinates

$$x = \rho \cos \theta \,, \tag{30.12}$$

$$y = \rho \sin \theta \,.$$

It follows that

$$V = \int_{z_0}^{z_1} n(\rho, z) \sqrt{1 + \dot{\rho}^2 + \rho^2 \dot{\theta}^2} \, dz \,. \tag{30.13}$$

The variation with respect to θ yields

$$\frac{d}{dz} \left(\frac{n \rho^2 \dot{\theta}}{\sqrt{1 + \dot{\rho}^2 + \rho^2 \dot{\theta}^2}} \right) = 0 \tag{30.14}$$

and hence

$$\frac{n\rho^2\dot\theta}{\sqrt{1 + \dot\rho^2 + \rho^2\dot\theta^2}} = h, \tag{30.15}$$

where h is a constant.

We can interpret this result as follows: We introduce the optical direction cosines

$$p = n\frac{\dot x}{\sqrt{1 + \dot x^2 + \dot y^2}}, \qquad q = n\frac{\dot y}{\sqrt{1 + \dot x^2 + \dot y^2}}. \tag{30.16}$$

Then it follows that

$$xq - yp = n\frac{x\dot y - y\dot x}{\sqrt{1 + \dot x^2 + \dot y^2}} \tag{30.17}$$

and with the aid of (30.12)

$$xq - yp = \frac{n\rho^2\dot\theta}{\sqrt{1 + \dot\rho^2 + \rho^2\dot\theta^2}} = h. \tag{30.18}$$

This means: The expression xq - yp is a constant along a given ray.

30.2 We can use the above result to transform Fermat's problem into a problem of variation for the function $\rho(z)$ alone. We proceed in a manner similar to that used before by applying Friedrich's method of transformation. The problem is to find two functions $\rho(z)$ and $\theta(z)$ with given boundary values

$$\rho(z_0) = \rho_0; \qquad\qquad \rho(z_1) = \rho_1$$
$$\theta(z_0) = \theta_0; \qquad\qquad \theta(z_1) = \theta_1 \tag{30.21}$$

such that the integral (30.13) assumes an extremum. We eliminate the boundary conditions for $\theta(z)$ by introducing Lagrangian multipliers α, β. We also replace $\dot\theta$ in (30.13) by an independent function u(z) and add a new integral to (30.13) with h(s) ($\dot\theta$ - u) as integrand, where h(s) is another multiplier. The result is

$$V = \int_{z_0}^{z} \left[n\sqrt{1 + \dot\rho^2 + \rho^2 u^2} + h(s)(\dot\theta - u)\right] dz$$
$$+ \alpha\left(\theta(z_0) - \theta_0\right) - \beta\left(\theta(z_1) - \theta_1\right). \tag{30.22}$$

Euler's equations with respect to $u(z)$, $\theta(z)$, $h(z)$ and the constants α, β lead to the conditions

$$\frac{n\rho^2 u}{\sqrt{1 + \dot\rho^2 + \rho^2 u}} - h = 0 ; \quad \dot h = 0 ; \quad \begin{aligned} h(z_0) &= \alpha \\ h(z_1) &= \beta \end{aligned} \qquad (30.23)$$

and

$$\dot\theta - u = 0 ; \quad \theta(z_0) = \theta_0 ; \quad \theta(z_1) = \theta_1 . \qquad (30.24)$$

Instead of imposing the conditions (30.24) upon (30.22) which would lead us back to the original problem (30.13), we impose (30.23) and obtain an equivalent problem V with $\rho(z)$ and $h(z)$ as variable functions. Since $\dot h = 0$ implies that $h(z)$ is a constant, we shall find a problem of variation for a function $\rho(z)$ and a constant h.

Let us first consider the integral

$$\int_{z_0}^{z_1} \left[n\sqrt{1 + \dot\rho^2 + \rho^2 u^2} - h\,u(z) \right] dz . \qquad (30.25)$$

By introducing h from the first equation (30.23), we obtain the integral

$$\int_{z_0}^{z_1} \frac{n(1 + \dot\rho^2)}{\sqrt{1 + \dot\rho^2 + \rho^2 u^2}} \, dz \qquad (30.251)$$

in which u has to be expressed as a function of $\dot\rho$ and h with the aid of (30.23). We find

$$u = \frac{h}{\rho^2} \frac{\sqrt{1 + \dot\rho^2}}{\sqrt{n^2 - \dfrac{h^2}{\rho^2}}} ; \quad \sqrt{1 + \dot\rho^2 + \rho^2 u^2} = \frac{n\sqrt{1 + \dot\rho^2}}{\sqrt{n^2 - \dfrac{h^2}{\rho^2}}} , \qquad (30.26)$$

and hence

$$\int_{z_0}^{z_1} \left[n\sqrt{1 + \dot\rho^2 + \rho^2 u^2} - hu \right] dz = \int_{z_0}^{z_1} \sqrt{n^2 - \frac{h^2}{\rho^2}} \sqrt{1 + \dot\rho^2} \, dz . \qquad (30.27)$$

From the elimination of θ in (30.22) by partial integration, we obtain the problem of variation

$$V = \int_{z_0}^{z_1} \sqrt{n^2 - \frac{h^2}{\rho^2}} \sqrt{1 + \dot\rho^2} \, dz + h(\theta_1 - \theta_0) \qquad (30.28)$$

with

$$\rho(z_0) = \rho_0 ; \qquad\qquad \rho(z_1) = \rho_1$$

as the only boundary conditions.

The variation with respect to h yields

$$\theta_1 - \theta_0 = h \int_{z_0}^{z_1} \frac{\sqrt{1 + \dot\rho^2}}{\rho^2 \sqrt{n^2 - \dfrac{h^2}{\rho^2}}} \, dz \, , \tag{30.281}$$

and the variation with respect to $\rho(z)$ yields

$$\frac{d}{dz}\left(\frac{m\dot\rho}{\sqrt{1 + \dot\rho^2}}\right) - m_\rho \sqrt{1 + \dot\rho^2} = 0 \tag{30.282}$$

where

$$m(\rho,z) = \sqrt{n^2(\rho,z) - \frac{h^2}{\rho^2}} \, . \tag{30.29}$$

30.3 We can interpret this result as follows: Skew rays in a system of rotational symmetry can be treated as primary rays (h = 0) if, in the ρz-plane, the index of refraction $n(\rho,z)$ is replaced by $m(\rho,z)$ given by (30.29).

Let us consider the following problem: To find a ray which intersects the plane $z = z_0$ at the point

$$x_0 = \rho_0 \cos \theta_0 \, , \qquad\qquad y_0 = \rho_0 \sin \theta_0 \tag{30.31}$$

with a direction p_0, q_0. First we determine the quantity by

$$h = x_0 q_0 - y_0 p_0 \, . \tag{30.32}$$

The problem remains to find $\rho(z)$ as a solution of 30.282 with the initial values

$$\rho(z_0) = \rho_0 \, ,$$

$$\dot\rho(z_0) = \frac{x_0 p_0 + y_0 q_0}{\rho_0 \sqrt{n_0^2 - p_0^2 - q_0^2}} \, . \tag{30.33}$$

This last condition follows from

$$\rho \dot{\rho} = x \dot{x} + y \dot{y} \qquad (30.34)$$

with the aid of the definition (30.16) of p and q.

Finally the functions $x(z)$ and $y(z)$ of the light ray are obtained in the form

$$x = \rho(z) \cos \theta(z)$$

$$y = \rho(z) \sin \theta(z) \qquad (30.35)$$

where $\theta(z)$ is given by the integral

$$\theta(z) = \theta_0 + h \int_{z_0}^{z} \frac{\sqrt{1 + \dot{\rho}^2}}{\rho^2 m} \, dz \quad . \qquad (30.36)$$

Our results allow us to formulate the condition for perfect optical instruments more precisely than at the end of §29. Let us assume that the line element $ds^2 = n^2(\rho, z)(dz^2 + d\rho^2)$ is that of a perfect optical instrument in the ρ, z-plane. In order to have a perfect instrument of revolution we must also require that all line elements $ds^2 = \left(n^2 - \frac{h^2}{\rho^2}\right)(dz^2 + d\rho^2)$ with arbitrary h be perfect with the same location of conjugate points on $z = z_0$ and $z = z_1$. Furthermore, the integral (30.281) must have a constant value for all values of the parameter h.

30.4 <u>The characteristic functions</u>. We have found in examples that Hamilton's characteristic functions for systems of rotational symmetry depend only upon three combinations which are invariants of rotation about the z-axis. Let $F(a_0 b_0; a_1 b_1; z_0 z_1)$ be one of the four characteristics V, W, W*, and T. The variables a_0, b_0; a_1, b_1 will represent the two pairs of the variables x_0, y_0; p_0, q_0; x_1, y_1; p_1, q_1 on which the particular function depends. We shall prove that F depends only on the combinations

$$u = a_0^2 + b_0^2 \; ,$$

$$v = a_1^2 + b_1^2 \; , \qquad (30.41)$$

$$w = 2(a_0 a_1 + b_0 b_1) \; .$$

Let us first consider the function $W(x_0, y_0, p_1, q_1; z_0, z_1)$ in which case

$$u = x_0^2 + y_0^2, \qquad v = p_1^2 + q_1^2 ,$$

$$w = 2(x_0 p_1 + y_0 q_1) \; . \qquad (30.42)$$

We can characterize W as a solution of the partial differential equation of first order:

$$\frac{\partial W}{\partial z_0} = -\sqrt{n^2(x_0, y_0, z_0) - W_{x_0}^2 - W_{y_0}^2} \ , \tag{30.43}$$

which satisfies the boundary condition at $z_0 = z_1$

$$W(x_0, y_0; p_1, q_1; z_1, z_1) = -(x_0 p_1 + y_0 q_1) = -\frac{1}{2} w \ . \tag{30.44}$$

We make the assumption that this boundary value problem has a unique solution. This assumption is certainly true for the function $n(u,z)$ in the neighborhood of the boundary plane z_1 under very general conditions. This follows from the general theory of partial differential equations of the first order. Let us now introduce in (30.43) a function

$$W = \Omega(u,w;z_0) \ . \tag{30.45}$$

We obtain the equation

$$\frac{\partial \Omega}{\partial z_0} = -\sqrt{n^2(u,z) - 4(u \Omega_u^2 + v \Omega_w^2 + w \Omega_u \Omega_w)} \tag{30.46}$$

by using the fact that $n = n(x,y,z)$ is a function of $u = x_0^2 + y_0^2$ and z. The quantity $v = p_1^2 + q_1^2$ in this equation is a constant parameter. Let us again assume that the solutions of this equation are uniquely determined by the boundary values on $z_0 = z_1$. Let $\Omega(u,w,z_0)$ be the solution with the boundary values $\Omega = -\frac{1}{2} w$. This is a function of u,w,z_0 and the parameters v and z_1.

On the other hand the function $W = \Omega(u,v,w;z_0,z_1)$ which is obtained by introducing the expressions (30.42) is a solution of the problem (30.43) and (30.44) and therefore must be the desired characteristic W. This, however, proves our statement that W depends only on u,v,w and z_0,z_1.

This result gives us another proof of the invariance of the expression $xq - yp$ along a light ray. Indeed, we have

$$x_1 = -\frac{\partial W}{\partial p_1} = -2(x_0 W_w + W_v p_1) \ , \qquad p_0 = -\frac{\partial W}{\partial x_0} = -2(W_u x_0 + W_w p_1) \ ,$$

$$y_1 = -\frac{\partial W}{\partial q_1} = -2(y_0 W_w + W_v q_1) \ , \qquad q_0 = -\frac{\partial W}{\partial y_0} = -2(W_u y_0 + W_w q_1) \ .$$

$$\tag{30.46}$$

It follows that

$$x_1q_1 - y_1p_1 = -2(x_0q_1 - y_0p_1)W_w \qquad x_0q_0 - y_0p_0 = -2(x_0q_1 - y_0p_1)W_w$$

and hence

$$x_1q_1 - y_1p_1 = x_0q_0 - y_0p_0 . \qquad (30.47)$$

30.5 The characteristic $V(x_0,y_0; x_1,y_1; z_0,z_1)$ can be obtained from W by the Legendre transformation

$$x_1 = -\frac{\partial W}{\partial p_1} ,$$
$$V - W = x_1p_1 + y_1q_1 . \qquad (30.51)$$
$$y_1 = -\frac{\partial W}{\partial q_1} ,$$

We use these relations to demonstrate that V depends only on

$$u' = u = x_0^2 + y_0^2 ,$$
$$v' = x_1^2 + y_1^2 , \qquad (30.52)$$
$$w' = 2(x_0x_1 + y_0y_1) .$$

We have

$$x_1 = -2(x_0W_w + p_1W_v) ,$$
$$y_1 = -2(y_0W_w + q_1W_v) , \qquad (30.53)$$
$$V - W = -(wW_w + 2vW_v) .$$

From the first equations (30.53), we obtain the following relations:

$$v' = 4(uW_w^2 + vW_v^2 + wW_vW_w) ,$$
$$\frac{1}{2}w' = -2(uW_w + \frac{1}{2}wW_v) . \qquad (30.54)$$

These equations allow us to express v and w as functions of v', w', and $u = u'$. We introduce these expressions in

$$V = W - wW_w - 2vW_v \qquad (30.55)$$

and obtain V as a function of u', v', w'; z_0, z_1. We have to assume of course that the eliminations in (30.54) can be made.

Similar considerations can be applied to the case of the remaining characteristics W^* and T. We summarize the result as follows:

Point characteristic: $V = V(u,v,w;z_0,z_1)$;

$u = x_0^2 + y_0^2$; $v = x_1^2 + y_1^2$; $w = 2(x_0 x_1 + y_0 y_1)$.

Mixed Characteristic: $W = W(u,v,w;z_0,z_1)$;

$u = x_0^2 + y_0^2$, $v = p_1^2 + q_1^2$, $w = 2(x_0 p_1 + y_0 q_1)$.

Mixed Characteristic: $W^* = W^*(u,v,w;z_0,z_1)$;

$u = p_0^2 + q_0^2$, $v = x_1^2 + y_1^2$, $w = 2(p_0 x_1 + q_0 y_1)$.

Angular Characteristic: $T = T(u,v,w;z_0,z_1)$;

$u = p_0^2 + q_0^2$, $v = p_1^2 + q_1^2$, $w = 2(p_0 p_1 + q_0 q_1)$.

Our results are quite clear from a geometrical point of view. In the case of the point characteristic V, for example, it is easy to see that the optical path between two points (x_0,y_0) and (x_1,y_1) depends only on the distances $\rho_0 = \sqrt{u}$ and $\rho_1 = \sqrt{v}$ of the two points from the axis and on the absolute value $|\theta_1 - \theta_0|$ of the angular difference. This last statement is equivalent to the relationship

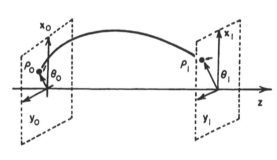

Figure 115

$$\cos(\theta_1 - \theta_0) = \frac{w}{2\sqrt{uv}} ,$$

(30.56)

so that V is determined by the three quantities u,v,w. In other words: The value of the function $V(x_0,y_0,x_1,y_1)$ is unchanged if the coordinates (x_0,y_0) and (x_1,y_1) are submitted to simultaneous rotations or reflections on planes through the z-axis. Therefore it can only depend on the three invariants of these transformations u,v,w.

§31. SPHERICAL ABERRATION AND COMA. CONDITION FOR COMA FREE INSTRUMENTS.

31.1 In this section we shall investigate the image of a small part of the object plane located about the axial point $x_0 = y_0 = 0$. The instrument is

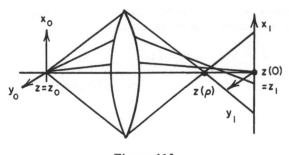

Figure 116

assumed to be symmetric with respect to the z-axis. It is a remarkable and important fact that it is possible to determine the image of such a surface element from the one bundle of rays which originates at the point $x_0 = y_0 = 0$. Let us first consider this bundle, the so called <u>axial bundle</u>. The intersections x_1, y_1 of a ray of this bundle with the image plane are functions of its optical direction cosines p_1, q_1 after refraction. These two functions are given by the derivatives

$$x_1 = - \frac{\partial W_0}{\partial p_1} \; ;$$

$$y_1 = - \frac{\partial W_0}{\partial q_1}$$

(31.11)

of the mixed characteristic $W_0 = W(0,0; p_1,q_1; z_0,z_1)$ belonging to the point $x_0 = y_0 = 0$. Since the instrument is symmetric with respect to the z-axis, we know that $W_0(p_1,q_1)$ depends only on the expression

$$\rho = \sqrt{p_1^2 + q_1^2} = \sqrt{v} \; .$$

(31.12)

We introduce

$$p_1 = \rho \cos \varphi$$

$$q_1 = \rho \sin \varphi$$

(31.121)

and thus obtain from (31.11) the relations

$$x_1 = - W_0'(\rho) \cos \varphi \, ,$$

$$y_1 = - W_0'(\rho) \sin \varphi \, .$$

(31.13)

The function

$$\ell(\rho) = W_0'(\rho)$$

(31.14)

is called the <u>lateral spherical aberration</u> of the bundle. It determines the radius of the circle in which the image plane is intersected by rays of the axial bundle which include an angle α_1 with the axis, where α_1 is given by

$$\sin \alpha_1 = \frac{1}{n_1} \sqrt{p_1^2 + q_1^2} = \frac{1}{n_1} \rho \ . \tag{31.15}$$

The function $\ell(\rho)$ can be found by tracing rays through the system. We remark that $W_0(\rho)$ can be obtained by integration if $\ell(\rho)$ is known, namely,

$$W_0(\rho) = W_0(0) + \int_0^\rho \ell(\rho) \, d\rho \ . \tag{31.16}$$

Let $Z(\rho)$ be the intersection of the ray with the z-axis. It is given by the equation

$$Z = -\frac{r_1}{p_1} x_1 = -\frac{r_1}{q_1} y_1 \ ,$$

or, from (31.13), by

$$Z(\rho) - z_1 = \frac{\ell(\rho)}{\rho} \sqrt{n^2 - \rho^2} \ . \tag{31.17}$$

We assume that $Z(\rho)$ reaches a finite limit $Z(0)$ if $\rho \to 0$ and that the image plane is chosen at this point so that $Z(0) = z_1$.

In general, we call the difference

$$L(\rho) = Z(\rho) - Z(0) \tag{31.18}$$

the <u>longitudinal spherical aberration</u>. It is related to the lateral spherical aberration $\ell(\rho)$ by the equation (31.17), i.e., by

$$L(\rho) = \frac{\sqrt{n^2 - \rho^2}}{\rho} \ell(\rho) \ . \tag{31.19}$$

31.2 Let (p_0, q_0) and (p_1, q_1) be the optical direction cosines of a ray of the axial bundle before and after refraction. Then we have the relation

$$\frac{p_0}{p_1} = \frac{q_0}{q_1} \tag{31.21}$$

which follows from the symmetry of the ray bundle with respect to the z-axis. This ratio determines another function of ρ namely

$$M(\rho) = \frac{p_0}{p_1} = \frac{q_0}{q_1} \tag{31.22}$$

which we shall call the <u>zonal magnification</u> of the bundle. In general this function converges to a finite limit $M(0) = M_0$ which, as we shall see in the next chapter, is equal to the Gaussian magnification of the object plane. $M(\rho)$ and the difference

$$\Delta M = M(\rho) - M_0 \tag{31.23}$$

are explicitly known when the rays of the axial bundle have been traced.

31.3 We can now demonstrate that the two functions $\ell(\rho)$ and $M(\rho)$ which are given by the axial bundle of rays through $x_0 = y_0 = 0$ allow us to determine not only the image of the point $x_0 = y_0 = 0$ but also the image of points $x_0, y_0 \neq 0$ near the axis. Let us consider the mixed characteristic $W(x_0, y_0; p_1, q_1)$ for the two planes $z = z_0$ and $z = z_1$. We develop this function in a Taylor series with respect to x_0 and y_0 with coefficients depending on p_1 and q_1 and disregard all terms which are non-linear in x_0 and y_0. Since W depends only on $u = x_0^2 + y_0^2$, $v = p_1^2 + q_1^2 = \rho^2$, and $w = 2(x_0 p_1 + y_0 q_1)$, we develop with respect to u and w and disregard all terms involving u, and all powers of w higher than the first power. It follows that

$$W = W_0(\rho) + 2(x_0 p_1 + y_0 q_1) W_1(\rho) + \dots . \tag{31.31}$$

The function W_0 obviously is identical with (31.16) given by the lateral spherical aberration of the axial bundle.

We use the general relation

$$p_0 = -\frac{\partial W}{\partial x_0} ; \qquad q_0 = -\frac{\partial W}{\partial y_0} , \tag{31.32}$$

in order to find $W_1(\rho)$. It follows that

$$p_0 = -2 p_1 W_1(\rho) + \dots ; \qquad q_0 = -2 q_1 W_1(\rho) + \dots$$

where the dots indicate terms which are homogeneous functions of x_0 and y_0 of at least first order. Letting $x_0 = y_0 = 0$, we obtain

$$-2 W_1(\rho) = \frac{p_0}{p_1} = \frac{q_0}{q_1} = M(\rho) \tag{31.33}$$

where $M(\rho)$ is the zonal magnification defined for the axial bundle. We thus have the result

$$W = W_0(\rho) - (x_0 p_1 + y_0 q_1) M(\rho) + \dots \tag{31.34}$$

which allows us to investigate the images of points x_0, y_0 in the neighborhood of the axis, in other words the image of a surface element about the point $x_0 = y_0 = 0$.

31.4 The ray intersection with the image plane is given by

$$x_1 = -\ell(\rho) \cos \varphi + \frac{\partial}{\partial p_1} \left[(x_0 p_1 + y_0 q_1) M(\rho) \right]$$

$$y_1 = -\ell(\rho) \sin \varphi + \frac{\partial}{\partial q_1} \left[(x_0 p_1 + y_0 q_1) M(\rho) \right] \ .$$

(31.41)

Without loss of generality we may assume $y_0 = 0$; hence in this case

$$x_1 = -\ell(\rho) \cos \varphi + x_0 \frac{\partial}{\partial p_1} (p_1 M)$$

$$y_1 = -\ell(\rho) \sin \varphi + x_0 p_1 \frac{\partial M}{\partial q_1} \ .$$

(31.42)

A perfect image of the point $(x_0, 0)$ would be obtained if $x_1 = M_0 x_0$ and $y_1 = 0$. The aberrations from this ideal are given by the differences $\Delta x_1 = x_1 - M_0 x_0$; $\Delta y_1 - y_1 - M_0 y_0 = y_1$. Hence we obtain

$$\Delta x_1 = -\ell(\rho) \cos \varphi + M_0 x_0 \frac{\partial}{\partial p_1} (p_1 \psi)$$

$$\Delta y_1 = -\ell(\rho) \sin \varphi + M_0 x_0 p_1 \frac{\partial \psi}{\partial q_1}$$

(31.43)

where $\psi(\rho)$ is defined by

$$\psi(\rho) = \frac{1}{M_0} \left(M(\rho) - M(0) \right) = \frac{\Delta M}{M} \ .$$

(31.44)

We readily find from (31.43) that

$$\Delta x_1 = -\ell(\rho) \cos \varphi + M_0 x_0 (\psi + \rho \psi' \cos^2 \varphi)$$

$$\Delta y_1 = -\ell(\rho) \sin \varphi + M_0 x_0 \rho \psi' \sin \varphi \cos \varphi$$

or

$$\Delta x_1 = -\ell(\rho) \cos \varphi + m + R \cos 2\varphi$$

$$\Delta y_1 = -\ell(\rho) \sin \varphi \qquad + R \sin 2\varphi$$

(31.45)

where $m(\rho)$ and $R(\rho)$ are the functions

$$m(\rho) = M_0 x_0 \left(\psi + \frac{1}{2} \rho \psi' \right)$$

$$R(\rho) = M_0 x_0 \frac{1}{2} \rho \psi' \ .$$

(31.46)

Let us first assume that $\ell(\rho) = 0$ so that the bundle is free from spherical aberration. We see from (31.45) that this alone is not sufficient to guarantee that even a small surface element about the center is sharply imaged. However this is the case if and only if the function ψ is zero. We have thus obtained the Sine condition of Abbe: A surface element (x_0, y_0) of the object plane is imaged sharply if and only if the condition

$$M(\rho) - M(0) = \frac{n_0 \sin \alpha_0}{n_1 \sin \alpha_1} - M_0 = 0 \qquad (31.47)$$

is satisfied by the rays of the axial bundle.

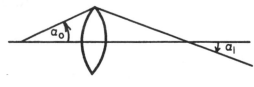

Figure 117

Let us now investigate what type of aberration is introduced if the condition (31.47) is not satisfied. We still assume $\ell(\rho) = 0$. We consider the rays which form the surface of a cone of angular opening $\rho = n_1 |\sin \alpha_1|$. The rays of this cone intersect the image plane in the circle

$$\Delta x_1 = m + R \cos 2\varphi$$
$$\Delta y_1 = \qquad R \sin 2\varphi \qquad (31.48)$$

of center $m(\rho)$ and radius $R(\rho)$. We notice that this circle is described twice if φ goes from 0 to 2π. In fact rays which are at opposite ends of the cone (at φ and $\varphi + \pi$) have the same intersection point. Since m and R vary with ρ we obtain different circles for different bundles ρ. By superimposing these different circles, we get the light spot which is the image of the point $(x_0, 0)$.

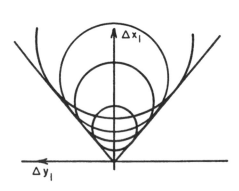

Figure 118

In general a figure of the type shown in Figure 118 is produced. In view of the characteristic shape of this figure, this aberration is known as Coma.

For small apertures the function ψ can be replaced by the quadratic function

$$\psi = B \rho^2 .$$

We find in this case

$$m = 2 B M_0 x_0 \rho^2$$
$$R = B M_0 x_0 \rho^2 \qquad (31.49)$$

i.e., $m = 2R$. The envelope of the circles in Figure 118 is given by two straight lines which include an angle of 30° with the Δx_1 direction.

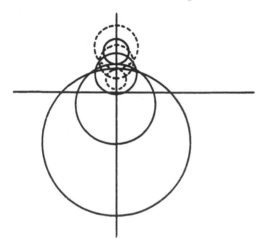

Coma flare in case $\psi = \rho^2 - \rho^4$

Figure 119

For larger apertures this simple relation is not valid any more as can be seen in the example shown in Figure 119 for the case $\psi = \rho^2 - \rho^4$. The relations (31.46) allow us to construct the comatic pattern rigorously in this case.

In the presence of spherical aberration, i.e., if $\ell(\rho) \neq 0$, we obtain a more complicated set of curves. For a given ρ these curves are represented by the equations (31.45). We can construct these curves as follows: We consider first a circle of radius $\ell(\rho)$ and center $m(\rho)$. The center P of another circle of radius R moves once around the circumference of the first circle while the second circle rotates twice around P. A point Q on the second circle describes an epicycle which determines the intersection curve (31.45). Its shape depends, of course, on the relative size of the quantities ℓ and R.

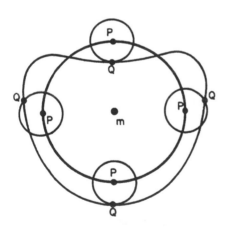

Figure 120

31.5 A generalization of Abbe's sine condition is obtained by the following consideration. The axial bundle of rays is a symmetrical bundle with the z-axis as the axis of symmetry. An oblique bundle originating at a point $(x_0,0)$ of the object plane has in general only a plane of symmetry, the xz-plane, but not an axis of symmetry. The aberration caused by this asymmetry is called the <u>coma of the oblique bundle</u> if the point $(x_0,0)$ lies in the immediate neighborhood of the point $x_0 = y_0 = 0$.

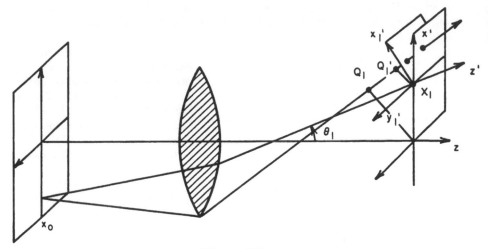

Figure 121

If the total bundle from $(x_0,0)$ after refraction is symmetrical at all then there must exist a ray in the xz-plane such that this ray is an axis of revolution for the bundle. In this case we can introduce a new coordinate system with a z_1'-axis given by the above ray and an $x_1'y_1'$-plane normal to it. The origin of this system is the point $(X_1,0)$ where the above ray intersects the image plane. Let p_1',q_1',r_1' be the optical direction cosines of a ray in this new coordinate system and

$$x_1' = x_1'(p_1',q_1')$$

$$y_1' = y_1'(p_1',q_1')$$

(31.51)

its intersection with the new $x_1'y_1'$-plane.

Then it is clear that the ray bundle can be symmetrical to the z_1'-axis if and only if the functions (31.51) are odd, i.e., do not contain any terms of even power of p_1' and q_1'.

This consideration leads us to the following procedure of investigating the asymmetry of an oblique bundle. We select a ray from $(x_0,0)$ which lies in the primary or the xz-plane. Let θ_1 be the angle of this ray with the z-axis and $(X_1,0)$ its intersection with the image plane. We assume that both X_1 and θ_1 are functions of x_0 such that $X_1 \to 0$ and $\theta_1 \to 0$ if $x_0 \to 0$, in order to insure that the selected rays tend to the position of the z-axis if $x_0 \to 0$. For small values of x_0 these functions have the form

$$X_1 = M_0 x_0 + \ldots , \qquad \theta_1 = \mu x_0 + \ldots .$$

(31.52)

The constant M_0 is the Gaussian magnification because we have chosen the plane $z = z_1$ at the point $z_1 = Z(0)$ where narrow bundles from the point $x_0 = y_0 = 0$ come to a focus. The constant μ is determined by the relation

$$\frac{1}{\mu} = \frac{1}{M_0} \lim_{x_0 \to 0} \frac{X_1}{\theta_1} = -\frac{1}{M_0} \lim_{x_0 \to 0} Z^* \tag{31.53}$$

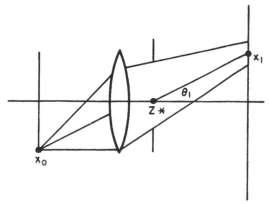

where Z^* is the point at which the selected ray from x_0 intersects the z-axis after refraction. In optical practice one calls this special ray the principal ray of the oblique bundle and Z^* the pupil point. In general this point is determined by the center of a diaphragm for which the principal ray is a ray in the central part of the bundle.

Figure 122

We now introduce a new coordinate system with $(X_1,0)$ as origin and the principal ray as z'-axis. The y'-axis of this system is parallel to the original y-axis; the other axes are rotated by the angle θ_1 with respect to the x and z axes. We express the coordinates x_1' and y_1' of the point of intersection with the x'y'-plane as functions of p_1' and q_1' and determine the even terms in these functions. These terms then determine the asymmetry of the bundle.

31.6 In order to carry out the above program, we make use of the fact that the relations

$$x_1 = -\frac{\partial W}{\partial p_1}, \qquad y_1 = -\frac{\partial W}{\partial q_1}, \qquad r_1 = \frac{\partial W}{\partial z_1} \tag{31.61}$$

are invariant with respect to orthogonal transformations of the coordinate system (x,y,z) in the image space. This is to be understood in the following way. Let x_1', y_1' be the intersection of a ray with the plane $z' = z_1'$ of a new coordinate system and W' the optical path from (x_0,y_0) to the base point Q_1' of the perpendicular dropped from the point $x_1' = y_1' = 0$ of the plane $z' = z_1'$ onto the ray. Then x_1',y_1', are again given by the relations

$$x_1' = -\frac{\partial W'}{\partial p_1'}; \qquad y_1' = -\frac{\partial W'}{\partial q_1'} \tag{31.62}$$

and we have

$$r_1' = \frac{\partial W'}{\partial z_1'} \tag{31.621}$$

p_1', q_1', r_1' are the optical direction cosines of the ray in the new coordinate system.

For the proof of this theorem let us first consider the point characteristic $V(x_0,y_0,z_0; x_1,y_1,z_1)$ for a fixed

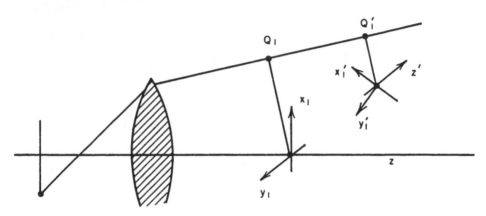

Figure 123

point x_0, y_0, z_0. We have

$$dV = p_1 dx_1 + q_1 dy_1 + r_1 dz_1 \quad . \tag{31.63}$$

By introducing new Cartesian coordinates, $x_1',y_1',z_1',p_1',q_1',r_1'$, we obtain immediately

$$dV' = p_1' dx_1' + q_1' dy_1' + r_1' dz_1' \tag{31.64}$$

since the expression $p_1 dx_1 + q_1 dy_1 + r_1 dz_1$ is an invariant under orthogonal transformation. We introduce the mixed characteristic W' as in §20.1 by the relation

$$dW' = d(V' - x_1'p_1' - y_1'q_1') = - x_1'dp_1' - y_1'dq_1' + r_1'dz_1' \tag{31.65}$$

and this gives the equations (31.62) and (31.621). $W' = V' - x_1'p_1' - y_1'q_1'$ can be interpreted as the optical path to the base point Q_1' of the perpendicular from the origin of the plane $z' = z_1'$.

31.7 From Figure 123 it follows that

$$W' = W + X_1 p_1 .$$ (31.71)

Let us develop W' in a Taylor series of powers of x_0. Since $X_1 = M_0 x_0 + \ldots$ and by (31.34):

$$W = W_0(\rho) - x_0 p_1 M(\rho)$$

we obtain

$$W' = W_0(\rho) - x_0 p_1 \left(M(\rho) - M_0 \right)$$ (31.72)

in which new direction cosines p_1' and q_1' are to be introduced by the relations:

$$p_1 = p_1' \cos \theta_1 + r_1' \sin \theta_1$$

$$r_1 = - p_1' \sin \theta_1 + r_1' \cos \theta_1$$ (31.73)

$$q_1 = q_1' .$$

In the case of small angles θ_1 we have

$$p_1 = p_1' + r_1' \theta_1$$

$$q = q_1'$$ (31.74)

$$r_1 = - p_1' \theta_1 + r_1'$$

or, from (31.52):

$$p_1 = p_1' + \mu r_1' x_0$$

$$q_1 = q_1'$$ (31.75)

$$r_1 = - \mu p_1' x_0 + r_1' .$$

We introduce these expressions in (31.72) and neglect powers of x_0 higher than the first. It follows that

$$W' = W_0(\rho') + \mu r_1' x_0 \frac{\partial W_0(\rho')}{\partial p_1'} - x_0 p_1' \left(M(\rho') - M_0 \right)$$ (31.76)

where $\rho' = \sqrt{p_1'^2 + q_1'^2} .$

From

$$\frac{\partial W_0(\rho')}{\partial p_1'} = \frac{p_1'}{\rho'} W_0'(\rho') = \frac{p_1'}{\rho'} \ell(\rho') \qquad (31.77)$$

we obtain

$$r_1' \frac{\partial W_0(\rho')}{\partial p_1'} = p_1' \frac{r_1'}{\rho'} \ell(\rho') = p_1'L(\rho') \qquad (31.771)$$

and hence

$$W' = W_0(\rho') + x_0p_1' \left[M(\rho') - M_0 - \mu L(\rho')\right] . \qquad (31.78)$$

31.8 We now define the function

$$\phi(\rho') = \frac{1}{M_0} \left[\Delta M - \mu L(\rho')\right] \qquad (31.81)$$

which is a generalization of the definition (31.44) of $\psi(\rho)$. The characteristic W' becomes

$$W' = W_0(\rho') - M_0x_0p_1'\phi(\rho')$$

and gives the ray intersection with the x_1',y_1' plane by differentiation with respect to p_1' and q_1'. It follows that

$$x_1' = -\ell(\rho') \cos \varphi + m + R \cos 2\varphi$$
$$\qquad (31.82)$$
$$y_1' = -\ell(\rho') \sin \varphi + R \sin 2\varphi$$

where $m(\rho')$ and $R(\rho')$ are functions similar to (31.46) namely

$$m = M_0x_0\left(\phi + \frac{1}{2}\rho'\phi'\right)$$
$$\qquad (31.83)$$
$$R = M_0x_0 \frac{1}{2}\rho'\phi' .$$

The curves of intersection with the $x_1'y_1'$ plane represented by (31.82) for a given value of ρ' are of the same type as those discussed in (31.4). They are symmetrical if and only if

$$\phi = \frac{1}{M_0} \left(\Delta M - \mu L(\rho')\right) \equiv 0 . \qquad (31.84)$$

This is the condition of <u>Staeble-Lihotzki</u>: <u>An oblique bundle of rays can be</u> <u>symmetric with respect to a given principal ray if the expression (31.84) is</u> <u>identically zero.</u>

Since both L and ΔM can be found simply by tracing the rays from the axial point $x_0 = y_0 = 0$ this is one of the most valuable criteria in optical design. If $\phi(\rho')$ is not zero we are in a position to construct the intersection curves, and hence the light distribution, with the aid of (31.82) and (31.83).

If the principal ray is not given, i.e., if the question is whether there exists a ray to which the bundle is rotationally symmetric, we can find directly from (31.84) that: A bundle from a point $(x_0,0)$ near the z-axis possesses an axis of revolution if and only if the two functions ΔM and L of the axial bundle are linearly dependent.

In this case there exists a constant μ such that $\phi \equiv 0$.

We remark explicitly that approximations in the above equation have been made only with respect to the variables x_0, y_0 but not with respect to p_1 and q_1. In other words our result holds for instruments in which a small field is imaged by a large bundle of rays. Microscopes are instruments of this type.

31.9 The aplanatic points of a sphere are perfect conjugate points in which Abbe's sine condition is satisfied. In fact, from (23.42) and (23.61) it follows that

$$\frac{n_0}{n_1} = \frac{r_1}{r_0} \tag{31.91}$$

and from Fig. 76:

$$\frac{r_1}{r_0} = \frac{\sin \alpha_0}{\sin \alpha_1} \tag{31.92}$$

i.e.,

$$M = \frac{n_0}{n_1} \frac{\sin \alpha_0}{\sin \alpha_1} = \frac{n_0^2}{n_1^2} \tag{31.93}$$

and $\Delta M = 0$.

By using aplanatic lenses in the front part of a microscope objective (Fig. 80) we can thus decrease the angular opening of the axial bundle without introducing spherical aberration or coma.

In general, two perfect conjugate points on the axis of revolution of an optical medium are called aplanatic if the sine condition (31.47) is satisfied. For example, the conjugate points in the case of Maxwell's fisheye are aplanatic.

It is possible to obtain optical instruments consisting of two aspheric surfaces such that two given points P_0 and Q_0 are aplanatic points of the instrument. We illustrate this by describing a graphical construction of two aspheric mirrors of this type.

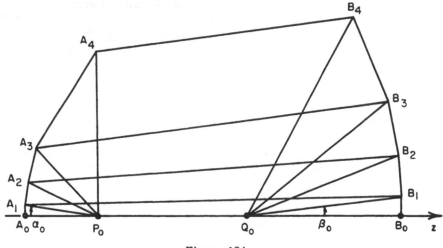

Figure 124

Let A_0 and B_0 be the vertices of the two mirrors. First we construct a set of radial lines through P_0 and Q_0 such that the sines of the angles α_y and β_y with the axis form arithmetic progressions, $\sin \alpha_y = \gamma h$; $\sin \beta_y = \gamma k$. We draw two lines $A_0 A_1$ and $B_0 B_1$ normal to the z-axis and connect A_1 with B_1. We now construct a line through A_1 normal to the bisecting line of the angle $P_0 A_1 B_1$. Let A_2 be its intersection with the second radial line through P_0. Similarly, a line is drawn through B_1 normal to the bisecting line of the angle $Q_0 B_1 A_1$ giving the point B_2. We connect A_2 and B_2 and repeat the procedure. The results are two curves consisting of a number of straight sections. These curves are approximations to two aspheric curves which are obtained in the limit for $h \rightarrow 0$; $k \rightarrow 0$ where h/k is kept constant.[†]

A similar construction can be carried out for two refracting aspheric surfaces.

31.95 The case where the object is at infinity can be treated in a manner similar to the above by developing the angular characteristic $T(p_0, q_0; p_1, q_1)$ in powers of p_0 and q_0. By neglecting powers of higher than first order, we

---

[†]In the case for the point P_0 at infinity the analytic solution of the problem is given by Schwarzschild; Unters. zur Geom. Optic II Abb. der Ges. d. Wiss. Goettingen

obtain an angular characteristic which allows us to investigate the images of parallel incident bundles which subtend a small angle with the z–axis. High speed photographic objectives of small angular field are examples of this case.

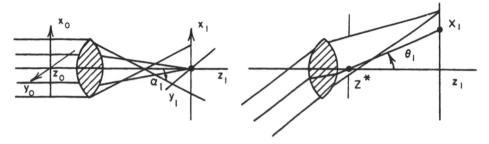

Figure 125

We give without proof the following results which are modifications of the above:

Let us first consider the axial bundle of rays parallel to the z–axis. We choose a reference plane at an arbitrary point $z = z_0$ of the axis. Lateral and longitudinal spherical aberration, i.e., $\ell(\rho)$ and $L(\rho)$ where $\rho = \sqrt{p_1^2 + q_1^2}$ are defined as before. However, instead of $M(\rho)$ we now consider the function

$$f(\rho) = -\frac{x_0}{p_1} = -\frac{y_0}{q_1} . \qquad (31.951)$$

This function is called the zonal focal length of the axial bundle. The value $f(0) = f_0$ defines the equivalent focal length of the optical system.

With the aid of the two functions $\ell(\rho)$ and $f(\rho)$ of the axial bundle, we obtain the angular characteristic T for small inclinations $(p_0, q_0 = 0)$ in the form

$$T = T_0(\rho) - f_0 p_0 \psi(\rho) \qquad (31.952)$$

where

$$\psi(\rho) = \frac{1}{f_0} \Delta f = \frac{1}{f_0} \left(f(\rho) - f_0 \right) . \qquad (31.953)$$

The aberrations

$$\Delta x_1 = x_1 - f_0 \frac{p_0}{\sqrt{n_0^2 - p_0^2 - q_0^2}}$$

(31.954)

$$\Delta y_1 = y_1 - f_0 \frac{q_0}{\sqrt{n_0^2 - p_0^2 - q_0^2}}$$

of an oblique bundle of small inclination p_0 and $q_0 = 0$ are then given by

$$\Delta x_1 = -\ell(\rho) \cos \varphi + m + R \cos 2\varphi$$

(31.955)

$$\Delta y_1 = -\ell(\rho) \sin \varphi \qquad + R \sin 2\varphi$$

in which

$$m = f_0 p_0 \left(\psi + \frac{1}{2} \rho \psi' \right)$$

(31.956)

$$R = f_0 p_0 \frac{1}{2} \rho \psi' .$$

If $\ell(\rho) = 0$, i.e., if the axial bundle is free of spherical aberration then we conclude: Oblique bundles of small inclination can give sharp images if and only if the condition $\psi(\rho) = 0$ is satisified by the axial bundle.

This is Abbe's sine condition for an infinitely distant object

$$\frac{x_0}{n_1 \sin \alpha_1} = -f_0 .$$

(31.957)

The question of rotational symmetry of the refracted bundle leads to a modification of the condition of Staeble-Lihotzki. We consider a principal ray of the oblique bundle which, after refraction, intersects the z-axis at the exit pupil point Z^*. Let θ_1 be the angle it makes with the axis, and X_1 its intersection with the image plane. X_1 and θ_1 are functions of p_0 which for small values of p_0 have the form

$$X_1 = f_0 p_0 + \ldots$$

(31.958)

$$\theta_1 = \mu p_0 + \ldots .$$

As before, we find

$$\frac{1}{\mu} = \frac{1}{f_0} \lim_{p_0 \to 0} \frac{X_1}{\theta_1} = -\frac{1}{f_0} \lim_{p_0 \to 0} Z^* .$$

By introducing a new coordinate system with the principal ray as the z'-axis, we obtain for the intersection with the x_1',y_1'-plane the formulae (31.82) and (31.83) with a modified definition of ϕ, namely,

$$\phi = \frac{1}{f_0}\left(\Delta f - \mu L(\rho')\right) . \tag{31.959}$$

Hence, the refracted bundle is symmetric to the principal ray if and only if $\Phi \equiv 0$. It possesses an axis of revolution if and only if the two functions Δf and $L(\rho)$ of the axial bundle are linearly dependent.

§32. THE CONDENSER PROBLEM.

32.1 In the problem of designing condenser systems one is led to a condition similar to the condition (31.84) of Staeble-Lihotzki.

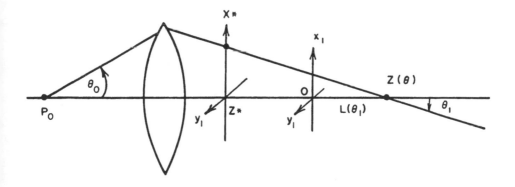

Figure 126

A condenser is a lens system which transforms the rays diverging from a filament, P_0, into a convergent bundle. It is required that the illumination of a section of a given plane $Z*$ shall be uniform and that the convergence is as good as possible. The condition of uniform illumination is the important one. The other condition actually requires only that, if possible, all light of the condensed bundle enters an objective which is placed in the region of convergence. For objectives with small apertures the spherical aberration of the condenser must be small; this is not necessary in the case of objectives with large apertues where a large amount of spherical aberration can be allowed without loss of efficiency. The main problem, therefore is to find a lens system which illuminates a section of the plane $z = Z*$ uniformly.

We analyze this problem in the case of small filaments. We can characterize the light emission from such a filament by a function

$$I = I(\theta_0, \varphi_0) \tag{32.11}$$

which determines the energy flux in the solid angle $\sin \theta_0 d\theta_0 d\varphi_0$ by

$$dF = I(\theta_0, \varphi_0) \sin \theta_0 d\theta_0 d\varphi_0 \quad . \tag{32.12}$$

The function $I(\theta_0, \varphi_0)$ is called the <u>intensity</u> of the filament.

We now consider the axial bundle of rays from the point $x_0 = y_0 = 0$ of the filament. Let $Z(\theta_1)$ be the point where the z-axis is intersected by a ray which subtends an angle θ_1 with the axis after refraction. In order to find $Z(\theta_1)$ it is sufficient to trace rays in the xz-plane. We can assign positive and negative values to θ_1 in this plane by using the customary convention of analytic geometry. Let us choose the Gaussian point $z = Z(0)$ as the zero point of the z-axis so that the spherical aberration of the bundle is given by

$$L(\theta_1) = Z(\theta_1) \quad . \tag{32.13}$$

A ray of this bundle intersects the plane $z = Z*$ at the point

$$X* = (Z* - L) \tan \theta_1 \cos \varphi_0$$
$$Y* = (Z* - L) \tan \theta_1 \sin \varphi_0 \quad . \tag{32.14}$$

We introduce the constant

$$\mu = -\frac{M_0}{Z*} \tag{32.15}$$

in analogy to the definition (31.53). We introduce the function

$$G(\theta_1) = (M_0 + \mu L) \tan \theta_1 \tag{32.16}$$

and write (31.14) in the form

$$X* = \frac{1}{\mu} G(\theta_1) \cos \varphi_0$$
$$Y* = \frac{1}{\mu} G(\theta_1) \sin \varphi_0 \quad . \tag{32.17}$$

If $E(X*, Y*)$ is the <u>illumination</u> of the surface $z = Z*$ then the flux through the surface element $dX* dY*$ is given by

$$dF = E \ dX* dY* = \frac{1}{\mu^2} E G G' \ d\theta_1 d\varphi_0 \quad .$$

This flux, on the other hand, must be equal to the flux (32.12) through the corresponding solid angle $\sin \theta_0 d\theta_0 d\varphi_0$ of the incident bundle.

Hence, we have the equation

$$\frac{1}{\mu^2} E G G' d\theta_1 = I(\theta_0,\varphi_0) \sin \theta_0 d\theta_0 \tag{32.18}$$

or

$$E = \mu^2 I(\theta_0,\varphi_0) \frac{\sin \theta_0 d\theta_0}{GG' d\theta_1} . \tag{32.19}$$

32.2 One easily verifies from (32.18) that the illumination at the center of the plane $z = Z^*$ is determined by

$$E(0,0) = \mu^2 I(0) ,$$

i.e., by the quantity

$$\mu^2 = \frac{M_0^2}{Z^{*2}} . \tag{32.21}$$

It increases with the square of the magnification of the filament and decreases inversely as the square of the distance of the slide from the Gaussian image point. The relative illumination $J^* = \dfrac{E}{E(0,0)}$ and the relative intensity $J = \dfrac{I}{I(0)}$ are consequently related by the equation

$$J^* = J(\theta_0,\varphi_0) \frac{\sin \theta_0 d\theta_0}{GG' d\theta_1} . \tag{32.22}$$

We can use this equation to determine J^* numerically if $J(\theta_0,\varphi_0)$ is known.

32.3 The illumination of the plane is uniform if $J^* = 1$. Equation (32.22) shows that this is possible only if $J(\theta_0,\varphi_0)$ is independent of φ_0. If this is the case, we find from (32.22)

$$G^2(\theta_1) = 2 \int_0^{\theta_0} J(\theta_0) \sin \theta_0 d\theta_0 \tag{32.31}$$

or with the aid of (32.16)

$$\frac{1}{\tan \theta_1} \sqrt{2 \int_0^{\theta_0} J(\theta_0) \sin \theta_0 d\theta_0} - M_0 - \mu L(\theta_1) = 0. \tag{32.32}$$

In analogy to (31.84), let us introduce the notation

$$\phi = \frac{1}{M_0} \left[\frac{1}{\tan \theta_1} \sqrt{2 \int_0^{\theta_0} J(\theta_0) \sin \theta_0 d\theta_0} - M_0 - \mu L(\theta_1) \right] .$$

(32.33)

We then find that the <u>necessary and sufficient condition for uniform illumination is that the rays of the axial bundle must satisfy the equation</u> $\phi \equiv 0$.

In case the filament is a point source we have $J(\theta_0) = 1$ and hence

$$\phi = \frac{1}{M_0} \left(\frac{2 \sin \dfrac{\theta_0}{2}}{\tan \theta_1} - M_0 - \mu L(\theta_1) \right) .$$

(32.34)

If the filament emits light according to Lambert's law

$$J(\theta_0) = \cos \theta_0$$

we obtain

$$\phi = \frac{1}{M_0} \left(\frac{\sin \theta_0}{\tan \theta_1} - M_0 - \mu L(\theta_1) \right) .$$

(32.35)

Neither of these two conditions can be satisfied if the condenser is coma free, i.e., if

$$\phi = \frac{1}{M_0} \left(\frac{\sin \theta_0}{\sin \theta_1} - M_0 - \mu L(\theta_1) \right) = 0 .$$

(32.36)

It follows that <u>uniform illumination can only be obtained if the condenser has coma</u>.

The position, Z^*, of the plane appears in (32.33) and in the formula (32.22) for the relative illumination, J^*, only in form of the combination $\mu L(\theta_1) = -\dfrac{M_0}{Z^*} L(\theta_1)$. It follows <u>that J^* is independent of the position of the plane if the condenser has no spherical aberration.</u>

If for such condensers the illumination is uniform for one position of the slide then it is uniform for all poisitions.

32.4   In general it will be impossible to satisfy the condition $\phi \equiv 0$ with a finite number of lenses with spherical surfaces. It is, however, possible to design the condenser in such a manner that $\phi(\theta_1) = 0$ for a certain angle $\theta_1 = \alpha$.

The illumination J^*, of course, is not uniformly equal to one but we can show that the departure from one is zero in the average, the average being taken over the section of the plane which is determined by the rays of angle $\theta_1 = \alpha$. Indeed, by (31.22) it follows that

$$(J^* - 1)GG'd\theta_1 = J(\theta_0) \sin \theta_0 d\theta_0 - GG'd\theta_1 . \tag{32.41}$$

Hence with the aid of

$$dx^* dy^* = \frac{1}{\mu^2} GG'd\theta_1 d\varphi_0$$

we find the relation

$$\mu^2 \iint_{\theta_1 \leqq \alpha} (J^* - 1)dx^* dy^* = 2\pi \left[\int_0^{\theta_0} J(\theta_0) \sin \theta_0 d\theta_0 - \frac{1}{2} G^2(\alpha) \right] . \tag{32.42}$$

The right side is zero, if $\phi(\alpha) = 0$. Therefore, in this case:

$$\iint_{\theta_1 \leqq \alpha} (J^* - 1) dx^* dy^* = 0.$$

This indicates that the best compromise will be obtained if $\phi(\alpha) = 0$ at the maximum angle $\theta_1 = \alpha$ corresponding to the boundary of the circular section of the plane $z = Z^*$ which is to be used.

CHAPTER IV

FIRST ORDER OPTICS

§33. THE FIRST ORDER PROBLEM IN GENERAL.

33.1   Let us consider an optical medium which is homogeneous outside the domain $\ell_0 \leqq z \leqq \ell_1$ of the xyz-space. Thus we

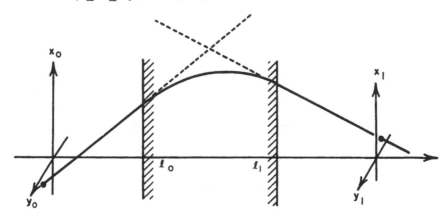

Figure 127

assume

$$n(x,y,z) = n_0; \ z < \ell_0,$$

$$n(x,y,z) = n_1; \ z > \ell_1.$$

and call the space $z < \ell_0$ the <u>object space</u> and the space $z > \ell_1$ the <u>image space</u>. The light rays in the object and in the image space are sections of straight lines. The optical instrument, i.e., the medium between ℓ_0 and ℓ_1 determines a transformation of the straight lines of the object space into the straight lines of the image space.

In view of this fact it is convenient to consider not only the sections of these straight lines which as light rays have physical reality, but also their total extensions. We call these extensions the <u>virtual part</u>, and the original sections the <u>real part</u> of the rays. A similar extension can be made for the object and image space. The half space $z > \ell_0$ is the virtual object space, the half space $z < \ell_1$ the virtual image space. A bundle of rays is said to

216

belong to a real object point, if it originates at a point of the real part of the object space, to a virtual object point if the rays converge to a point of the virtual part. In a similar manner we define real and virtual image points in the image space as the points of intersection of real or virtual parts of a bundle of rays, respectively.

We now take two reference planes $z = z_0$ and $z = z_1$ in object and image space, respectively. These planes can be chosen in the real or virtual parts of their spaces. The transformation of the object and image rays is then given by four functions:

$$x_1 = x_1 (x_0, y_0, p_0, q_0)$$

$$y_1 = y_1 (x_0, y_0, p_0, q_0)$$

$$\quad\quad\quad\quad\quad\quad\quad\quad (33.11)$$

$$p_1 = p_1 (x_0, y_0, p_0, q_0)$$

$$q_1 = q_1 (x_0, y_0, p_0, q_0)$$

which determine the direction p_1, q_1 of an image ray and its intersection with the plane $z = z_1$ as functions of the corresponding coordinates of the object ray. We have seen that this transformation (33.11) is a <u>canonical transformation</u> defined by the condition that the differential expression

$$p_1 dx_1 + q_1 dy_1 - p_0 dx_0 - q_0 dy_0 \quad\quad\quad\quad (33.12)$$

is a total differential.

This result was a consequence of the more general theorem: Let $V (x_0, y_0, z_0; x_1, y_1, z_1)$ be the optical distance of two points (x_0, y_0, z_0) and (x_1, y_1, z_1) of object and image space, respectively. Then we have

$$dV = p_1 dx_1 + q_1 dy_1 + r_1 dz_1 - p_0 dx_0 - q_0 dy_0 - r_0 dz_0 \quad\quad (33.13)$$

where p_0, q_0, r_0 are the optical direction cosines of the object ray and p_1, q_1, r_1 those of the image ray. In the special case that z_0 and z_1 are constants we obtain the statement (33.12).

33.2 The differential expression (33.13) remains invariant if other Cartesian coordinate systems are introduced, either in object or image space or in both. This implies the invariance of the statement (33.12), thus: Let x_0, y_0 be the point of intersection of an object ray with a plane, $z_0 =$ a constant, of an arbitrary Cartesian coordinate system and p_0, q_0 its optical direction cosines in this system. Let x_1, y_1, p_1, q_1 be analogous coordinates in the image space with respect to another arbitrary coordinate system. Then the functions (33.11) must satisfy the relation

$$dV = p_1 dx_1 + q_1 dy_1 - p_0 dx_0 - q_0 dy_0. \quad\quad\quad (33.21)$$

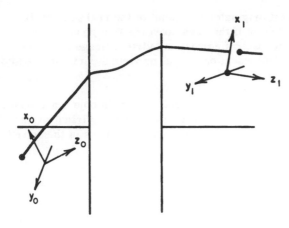

Figure 128

We have already made use of this fact in § 31 and have seen that a similar invariant character is connected with the other Hamiltonian characteristics W, W*, T and their differentials.

33.3 The condition (33.21) requires that the partial derivatives of the functions (33.11) must satisfy certain relations. For the derivation of these conditions, let us consider general canonical transformation, i.e., transformations

$$X_i = X_i (x_1 \ldots, x_n; p_1, \ldots, p_n)$$
$$P_i = P_i (x_1 \ldots, x_n; p_1, \ldots, p_n)$$

$$i = 1, \ldots, n \qquad (33.31)$$

which transform a pair of conjugate vectors $x = (x_1, \ldots, x_n)$ and $p = (p_1, \ldots, p_n)$ into the vectors $X = (X_1, \ldots, X_n)$, $P = (P_1, \ldots, P_n)$ so that the condition

$$\sum_{i=1}^{n} (P_i \, dX_i - p_i \, dx_i) = dV \qquad (33.32)$$

is satisfied.

We write the differentials of the functions (33.31) in the form

$$dX = A \, dx + B \, dp$$
$$dP = C \, dx + D \, dp$$

$$(33.33)$$

where A, B, C, D are the matrices

$$A_{ik} = \frac{\partial X_i}{\partial x_k} \qquad\qquad B_{ik} = \frac{\partial X_i}{\partial p_k}$$

$$C_{ik} = \frac{\partial P_i}{\partial x_k} \qquad\qquad D_{ik} = \frac{\partial P_i}{\partial p_k} \qquad (33.34)$$

Our aim is to show that (33.32) implies that these matrices must satisfy the relations

$$A'C - C'A = 0$$

$$B'D - D'B = 0 \tag{33.35}$$

$$D'A - B'C = 1$$

where A', B', C', D' are the transposed matrices of A, B, C, D.

From (33.32) if follows that

$$\sum_{i=1}^{n} P_i \left(\sum_{k=1}^{n} \frac{\partial X_i}{\partial x_k} dx_k + \frac{\partial X_i}{\partial p_k} dp_k \right) - \sum_{i=1}^{n} p_i dx_i = dV$$

or

$$\sum_{k=1}^{n} \left(\sum_{i=1}^{n} P_i \frac{\partial X_i}{\partial x_k} - p_k \right) dx_k + \sum_{k=1}^{n} \left(\sum_{i=1}^{n} P_i \frac{\partial X_i}{\partial p_k} \right) dp_k = dV. \tag{33.36}$$

By temporarily introducing the notation

$$R_k = \sum_i P_i \frac{\partial X_i}{\partial x_k} - p_k ,$$

$$\tag{33.37}$$

$$S_k = \sum_i P_i \frac{\partial X_i}{\partial p_k}$$

we obtain from (33.36) the relations

$$\frac{\partial R_k}{\partial x_\nu} - \frac{\partial R_\nu}{\partial x_k} = 0 ,$$

$$\frac{\partial R_k}{\partial p_\nu} - \frac{\partial S_\nu}{\partial x_k} = 0 ,$$

$$\frac{\partial S_k}{\partial p_\nu} - \frac{\partial S_\nu}{\partial p_k} = 0 ,$$

which leads to

$$\sum_i \left(\frac{\partial P_i}{\partial x_\nu} \frac{\partial X_i}{\partial x_k} - \frac{\partial P_i}{\partial x_k} \frac{\partial X_i}{\partial x_\nu} \right) = 0$$

$$\sum_i \left(\frac{\partial P_i}{\partial p_\nu} \frac{\partial X_i}{\partial x_k} - \frac{\partial P_i}{\partial x_k} \frac{\partial X_i}{\partial p_\nu} \right) = \begin{cases} 0; & k \neq \nu \\ 1; & k = \nu \end{cases} \tag{33.38}$$

$$\sum_i \left(\frac{\partial P_i}{\partial p_\nu} \frac{\partial X_i}{\partial p_k} - \frac{\partial P_i}{\partial p_k} \frac{\partial X_i}{\partial p_\nu} \right) = 0 \ .$$

If we use the notation of the <u>Lagrangian Brackets</u>:

$$[u,v] = \sum_{i=1}^{n} \frac{\partial (X_i, P_i)}{\partial (u,v)} \ , \tag{33.381}$$

we can express the above equations in the form

$$[x_k, x_\nu] = 0$$

$$[x_k, p_\nu] = \begin{cases} 0, & k \neq \nu \\ 1, & k = \nu \end{cases} \tag{33.382}$$

$$[p_k, p_\nu] = 0 \ .$$

If, however, the notation (33.34) is introduced, we obtain

$$\sum_i (C_{i\nu} A_{ik} - C_{ik} A_{i\nu}) = 0$$

$$\sum_i (D_{i\nu} A_{ik} - B_{ik} C_{i\nu}) = \begin{cases} 0, & i \neq \nu \\ 1, & i = \nu \end{cases} \tag{33.383}$$

$$\sum_i (D_{i\nu} B_{ik} - D_{ik} B_{i\nu}) = 0 \ ,$$

i.e., the relations (33.35).

We therefore recognize that the matrix conditions (33.35) are nothing but another form of writing Lagrange's conditions (33.382).

We can use the relations (33.35) in order to invert the linear equations (33.33). In fact, it readily follows that:

$$dx = D' \, dX - B' \, dP,$$

$$dy = -C' \, dX + A' \, dP.$$

(33.39)

This shows that the determinant

$$\Delta = \begin{vmatrix} A & B \\ C & D \end{vmatrix}$$

(33.391)

of the linear equations (33.33) cannot be zero under any circumstances. One can prove, incidentally, that $\Delta = 1$.

33.4 We now consider a single light ray which passes through the optical instrument. We choose two coordinate systems in object and image space such that the z_0 and z_1 axes coincide with the object and the image ray, respectively. This ray shall be called the prin-cipal ray in the following: The problem of first order optics is to investigate the properties of the four parameter manifold of rays which lie in the immediate neighborhood of a given ray, i.e., of the principal ray. In this neighborhood we can replace the general formulae (33.11) by linear relations

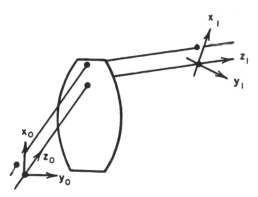

Figure 129

$$X_1 = A \, X_0 + B \, P_0$$

$$P_1 = C \, X_0 + D \, P_0$$

(33.41)

where X_0, P_0; X_1, P_1 are the vectors (x_0, y_0); (p_0, q_0); (x_1, y_1) and (p_1, q_1), respectively, and A,B,C,D are the matrices

$$A = \begin{pmatrix} A_{11} & A_{12} \\ A_{21} & A_{22} \end{pmatrix}, \ B = \begin{pmatrix} B_{11} & B_{12} \\ B_{21} & B_{22} \end{pmatrix}, \ C = \begin{pmatrix} C_{11} & C_{12} \\ C_{21} & C_{22} \end{pmatrix}, \ D = \begin{pmatrix} D_{11} & D_{12} \\ D_{21} & D_{22} \end{pmatrix} .$$

(33.42)

The mathematical problem is to study the linear transformations (33.41) where the matrices A,B,C,D satisfy the conditions (33.35) and to interpret the results optically as to the formation of images by the narrow bundle.

The easiest mathematical approach to this problem is given by Hamilton's method of characteristic functions. However, this advantage involves a certain loss of generality as we have to assume that not all of the determinants $|A|$, $|B|$, $|C|$, $|D|$ of the matrices (33.42) are zero. In fact, the conditions (19.25), (20.11), (20.35) and (20.41), necessary for the existence of the corresponding characteristic function, lead to the following conditions for the use of characteristic functions in our present problem:

$$\text{Point characteristic} \quad V \quad : \quad |B| \neq 0 \, ,$$

$$\text{Mixed Characteristics} \quad W \quad : \quad |D| \neq 0 \, ,$$

$$W* \quad : \quad |A| \neq 0 \, , \tag{33.43}$$

$$\text{Angular Characteristics} \quad T \quad : \quad |C| \neq 0 \, .$$

It is possible to give examples where all determinants are zero and yet the matrices A, B, C, D satisfy the conditions (33.35)[†] .

33.5 By excluding these exceptional cases we shall, however, make use of Hamilton's method in deriving the essential results of first order optics. Let us, for example, assume that $|C| \neq 0$[††]. It is then possible to express X_0 and X_1 in (33.41) as functions of P_0 and P_1, namely

$$X_0 = - (C^{-1}D)P_0 + C^{-1}P_1 \, ,$$

$$X_1 = (B - AC^{-1}D)P_0 + (AC^{-1})P_1 \, . \tag{33.51}$$

We write this in the form

$$X_0 = A*P_0 + B*P_1$$

$$- X_1 = C*P_0 + D*P_1 \tag{33.52}$$

where

$$A* = - C^{-1}D, \qquad B* = C^{-1} \, ,$$

$$C* = AC^{-1}D - B, \quad D* = - AC^{-1} \tag{33.53}$$

[†]Caratheodory, Geometrische Optik, pp. 51-57, Erg. der Math. and ihrer Grenzgebiete, Bd IV, Berlin 1937.

[††]The following considerations cannot be applied to the case of telescopic systems, in which C = 0, i.e., $P_1 = DP_0$.

With the aid of the relations (33.35) it is easy to show that A^* and D^* are symmetrical matrices and that

$$B^* = (C^*)' \ .$$

We are, however, in a position to obtain this result directly by deriving the equations (33.52) with the aid of the angular characteristic. We have seen that the function $T \ (p_0, \ p_1, \ q_0, \ q_1; \ z_0, \ z_1)$ has the form

$$T = T_0 \ (p_0, q_0; p_1, q_1) + z_1 \sqrt{n_1^2 - p_1^2 - q_1^2} - z_0 \sqrt{n_0^2 - p_0^2 - q_0^2}$$
$$(33.54)$$

in which T_0 is independent of z_0 and z_1.

We write, for small values of $p_0, \ q_0, \ p_1, \ q_1$:

$$T_0 - T_0 \ (0) = \frac{1}{2} \ (A_{11}p_0^2 + 2A_{12}p_0q_0 + A_{22}q_0^2)$$

$$+ \frac{1}{2} \ (C_{11}p_1^2 + 2C_{12}p_1q_1 + C_{22}q_1^2) \qquad (33.55)$$

$$- (F_{11}p_0p_1 + F_{12}p_0q_1 + F_{21}q_0p_1 + F_{22}q_0q_1)$$

and

$$z_1 \sqrt{n_1^2 - p_1^2 - q_1^2} = n_1z_1 - \frac{1}{2} \frac{z_1}{n_1} \ (p_1^2 + q_1^2) \ ,$$

$$z_0 \sqrt{n_0^2 - p_0^2 - q_0^2} = n_0z_0 - \frac{1}{2} \frac{z_0}{n_0} \ (p_0^2 + q_0^2) \ . \qquad (33.56)$$

The formulae

$$x_0 = \frac{\partial T}{\partial p_0} \ , \qquad - x_1 = \frac{\partial T}{\partial p_1} \ ,$$

$$y_0 = \frac{\partial T}{\partial q_0} \ , \qquad - y_1 = \frac{\partial T}{\partial q_1} \ , \qquad (33.57)$$

lead directly to the relations

$$x_0 = A_{11}p_0 + A_{12}q_0 - F_{11}p_1 - F_{12}q_1 + \frac{z_0}{n_0} \ p_0 \ ,$$

$$y_0 = A_{12}p_0 + A_{22}q_0 - F_{21}p_1 - F_{22}q_1 + \frac{z_0}{n_0} \ q_0 \ ,$$

$$- x_1 = - F_{11}p_0 - F_{21}q_0 + C_{11}p_1 + C_{12}q_1 - \frac{z_1}{n_1} \, p_1 \, ,$$

$$- y_1 = - F_{12}p_0 - F_{22}q_0 + C_{12}p_1 + C_{22}q_1 - \frac{z_1}{n_1} \, q_1 \, ,$$

or in matrix notation

$$X_0 = \left(A + \frac{z_0}{n_0} \, E \right) P_0 - F P_1$$

$$- X_1 = - F'P_0 + \left(C - \frac{z_1}{n_1} \, E \right) P_1$$

(33.581)

where A and C are the symmetrical matrices

$$A = \begin{pmatrix} A_{11} & A_{12} \\ A_{12} & A_{22} \end{pmatrix} ; \quad C = \begin{pmatrix} C_{11} & C_{12} \\ C_{12} & C_{22} \end{pmatrix} ,$$

(33.582)

E the unit matrix

$$E = \begin{pmatrix} 1 & 0 \\ 0 & 1 \end{pmatrix}$$

(33.583)

and F a matrix

$$F = \begin{pmatrix} F_{11} & F_{12} \\ F_{21} & F_{22} \end{pmatrix}$$

(33.584)

which, in general, is not symmetric.

The equations (33.581), however, express our above statement. We note in passing that one can use these equations in order to prove the matrix conditions (33.35) independently of the former derivations for systems in which $|C| \neq 0$.

33.6 We know that, by a rotation of the coordinate systems about the z_0 and z_1 axes, it is possible to transform the two quadratic forms

$$A \, [p_0, q_0] = A_{11}p_0^2 + 2A_{12}p_0q_0 + A_{22}q_0^2 \, ,$$

$$C \, [p_1, q_1] = C_{11}p_1^2 + 2C_{12}p_1q_1 + C_{22}q_1^2$$

(33.61)

into the normal form

$$A \left[p_0, q_0 \right] = A_1 p_0^2 + A_2 q_0^2 ,$$

(33.62)

$$C \left[p_1, q_1 \right] = C_1 p_1^2 + C_2 q_1^2 .$$

Without loss of generality we can therefore assume that A and C have this form in our chosen coordinate system.

Hence our problem is to investigate linear transformations of the general type:

$$x_0 = \left(A_1 + \frac{z_0}{n_0} \right) p_0 - F_{11} p_1 - F_{12} q_1$$

$$y_0 = \left(A_2 + \frac{z_0}{n_0} \right) q_0 - F_{21} p_1 - F_{22} q_1$$

(33.63)

$$- x_1 = - F_{11} p_0 - F_{21} q_0 + \left(C_1 - \frac{z_1}{n_1} \right) p_1$$

$$- y_1 = - F_{12} p_0 - F_{22} q_0 + \left(C_2 - \frac{z_1}{n_1} \right) q_1 .$$

Finally we note that we can choose the origin $z_0 = 0$ and $z_1 = 0$ of the coordinate systems in such a manner that

$$A_1 = - A_2 = a ,$$

(33.64)

$$C_1 = - C_2 = - c .$$

With these choices there are six constants left:

$$a, c, \text{ and } \begin{pmatrix} F_{11} & F_{12} \\ F_{21} & F_{22} \end{pmatrix}$$

(33.65)

which determine the optical properties of the manifold of rays. The linear relations (33.63) become

$$x_0 = \left(a + \frac{z_0}{n_c} \right) p_0 - F_{11} p_1 - F_{12} q_1, \quad y_0 = \left(\frac{z_0}{n_0} - a \right) q_0 - F_{21} p_1 - F_{22} q_1$$

(33.66)

$$x_1 = F_{11} p_0 + F_{21} q_0 + \left(\frac{z_1}{n_1} + c \right) p_1, \quad y_1 = F_{12} p_0 + F_{22} q_0 + \left(\frac{z_1}{n_1} - c \right) q_1 .$$

The matrix F is of special significance; for reasons which will become obvious in the following pages we shall call it the <u>matrix of the focal lengths</u> of the system.

§34. GAUSSIAN OPTICS.

34.1 In the case when the principal ray of an infinitesimally narrow manifold of rays is the axis of a system of revolution we obtain Gauss' first order optics. For applications this is the most important case. We have seen in §29 that T is a function of

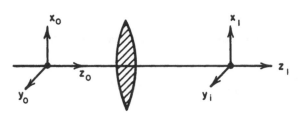

Figure 130

$$u = p_0^2 + q_0^2 ,$$

$$v = p_1^2 + q_1^2 , \qquad (34.11)$$

$$w = 2(p_0 p_1 + q_0 q_1) .$$

It follows that the matrices (33.58) have the form

$$A = \begin{pmatrix} a & 0 \\ 0 & a \end{pmatrix}; \quad F = \begin{pmatrix} f & 0 \\ 0 & f \end{pmatrix}; \quad C = \begin{pmatrix} c & 0 \\ 0 & c \end{pmatrix}. \qquad (34.12)$$

The equations (33.63) become

$$x_0 = \left(a + \frac{z_0}{n_0} \right) p_0 - f p_1 , \qquad y_0 = \left(a + \frac{z_0}{n_0} \right) q_0 - f q_1 ,$$

$$- x_1 = - f p_0 + \left(c - \frac{z_1}{n_1} \right) p_1 , \qquad - y_1 = - f q_0 + \left(c - \frac{z_1}{n_1} \right) q_1 , \qquad (34.13)$$

which shows that x_0, x_1, p_0, p_1 are related by the same equations as y_0, y_1, q_0, q_1. Hence it is sufficient to study only the first set of equations (34.13).

We are still at liberty to choose the origins, $z_0 = 0$ and $z_1 = 0$, of the coordinate systems at our convenience. Let us make this choice so that the determinant

$$\begin{vmatrix} a & -f \\ -f & c \end{vmatrix} = 0 . \qquad (34.14)$$

This obviously is always possible.

34.2 We first consider the images of points of the plane $z_0 = 0$ on the plane $z_1 = 0$. We have

$$x_0 = ap_0 - fp_1 ,$$

$$- x_1 = - fp_0 + cp_1 ,$$

(34.21)

from which follows

$$x_1 = - \frac{1}{f} \left(- cx_0 + (ac - f^2)p_0 \right) = \frac{c}{f} x_0 ,$$

$$p_1 = - \frac{1}{f} (x_0 - ap_0) .$$

(34.22)

This shows as a consequence of the condition (34.14), that the planes $z_0 = 0$ and $z_1 = 0$ are conjugate planes. All rays from a point x_0, y_0 of the plane $z_0 = 0$ intersect each other at the conjugate point

$$x_1 = \frac{c}{f} x_0 ,$$

$$y_1 = \frac{c}{f} y_0$$

(34.23)

of the plane $z_1 = 0$. Let us, therefore, introduce the notation

$$M_0 = \frac{c}{f} = \frac{f}{a}$$

(34.24)

as the <u>magnification of the plane</u> $z_0 = 0$. It follows from (34.13) that the coefficient of p_1 in the first equation (34.21) and of p_0 in the second has the value f independent of z_0 and z_1. This coefficient must therefore represent a physical property of the instrument which is invariant with respect to the choice of the reference planes. It is called the <u>equivalent focal length</u> of the system.

By introducing these notations in (34.21) and (34.22) we obtain

$$x_0 = f \left(\frac{p_0}{M_0} - p_1 \right) ,$$

$$- x_1 = f (- p_0 + M_0 p_1) ,$$

(34.25)

and

$$x_1 = M_0 x_0 , \qquad p_1 = - \frac{x_0}{f} + \frac{1}{M_0} p_0 ,$$

(34.26)

and, of course, similar equations for y_0, y_1; q_0, q_1.

34.3 We consider next another pair of reference planes z_0 and z_1. The equations (34.13) become

$$x_0 = \left(\frac{f}{M_0} + \frac{z_0}{n_0} \right) p_0 - f p_1 \ ,$$

$$- x_1 = - f p_0 + \left(M_0 f - \frac{z_1}{n_1} \right) p_1 \ . \tag{34.31}$$

The two planes z_0 and z_1 are conjugate if the determinant of these equations is zero. This leads to the <u>Lens equation</u>

$$\left(M_0 f - \frac{z_1}{n_1} \right) \left(\frac{f}{M_0} + \frac{z_0}{n_0} \right) = f^2 \ , \tag{34.32}$$

or

$$M_0 \frac{n_1}{z_1} - \frac{1}{M_0} \frac{n_0}{z_0} = \frac{1}{f} \ . \tag{34.33}$$

Equation (34.33) determines the position of any pair of conjugate planes relative to a given pair of conjugate planes.

The magnification, M, of the plane z_0 is obtained by writing (34.31) in a form analogous to (34.25):

$$x_0 = \frac{f}{M} p_0 - f p_1 \ ,$$

$$- x_1 = - f p_0 + M f p_1 \ , \tag{34.34}$$

and comparing coefficients with (34.31) it follows that

$$M = M_0 - \frac{1}{f} \frac{z_1}{n_1} = \frac{1}{\dfrac{1}{M_0} + \dfrac{1}{f} \dfrac{z_0}{n_0}} \ , \tag{34.35}$$

and hence

$$\frac{z_0}{n_0} = \frac{M_0 - M}{M_0 M} f \ ,$$

$$\frac{z_1}{n_1} = (M_0 - M) f \ . \tag{34.36}$$

The last equations give the positions of the object and image planes which belong to a given magnification M. From (34.36) we conclude the useful relation

$$M \cdot M_0 = \frac{z_1}{n_1} \bigg/ \frac{z_0}{n_0} \qquad\qquad (34.37)$$

between the magnifications and positions of two pairs of conjugate planes.

The coordination of the rays with z_0 and z_1 as reference planes is given by equations similar to (34.26), namely

$$x_1 = Mx_0 , \qquad\qquad\qquad y_1 = My_0 ,$$

$$\qquad\qquad\qquad\qquad\qquad\qquad\qquad\qquad\qquad\qquad (34.38)$$

$$p_1 = -\frac{x_0}{f} + \frac{1}{M} p_0 , \qquad q_1 = -\frac{y_0}{f} + \frac{1}{M} q_0 .$$

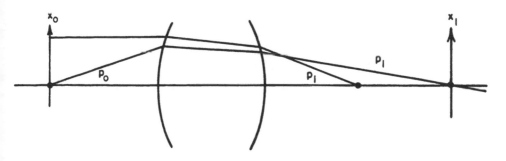

Figure 131

These equations indicate a simple method of determining the focal length f of the system and the magnification M of a plane z_0:

For any ray which is parallel to the z_0 axis in the object space the ratio $-\dfrac{x_0}{p_1}$ gives the focal length f of the system.

For any ray which originates at the point z_0; $x_0 = y_0 = 0$, the ratio $\dfrac{p_0}{p_1}$ determines the magnification M of the plane z_0.

34.4 It is customary to use as the original pair of conjugate planes the so-called unit planes of the system. This is the pair of conjugate planes for which $M_0 = 1$. The points $z_0 = 0$ and $z_1 = 0$ are then the unit points of the system. We obtain the special lens equation

$$\frac{n_1}{z_1} - \frac{n_0}{z_0} = \frac{1}{f} . \qquad\qquad (34.41)$$

The equations (34.36) become

$$\frac{z_0}{n_0} = \frac{1 - M}{M} f; \quad \frac{z_1}{n_1} = (1 - M)f \ , \qquad (34.42)$$

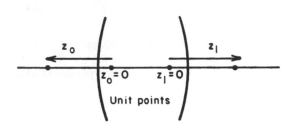

and the relation (34.37) becomes

$$M = \frac{z_1}{n_1} \bigg/ \frac{z_0}{n_0} \ . \qquad (34.43)$$

Figure 132

However, it is often more convenient to use the general equation (34.33) and determine the position of conjugate planes with respect to a known pair of magnification M_0. Let us, for example, consider the <u>nodal planes</u> of the system as planes $z_0 = z_1 = 0$. The axial points of these planes are called the <u>nodal points</u> of the instrument and are defined as follows: A ray of direction $\frac{p_0}{n_0}$ through the nodal point of the object space leaves the instrument with the same direction, i.e., for $x_0 = 0$ we have

$$\frac{p_0}{n_0} = \frac{p_1}{n_1} \quad \text{or} \quad \frac{p_0}{p_1} = M_0 = \frac{n_0}{n_1} \ . \qquad (34.44)$$

The Lens equation with respect to the nodal points becomes

$$\frac{n_0}{z_1} - \frac{n_1}{z_0} = \frac{1}{f} \ . \qquad (34.45)$$

Equation (34.37) yields

$$M = z_1 \ / \ z_0 \ . \qquad (34.46)$$

Unit points and Nodal points are called the <u>cardinal points</u> of the lens system.

In order to find the position of the cardinal points relative to each other, let us introduce $M = 1$ in (34.46). It follows that $z_1 = z_0$, and hence from (34.45)

$$z_1 = z_0 = (n_0 - n_1)f \qquad (34.47)$$

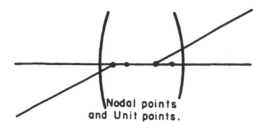

Nodal points
and Unit points.

Figure 133

is the distance of the unit points from the corresponding nodal points. This equation shows that unit points and nodal points coincide if object and image space have the same index of refraction.

34.5 <u>Focal points</u>; <u>Newton's Lens equation</u>. Every lens system of revolution has two focal points F_0 and F_1. An object ray through F_0 leaves the instrument parallel to the axis; an object ray parallel to the axis intersects it at F_1 after refraction. Let us determine the locations of these points relative to the unit points. Letting $z_0 = F_0$, $z_1 = \infty$, and $z_0 = \infty$, $z_1 = F_1$, we obtain from (34.41):

$$F_0 = - n_0 f; \quad F_1 = n_1 f .\tag{34.51}$$

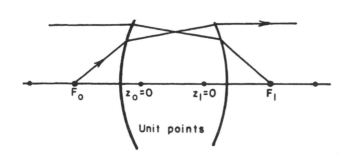

Figure 134

The equation (34.32) for $M_0 = 1$ assumes the form

$$\left(\frac{z_0}{n_0} + f\right)\left(f - \frac{z_1}{n_1}\right) = f^2;\tag{34.52}$$

or with the aid of (34.51):

$$\left(z_0 - F_0\right)\left(z_1 - F_1\right) = - n_0 n_1 f^2 .\tag{34.53}$$

Hence, if we determine the position of conjugate planes by their distances

$$Z_0 = z_0 - F_0 ,$$

$$Z_1 = z_1 - F_1 ,$$

from the respective focal points we find <u>Newton's lens equation</u>

$$Z_0 Z_1 = - n_0 n_1 f^2 .\tag{34.54}$$

The correlation between conjugate points thus can be characterized as an inversion on the points $\pm \sqrt{n_0 n_1}$ f.

We mention finally that the relations (34.42) also assume an extremely simple form if the distances Z_0 and Z_1 are introduced; we readily find:

$$\frac{Z_0}{n_0} = \frac{f}{M} \; ; \quad \frac{Z_1}{n_1} = - Mf \; . \tag{34.55}$$

34.6   An optical instrument of revolution can be considered as a perfect optical instrument in a first order approximation. Every real or virtual object point possesses a conjugate point in the image space, either in its real or its virtual part. The original transformation of object rays into image rays thus introduces a transformation of the points (x_0, y_0, z_0) into the points (x_1, y_1, z_1) of the image space. This transformation is given by the formulae $x_1 = Mx_0$, $y_1 = My_0$, if we introduce for M its expression in z_0. With the focal points as reference points on the z-axes we obtain from (34.54) and (34.55)

$$M = \frac{n_0 f}{Z_0} \; , \tag{34.61}$$

and hence the transformation formulae

$$x_1 = n_0 f \frac{x_0}{Z_0} \; ,$$

$$y_1 = n_0 f \frac{y_0}{Z_0} \; , \tag{34.62}$$

$$Z_1 = - \frac{n_0 n_1 f^2}{Z_0} \; .$$

The last equation follows from the lens equation (34.54).

The transformation (34.62) is a <u>collineation</u> since planes are transformed into planes, and consequently straight lines are transformed into straight lines since any straight line is formed by the intersection of two planes.

We can use this result to determine the image of a plane

$$x_0 = (z_0 - \alpha_0) \tan \theta_0 \tag{34.63}$$

which subtends an angle θ_0 with the optical axis but is not normal to it. The axial point $z_0 = \alpha_0$ is transformed into the point

$$x_1 = y_1 = 0; \quad \alpha_1 = - \frac{n_0 n_1 f^2}{\alpha_0} \; . \tag{34.64}$$

An infinitely distant point of the plane (34.63) becomes a finite point

$$X_1 = n_0 f \tan \theta_0 \; ,$$

$$Y_1 = 0 \; , \tag{34.65}$$

$$Z_1 = 0 \; ,$$

of the image space. The image surface of (34.63) must be plane and normal to the x_1, Z_1 plane; it goes through the points (34.64) and (34.65). The angle θ_1 subtended with the z-axis is given by

$$\tan \theta_1 = \frac{X_1}{Z_1 - \alpha_1} = \frac{\tan \theta_0}{n_1 f} \; \alpha_0 \; ,$$

or from (34.61) by

$$\frac{n_0 \tan \theta_0}{n_1 \tan \theta_1} = M \; . \tag{34.66}$$

Thus for very large values of $|M|$ we find θ_1 considerably decreased. If the oblique object plane is covered by the lines of a rectangular coordinate system we obtain two sets of lines on the image plane of which one is parallel to the y-axis. The lines of the other set form a plane bundle through the finite point (34.65) which is a point of the focal plane F_1.

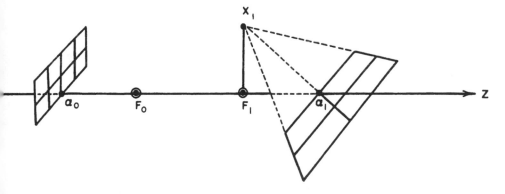

Figure 135

§35. ORTHOGONAL RAY SYSTEMS IN FIRST ORDER OPTICS.

35.1 A generalization of the Gaussian optics is obtained in the case of

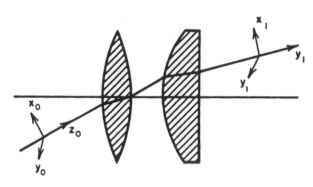

Figure 136

a bundle of rays about an oblique meridional ray in a system of revolution. We choose our coordinate systems such that the x_0- and x_1-axes lie in the plane of the ray. The angular characteristic T must then be an even function of q_0 and q_1 which excludes the presence of linear terms of q_0 and q_1 in the development (33.55). It follows that the matrices A,B, and C have the form

$$A = \begin{pmatrix} a_1 & 0 \\ 0 & a_2 \end{pmatrix}, \quad B = \begin{pmatrix} F & 0 \\ 0 & f \end{pmatrix}, \quad C = \begin{pmatrix} c_1 & 0 \\ 0 & c_2 \end{pmatrix} \tag{35.11}$$

and hence the equations:

$$x_0 = \left(a_1 + \frac{z_0}{n_0}\right)p_0 - Fp_1, \quad y_0 = \left(a_2 + \frac{z_0}{n_0}\right)q_0 - fq_1 ,$$

$$-x_1 = -Fp_0 + \left(c_1 - \frac{z_1}{n_1}\right)p_1, \quad -y_1 = -fq_0 + \left(c_2 - \frac{z_1}{n_1}\right)q_1 . \tag{35.12}$$

We still obtain two independent sets of equations for the coordinates (x_0,p_0,x_1, p_1) and (y_0,q_0,y_1,q_1). However, these equations are no longer identical as in the Gaussian case. First order ray systems of the type (35.11) are called orthogonal systems.

35.2 Let us first consider the two plane bundles of rays $q_0 = q_1 = 0$ and $p_0 = p_1 = 0$ which originate at the point $x_0 = y_0 = z_0 = 0$. We obtain

$$0 = a_1 p_0 - Fp_1 ,$$

$$-x_1 = -Fp_0 + \left(c_1 - \frac{z_1}{n_1}\right)p_1 \tag{35.21}$$

in the first case, and

$$0 = a_2 q_0 - f q_1 \ ,$$

$$- y_1 = - f q_0 + \left(c_2 - \frac{z_1}{n_1} \right) q_1$$

(35.22)

in the other case.

Every bundle clearly has a pair of conjugate points z_1 and z_2, which can be determined by letting the determinant of the equations (35.21) and (35.22) be zero. However, these two points z_1 and z_2 are not identical in general. They are known as the <u>primary or tangential</u> and <u>secondary or sagittal foci</u> of the point $z_0 = 0$. The difference $z_2 - z_1$ is called the

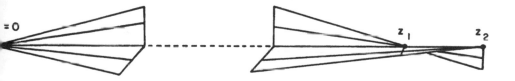

Figure 137

<u>astigmatic difference</u> of the bundle.

35.3 We consider next all the rays through the point $x_0 = y_0 = z_0 = 0$. Since $p_0 = + \dfrac{F}{a_1} \, p_1$ and $q_0 = + \dfrac{f}{a_2} \, q_1$ we obtain from (35.12):

$$- x_1 = \left(c_1 - \frac{F^2}{a_1} - \frac{z_1}{n_1} \right) p_1 \ ,$$

$$- y_1 = \left(c_2 - \frac{f^2}{a_2} - \frac{z_1}{n_1} \right) q_1 \ .$$

(35.31)

If we introduce $p_1 = \rho \cos \varphi$, $q_1 = \rho \sin \varphi$ we recognize that the intersections with the plane z_1 of rays which have a given inclination ρ form an ellipse. The axes of this ellipse are linear functions of z_1, the position of the reference plane. At a certain position we have a circle instead of an ellipse, namely if z_1 satisfies the condition

$$\left(c_2 - \frac{f^2}{a_2} \right) + \left(c_1 - \frac{F}{a_1} \right) = 2 \frac{z_1}{n_1} \ .$$

(35.32)

The other two positions z_1 and z_2 are distinguished by the fact that the ellipse degenerates into a section of a straight line. These two "focal lines" are normal to each other. The refracted bundle thus consists of all straight lines which intersect two line segments which are normal to each other.

Let us now assume that the point z_1 of (35.32) is the origin $z_1 = 0$ of our coordinate system, so that the coefficients of (35.12) satisfy the condition

$$c_2 - \frac{f^2}{a_2} = - \left(c_1 - \frac{F^2}{a_1{}^2} \right) = \lambda_0 . \tag{35.33}$$

We shall see presently that $2\lambda_0$ is closely related to the value which we have introduced above as the $\underline{\text{astigmatic difference}}$. By similar considerations as in 34.2 we are led to introduce the constants

$$M_0 = \frac{F}{a_1} , \qquad m_0 = \frac{f}{a_2} . \tag{35.34}$$

We shall call M_0 and m_0 the $\underline{\text{primary and secondary magnifications}}$ respectively and F and f the $\underline{\text{primary and secondary focal lengths}}$ of the system.

By expressing the coefficients in (35.12) in terms of λ_0 and the constants (35.34) we find

$$x_0 = \left(\frac{F}{M_0} + \frac{z_0}{n_0} \right) p_0 \qquad - Fp_1 ,$$
$$- x_1 = \qquad -Fp_0 \qquad + \left(M_0 F - \frac{z_1}{n_1} - \lambda_0 \right) p_1 ; \tag{35.35}$$

$$y_0 = \left(\frac{f}{m_0} + \frac{z_0}{n_0} \right) q_0 \qquad - fq_1 ,$$
$$- y_1 = \qquad -fq_0 \qquad + \left(m_0 f - \frac{z_1}{n_1} + \lambda_0 \right) q_1 . \tag{35.36}$$

35.4 Let us now investigate the images of the points of the plane $z_0 = 0$ at different positions of the plane z_1. From (35.35) and (35.36) it follows that

$$x_1 = M_0 x_0 + \left(\frac{z_1}{n_1} + \lambda_0 \right) p_1 ,$$
$$y_1 = m_0 y_0 + \left(\frac{z_1}{n_1} - \lambda_0 \right) q_1 . \tag{35.41}$$

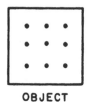

OBJECT

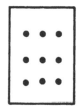

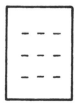

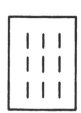

Figure 138

Let us use as a "test object" a system of points on a square plate.

We chose first $z_1 = 0$ and find

$$x_1 = M_0 x_0 + \lambda_0 p_1 \ ,$$
$$\tag{35.42}$$
$$y_1 = m_0 y_0 - \lambda_0 q_1 \ .$$

A given point x_0, y_0 thus is imaged in the form of a circular spot. The radius of this circle is proportional to λ_0 and its center is at the point $M_0 x_0$, $m_0 y_0$. The square assumes the form of a rectangle.

At the position $z_1 = - n_1 \lambda_0$ we have

$$x_1 = M_0 x_0 \ ,$$
$$\tag{35.43}$$
$$y_1 = m_0 y_0 - 2\lambda_0 q_1 \ ,$$

and thus we obtain as images horizontal sections of straight lines which are parallel to the y_1- axis. The location of the images is the same as in the former case.

Finally we obtain for $z_1 = + n_1 \lambda_0 = z_2$ the image functions

$$x_1 = M_0 x_0 + 2\lambda_0 p_1 \ ,$$
$$\tag{35.44}$$
$$y_1 = m_0 y_0 \ ,$$

which represent vertical sections of straight lines about centers which are located as above.

The difference $z_2 - z_1 = 2\lambda_0 n_1$ of the last two positions obviously is equal to the astigmatic difference of the bundle from $x_0 = y_0 = 0$. At other positions z_1 we find elliptical spots as images; we see that no plane z_1 exists where the definition is sharp.

35.5 The preceding considerations show that the images of points of the plane $z_0 = 0$ depend upon five first order coefficients M_0, m_0; F, f, and λ_0. The quantities F and f are independent of the choice of the reference plane $z_0 = 0$ and thus represent geometric characteristics of the whole manifold of rays. However, the quantities M_0, m_0, and λ_0 are different for another object plane and are thus functions of its position, z_0. The position z_1 of the plane where circular images are obtained also varies with z_0. We conclude the investigation of orthogonal systems by deriving these functions.

We know that the plane $z_1 = 0$ is the image plane of the plane $z_0 = 0$ in the sense that circular images are obtained. The equations (35.35) and (35.36) for $z_0 = z_1 = 0$ assume the form

$$x_0 = \frac{F}{M_0} \, p_0 - Fp_1 \, ,$$

$$- x_1 = - Fp_0 + (M_0 F - \lambda_0)p_1 \, ; \tag{35.51}$$

$$y_0 = \frac{f}{m_0} \, q_0 - fq_1 \, ,$$

$$- y_1 = - fq_0 + (m_0 f + \lambda_0)q_1 \, . \tag{35.52}$$

If in the case where $z_0 \neq 0$, the position of the image plane, understood in the above sense, is given by z_1, then the general equations (35.35) and (35.36) must assume the same form as (35.51) and (35.52) provided M_0, m_0, and λ_0 are replaced by the new quantities M, m and λ.

This leads to the relations

$$\frac{F}{M} = \frac{F}{M_0} + \frac{z_0}{n_0} \, ,$$

$$\frac{f}{m} = \frac{f}{m_0} + \frac{z_0}{n_0} \, ,$$

$$MF - \lambda = M_0 F - \frac{z_1}{n_1} - \lambda_0 \, , \tag{35.53}$$

$$mf + \lambda = m_0 f - \frac{z_1}{n_1} - \lambda_0 \, ,$$

i.e., four equations for the four unknown quantities M, m, λ, and z_1.

The first two equations give the functions M and m directly, namely

$$M = \frac{M_0}{1 + \frac{M_0}{n_0 F} z_0} \, , \qquad m = \frac{m_0}{1 + \frac{m_0}{n_0 f} z_0} \, . \tag{35.54}$$

We also obtain from these equations the relation

$$\frac{M_0 - M}{MM_0} F = \frac{m_0 - m}{mm_0} f .$$

(35.55)

The last two equations (35.53) lead to

$$\lambda - \lambda_0 = \frac{1}{2} (mm_0 - MM_0) \frac{z_0}{n_0} ,$$

$$\frac{z_1}{n_1} = \frac{1}{2} (MM_0 + mm_0) \frac{z_0}{n_0}$$

(35.56)

and hence, with the aid of (35.54) we find

$$2(\lambda - \lambda_0) = - \frac{M_0^2}{\dfrac{n_0}{z_0} + \dfrac{M_0}{F}} + \frac{m_0^2}{\dfrac{n_0}{z_0} + \dfrac{m_0}{f}} ,$$

$$2 \frac{z_1}{n_1} = \frac{M_0^2}{\dfrac{n_0}{z_0} + \dfrac{M_0}{F}} + \frac{m_0^2}{\dfrac{n_0}{z_0} + \dfrac{m_0}{f}} .$$

(35.57)

All these formulae are generalizations of the corresponding formulae in the case of Gaussian optics. The last equation (35.57) replaces the lens equation (34.33).

In the case of an infinite object point $z_0 = \infty$ we obtain

$$2(\lambda_\infty - \lambda_0) = - M_0 F + m_0 f ,$$

$$\frac{2}{n_1} z_1 (\infty) = M_0 F + m_0 f .$$

(35.58)

Since the reference plane $z_0 = 0$ is not distinct from other reference planes we conclude from (35.58) the following relations:

$$- MF + mf + 2\lambda = 2\lambda_\infty ,$$

$$MF + mf = \frac{2}{n_1} \left(z_1 (\infty) - z_1 \right)$$

(35.59)

which are valid for any object plane z_0.

By letting $\lambda = 0$ in (35.57) we obtain a quadratic equation for z_0. The solutions of this equation, if real, determine the so-called <u>stigmatic points</u> on the principal ray, i.e., points without astigmatism of the refracted bundle. The associated planes z_0 are imaged sharply but not necessarily undistorted.

§36. NON-ORTHOGONAL SYSTEMS.

36.1 It is not difficult to treat general first order manifolds of rays in a manner similar to the preceding special cases. Bundles of this type are obtained if the principal ray is a skew ray in a medium of rotational symmetry or, more generally, if it is a ray in an asymmetrical optical system such as the human eye. A complete theory of the image formation of such general systems of rays was first developed by A. Gullstrand.

For a detailed derivation of Gullstand's results we refer the reader to his original papers or to modern expositions of his results.[†] Here we will only demonstrate what type of image formation can be expected in general. Let us assume that the ray coordination in our manifold can be represented by the mixed characteristic

$$W = W(x_0, y_0, z_0; p_1, q_1, z_1)$$

for a certain choice of z_0 and z_1. We consider z_0 as fixed and z_1 as variable. The problem is to determine the images of a point x_0, y_0 of the object plane $z_0 = 0$ on different image planes z_1.

We develop W in a power series with respect to x_0, y_0, p_1, q_1 and consider only second order terms. By using the fact that W is a linear function of z_1 which has the form

$$W_0 = W(x_0, y_0; p_1, q_1) + z_1 \sqrt{n_1^2 - p_1^2 - q_1^2} , \qquad (36.11)$$

we write

$$W = \frac{1}{2}(A_1 x_0^2 + A_2 y_0^2) + \frac{1}{2}(C_1 p_1^2 + C_2 q_1^2) - \left[M_{11} x_0 p_1 + M_{12} y_0 p_1 \right.$$

$$\left. + M_{21} x_0 q_1 + M_{22} y_0 q_1 \right] - \frac{z_1}{2n_1}(p_1^2 + q_1^2) . \qquad (36.12)$$

That the quadratic forms $A[x_0, y_0]$ and $C[p_1, q_1]$ contain no mixed terms is the result of a suitable choice of the x_0, y_0 and x_1, y_1 axes of our coordinate systems on the ray.

[†] Herzberger, Strahlenoptik
  Caratheodory, Geometrische Optik.

From (36.12) we obtain the first order relations between the coordinates of the ray in the form

$$x_1 = M_{11}x_0 + M_{12}y_0 - \left(C_1 - \frac{z_1}{n_1} \right) p_1 ,$$

$$y_1 = M_{21}x_0 + M_{22}y_0 - \left(C_2 - \frac{z_1}{n_1} \right) q_1 ,$$

$$(36.13)$$

$$p_0 = - A_1 x_0 + M_{11}p_1 + M_{21}q_1 ,$$

$$q_0 = - A_2 y_0 + M_{12}p_1 + M_{22}q_1 .$$

36.2 We first consider the bundle of rays from the point $x_0 = y_0 = 0$. We obtain

$$- x_1 = \left(C_1 - \frac{z_1}{n_1} \right) p_1 ; \qquad p_0 = M_{11}p_1 + M_{21}q_1 ,$$

$$(36.21)$$

$$- y_1 = \left(C_2 - \frac{z_1}{n_1} \right) q_1 ; \qquad q_0 = M_{12}p_1 + M_{22}q_1 ,$$

which represents an astigmatic bundle in the image space of the type investigated in §35. The z_1 axis is intersected by the rays of two plane bundles, determined by $p_1 = 0$ and $q_1 = 0$.

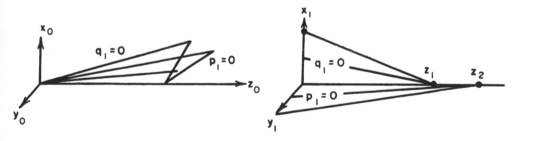

Figure 139

These two bundles are normal to each other. The focal points of these bundles are given by

$$q_1 = 0, \qquad z_1 = n_1 C_1 ; \qquad p_1 = 0, \qquad z_2 = n_1 C_2 . \qquad (36.23)$$

However, contrary to the case of orthogonal systems, we find that the corresponding plane bundles in the object plane are not normal to each other but are given by

$$p_0 = M_{11}p_1; \qquad q_0 = M_{12}p_1 \qquad \text{in the case } q_1 = 0 \; ;$$

$$p_0 = M_{21}q_1; \qquad q_0 = M_{22}q_1 \qquad \text{in the case } p_1 = 0 \; . \tag{36.24}$$

If they are normal to each other then

$$\frac{M_{11}}{M_{12}} = - \frac{M_{22}}{M_{21}} \tag{36.25}$$

which is not true in general.

36.3 We next consider the images of points x_0, y_0 of the plane $z_0 = 0$. We have by (36.13)

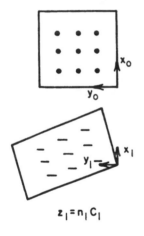

$$x_1 = M_{11}x_0 + M_{12}y_0 + \left(\frac{z_1}{n_1} - C_1 \right) p_1 \; ,$$

$$y_1 = M_{21}x_0 + M_{22}y_0 + \left(\frac{z_1}{n_1} - C_2 \right) q_1 \; . \tag{36.31}$$

Let us again use a square with a number of points as a test object. We choose first the position $z_1 = n_1 C_1$. It follows

$$x_1 = M_{11}x_0 + M_{12}y_0 \; ,$$

$$y_1 = M_{21}x_0 + M_{22}y_0 + (C_1 - C_2)q_1 \; . \tag{36.32}$$

The test points are imaged as sections of horizontal straight lines. The centers of these sections are located at points which are related to the object points by the linear relations

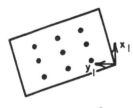

$$x_1 = M_{11}x_0 + M_{12}y_0 \; ,$$

$$y_1 = M_{21}x_0 + M_{22}y_0 \; . \tag{36.33}$$

The original square is transformed into a parallelogram, i.e., a sheared image of the object.

At $\dfrac{z_1}{n_1} = \dfrac{C_1 + C_2}{2}$ we obtain circular images with centers in the same locations as above.

Figure 140

Finally at $z_1 = n_1 C_2$ we obtain sections of vertical straight lines. At intermediate positions we obtain elliptical images.

36.4 The main difference from the orthogonal systems is the torsion of the projected image. In both cases sharp definition of the images of points is prohibited by astigmatism. But the optical projection of the object points differs principally in our last case. No torsion can be observed in orthogonal systems.

We observed that the image distortion (36.33) is closely related to the formulae (36.21), namely

$$p_0 = M_{11}p_1 + M_{21}q_1 \; ,$$

$$q_0 = M_{12}p_1 + M_{22}q_1$$

(36.41)

in the case of $x_0 = y_0 = 0$.

As a consequence of these equations we found that the plane bundles through $x_0 = y_0 = 0$ which correspond to the principal astigmatic bundles $p_1 = 0$ and $q_1 = 0$ in the image space are not orthogonal to each other. The non-orthogonality of these bundles and the torsion of the image thus are equivalent phenomena. For this reason Gullstrand uses the angle between these bundles as a criterion to determine the principal types of optical systems with respect to formation of images.

In a more complete investigation of non-orthogonal systems one has to determine how the constants which characterize the image formation change when the position of the object plane is varied. This involves a discussion of the general formulae (33.63) along similar lines to those we have carried out in §34 and §35.

§37. DIFFERENTIAL EQUATIONS OF FIRST ORDER OPTICS FOR SYSTEMS OF ROTATIONAL SYMMETRY.

37.1 In the preceding sections we have studied the different types of first order image formation which can be expected in general optical systems. All our results have been derived from the sole hypothesis that the transformation of object rays into image rays is a canonical transformation. In the important case of Gaussian optics we have seen that the image function of the ray bundle can be described by three essential constants; the focal length f of the system and the locations of the unit points. The manifold of rays in the neighborhood of the axis of an optical system of revolution determines a Gaussian system. In the case of more general first order systems, of course, a greater number of constants is needed to characterize the image formation.

In this section we shall be concerned with the problem of determining these constants explicitly. We will find that this leads to the problem of integrating certain linear differential equations or linear difference equations, depending upon whether the medium is continuous or discontinuous. We shall treat only the Gaussian case but we note that similar considerations can be carried out in general. We consider continuous media first.

37.2 The canonical variables x, y, p, q as functions of z satisfy the canonical equations

$$\dot{x} = H_p , \qquad \dot{p} = - H_x ,$$
$$\dot{y} = H_q , \qquad \dot{q} = - H_y ,$$

(37.21)

where the Hamiltonian H is given by the expression

$$H = -\sqrt{n^2 (x,y,z) - p^2 - q^2} .$$

(37.22)

We introduce $v = p^2 + q^2$ and $u = x^2 + y^2$ and use the fact that, in systems of revolution, n is a function of the two variables z and u.

From (37.21) it follows that

$$\dot{x} = 2H_v p , \qquad \dot{p} = - 2H_u x ,$$
$$\dot{y} = 2H_v q , \qquad \dot{q} = - 2H_u y .$$

(37.23)

We consider x, y, p, q as small quantities and develop H with respect to powers of u and v. Letting

$$n(u,z) = n_0 (z) + n_1 (z)u + \ldots$$

(37.24)

it follows that to a first order approximation

$$H = - n_0 - n_1 u + \frac{v}{2n_0} .$$

(37.25)

We now replace the original problem (37.22) by an "osculating" canonical problem in which the Hamiltonian (37.22) is replaced by the Hamiltonian (37.25). We again obtain a system of canonical equations:

$$\dot{x} = \frac{1}{n_0} p , \qquad \dot{p} = 2n_1 x ,$$
$$\dot{y} = \frac{1}{n_0} q , \qquad \dot{q} = 2n_1 y .$$

(37.26)

The solutions of this problem can be considered to be approximations of the exact light rays in the neighborhood of the axis. Therefore, these equations are called <u>paraxial equations</u> and their solutions <u>paraxial rays</u>. The solutions x, y, p, q of (37.26) are <u>not</u> small quantities, however, but finite. Only those solutions which lie close to the axis can be considered as approximations of the solutions of (37.21).

It is not difficult to find a geometric representation of these solutions. Let $x(z)$, $y(z)$ be a solution, represented by a curve in space. The quantities

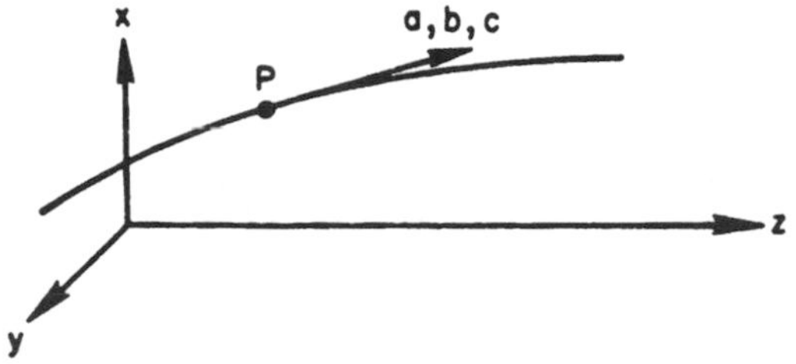

$$p = n_0(z) \frac{dx}{dz} ,$$

$$q = n_0(z) \frac{dy}{dz} ,$$

can then be interpreted by

Figure 141

$$p = n_0(z) \frac{a}{c} ,$$

$$q = n_0(z) \frac{b}{c} , \qquad (37.27)$$

where a,b,c are the direction cosines of the curve at a certain point P. For the exact solutions, however, we know that $p = na$, $q = nb$.

If the curve lies in the x,z-plane then $\frac{a}{c} = \tan \theta$, where θ is the angle made by the tangent to the curve and the z-axis. If, for example, the medium is homogeneous, with $n = 1$, then we have $p = \tan \theta$ for rays in the x,z-plane. The interpretation of p in the exact case is $p = \sin \theta$.

37.3 Since (37.26) is a system of canonical equations, all conclusions which we have drawn with regard to solutions of canonical equations apply to our paraxial rays. We can consider (37.26) as Euler's equations belonging to the problem of variation

$$V = \int_{z_0}^{z_1} (\dot{x}p + \dot{y}q - H)dz = \int_{z_0}^{z_1} \left[\dot{x}p + \dot{y}q + n_1 (x^2 + y^2) - \frac{1}{2n_0} (p^2 + q^2) \right] dz$$

$$+ \int_{z_0}^{z_1} n_0 dz \qquad (37.31)$$

for the functions x, y, p, q. By introducing $p = n_0 \dot{x}$ and $q = n_0 \dot{y}$ from (37.26) we obtain a problem of variation for x and y alone, namely

$$V = \int_{z_0}^{z_1} \left[\frac{1}{2}n_0 (\dot{x}^2 + \dot{y}^2) + n_1 (x^2 + y^2) \right] dz + \int_{z_0}^{z_1} n_0 dz \qquad (37.32)$$

which corresponds to the original Fermat problem. The last integral, of course, could be omitted since it is a constant.

The quantity V can be considered as a function of the points x_0, y_0, z_0 and x_1, y_1, z_1. Its differential is given by

$$dV = p_1 dx_1 + q_1 dy_1 - H_1 dz_1 - p_0 dx_0 - q_0 dy_0 + H_0 dz_0 . \qquad (37.33)$$

V satisfies two partial differential equations

$$\frac{\partial V}{\partial z_1} = - H_1 = n_0(z_1) + n_1(x_1^2 + y_1^2) - \frac{1}{2n_0(z_1)} (V_{x_1}^2 + V_{y_1}^2) ,$$

$$\frac{\partial V}{\partial z_0} = H_0 = - n_0(z_0) - n_1(x_0^2 + y_0^2) + \frac{1}{2n_0(z_0)} (V_{x_0}^2 + V_{y_0}^2) . \qquad (37.34)$$

We remark explicitly that the "wave fronts"

$$V (x_0, y_0, z_0; x, y, z) = C$$

about a point x_0, y_0, z_0 are "transversal" to the paraxial rays in accordance with the definition (21.262) but are no longer normal.

We finally mention that one can often make use of Jacobi's general theorem for solving the paraxial equations: This theorem states[†]: If $\psi(x, y, z; a, b)$ is a solution of the Hamiltonian differential equation

$$\frac{\partial \psi}{\partial z} + H\left(x, y, z; \frac{\partial \psi}{\partial x} , \frac{\partial \psi}{\partial y}\right) = 0 \qquad (37.35)$$

which depends on two parameters, a, b for which the Jacobian

$$\begin{vmatrix} \psi_{xa} & \psi_{xb} \\ \psi_{ya} & \psi_{yb} \end{vmatrix} \neq 0 , \qquad (37.36)$$

then the solution of the canonical equations is given by

$$\frac{\partial \psi}{\partial a} = \alpha, \quad \frac{\partial \psi}{\partial x} = p ,$$

$$\frac{\partial \psi}{\partial b} = \beta, \quad \frac{\partial \psi}{\partial y} = q . \qquad (37.37)$$

[†] Courant-Hilbert, Methoden der math. Physik, Vol. II, pp. 91-95.

In our case the equation (37.35) becomes

$$2n_0\psi_z + \psi_x{}^2 + \psi_y{}^2 = 2n_0{}^2 + 2n_0n_1 (x^2 + y^2) \ . \tag{37.38}$$

The term $2n_0{}^2$ on the right side can be omitted since the function

$$\psi - \int_0^z n_0 dz$$

gives the same result as ψ alone when substituted in (37.37). Thus we obtain the Hamiltonian equation

$$2n_0\psi_z + \psi_x{}^2 + \psi_y{}^2 = 2n_0n_1 (x^2 + y^2) \tag{37.39}$$

as equivalent to the paraxial equations (37.26).

   37.4 <u>Interpretation of the function $n_1(z)$</u>. The function $n_1(z)$, defined by the equation (37.24) has a simple physical meaning. Let us consider the surfaces $n(u,z) = C$ and assume that these surfaces intersect the optical axes at right angles. We thus exclude from this consideration media in which the

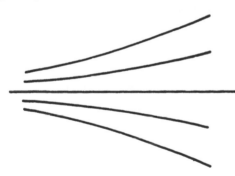

refracting surfaces $n = C$ have the z-axis as an asymptote. The refracting surface which intersects the axis at a point z_0 is given by

$$n(u,z) = n(0,z_0) = n_0(z_0) \ . \tag{37.41}$$

We develop the left side with respect to u and $z - z_0$ in order to find an equation of the surface in the neighborhood of its vertex z_0. It follows that

Figure 142

$$n_0(z_0) + \dot{n}_0(z_0)(z - z_0) + n_1(z_0)u = n_0(z_0) \ .$$

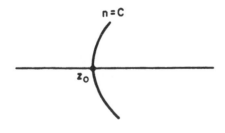

n = C

z_0

Hence

$$(z - z_0) = -\frac{n_1}{\dot{n}_0}(x^2 + y^2) \ . \tag{37.42}$$

This is the equation of a paraboloid which has the radius of curvature at its vertex,

Figure 143

$$R(z_0) = -\frac{\dot{n}_0}{2n_1} \ . \tag{37.43}$$

This is also the radius of curvature of the surface (37.41). Hence we obtain

$$2n_1 = - \frac{\dot{n}_0}{R} \ . \tag{37.44}$$

We remark that $R > 0$ means that the surface is convex towards the incident light.

The expression

$$D = \frac{\dot{n}_0}{R} \tag{37.45}$$

is called the <u>refracting power</u> of the surface or simply the <u>surface power</u>.

By using this notation in (37.26) we obtain the canonical equations

$$\dot{x} = \frac{1}{n} p \ , \quad \dot{p} = - Dx \ , $$

$$\dot{y} = \frac{1}{n} q \ , \quad \dot{q} = - Dy \ , \tag{37.46}$$

where $n = n(z)$ is the refractive index on the z-axis and $D(z) = \dfrac{\dot{n}(z)}{R(z)}$ the power of the refracting surface.

The equivalent Hamiltonian equation (37.39) assumes the form

$$2n\psi_z + \psi_x{}^2 + \psi_y{}^2 + nD(x^2 + y^2) = 0 \ . \tag{37.47}$$

The problem of solving this partial differential equation can be reduced to the corresponding problem for the equation

$$2n\psi_z + \psi_x{}^2 + nDx^2 = 0 \ . \tag{37.48}$$

In fact, if $\psi = \psi (x,z,a)$ is a solution of (37.48) which depends upon an arbitrary parameter a, then

$$\psi(x,y,z; a,b) = \psi(x,z,a) + \psi(y,z,b) \tag{37.49}$$

is a solution of (37.47) which depends on two arbitrary parameters a and b.

37.5 The functions $x(z)$, $p(z)$ and the functions $y(z)$, $q(z)$ satisfy the same differential equations. Therefore it is sufficient to study one of the two sets, for example, the equations

$$\dot{x} = \frac{1}{n} p \ , \quad \dot{p} = - Dx \ . \tag{37.51}$$

The general solutions of these equations can be expressed as a linear combination of two linearly independent particular solutions.

We choose these particular solutions as follows:

(1) The Axial Ray: Let z_0 be the position of an arbitrary reference plane

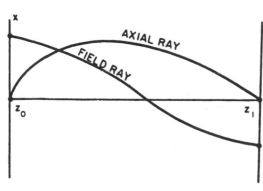

Figure 145

normal to the optic axis. We define the axial ray as the solution

$$x = h(z) \ ,$$
$$p = \vartheta(z) \tag{37.52}$$

of (37.51) which, at $z = z_0$ assumes the boundary values

$$h(z_0) = 0 \ ,$$
$$\vartheta(z_0) = 1 \ . \tag{37.521}$$

(2) The Field Ray: A second solution of (37.51) is chosen quite arbitrarily:

$$x = H(z) \ ,$$
$$p = \theta(z) \ . \tag{37.53}$$

For our present purpose we select this solution by the boundary values

$$H(z_0) = 1 \ ,$$
$$\theta(z_0) = 0 \ . \tag{37.531}$$

Thus the axial ray is a paraxial ray which originates at the point z_0 of the axis. The above field ray is a paraxial ray which leaves the object plane z_0 parallel to the axis.

With the aid of these two rays we can express any other paraxial ray in the form

$$x(z) = x_0 H(z) + p_0 h(z) \ , \qquad y(z) = y_0 H(z) + q_0 h(z) \ ,$$
$$p(z) = x_0 \theta(z) + p_0 \vartheta(z) \ , \qquad q(z) = y_0 \theta(z) + q_0 \vartheta(z) \ , \tag{37.54}$$

with arbitrary constants x_0, p_0, y_0, q_0. Indeed it follows from (37.521) and (37.531) that (37.54) is a ray which leaves the object plane z_0 at the point x_0, y_0 with the direction p_0, q_0.

Let us now assume that the axial ray $x = h(z)$ intersects the axis again at a point $z = z_1$, i.e., $h(z_1) = 0$. From (37.54) it follows that the coordinates $x_1 = x(z_1)$, $y_1 = y(z_1)$; $p_1 = p(z_1)$, $q_1 = q(z_1)$ of any other paraxial ray on the plane $z = z_1$ are given by

$$x_1 = H(z_1)x_0 , \qquad\qquad y_1 = H(z_1)y_0 ,$$

$$y = \theta(z_1)x_0 + \vartheta(z_1)p_0 , \qquad q = \theta(z_1)y_0 + \vartheta(z_1)q_0 . \tag{37.55}$$

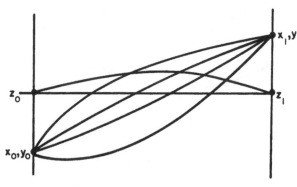

Figure 146

These equations demonstrate that the two planes $z = z_0$ and $z = z_1$ are conjugate planes: Every paraxial ray x_0 , y_0 of the object plane intersects the image plane $z = z_1$ at one and the same point

$$x_1 = Mx_0$$

$$y_1 = My_0 ,$$

where M is determined by the value $H(z_1)$ of the above field ray. By comparing the equations (37.55) with the equations which hold in general for Gaussian conjugate planes, namely

$$x_1 = Mx_0 , \tag{37.56}$$

$$p_1 = - \frac{x_0}{f} + \frac{1}{M} p_0 ,$$

we obtain the Gaussian constants M and f of the medium between z_0 and z_1 from the boundary values of the axial and field ray:

$$H(z_1) = M; \qquad \theta(z_1) = - \frac{1}{f} ; \qquad \vartheta(z_1) = \frac{1}{M} . \tag{37.561}$$

The above considerations allow us to formulate the following statement:
If an arbitrary paraxial ray $x = h(z)$, $p = \vartheta(z)$ intersects the z-axis at two different points z_0 and z_1, then the planes $z = z_0$ and $z = z_1$ are conjugate planes with magnification

$$M = \frac{\vartheta(z_0)}{\vartheta(z_1)} . \tag{37.57}$$

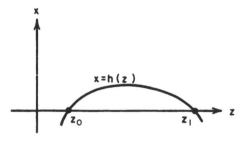

Figure 147

37.6 In many cases it is advantageous to choose the axial and field rays in a different manner. Let us assume that a diaphragm is placed at a point ζ of the optical axis between z_0 and z_1. We define the axial ray h, ϑ by the conditions

$$h(z_0) = 0$$

$$h(\zeta) = 1 \qquad (37.61)$$

and the field ray by

$$H(z_0) = 1 \ ,$$

$$H(\zeta) = 0 \ . \qquad (37.62)$$

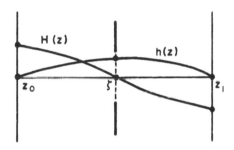

Figure 148

Hence the field ray goes through the center of the diaphragm. We now obtain the general solution of equations (37.46) in the form

$$x(z) = x_0 H(z) + \zeta h(z) \ , \qquad y(z) = y_0 H(z) + \eta h(z) \ ,$$

$$p(z) = x_0 \theta(z) + \zeta \vartheta(z) \ , \qquad q(z) = y_0 \theta(z) + \eta \vartheta(z) \ . \qquad (37.63)$$

These equations represent a paraxial ray which originates at a point x_0, y_0 of the object plane and intersects the plane of the diaphragm at the point ζ, η.

37.7 Lagrange's invariant. Two solutions x,p and X,P of the equations (36.61) are related to each other by a simple expression which is known as Lagrange's invariant. The solutions satisfy the equations

$$\dot{x} = \frac{1}{n} p \ , \qquad \dot{p} = - Dx \ ,$$

$$\dot{X} = \frac{1}{n} P \ , \qquad \dot{P} = - DX \ . \qquad (37.71)$$

We readily conclude from these equations that

$$P\dot{x} - p\dot{X} = 0 \ , \qquad x\dot{P} - X\dot{p} = 0 \ , \qquad (37.72)$$

and hence

$$\frac{d}{dz}(Xp - xP) = 0 .$$ (37.73)

It follows that the determinant

$$\Gamma = \begin{vmatrix} X & P \\ x & p \end{vmatrix}$$ (37.74)

of two arbitrary solutions has a constant value Γ. We call this determinant the _Lagrangian invariant_.

In the case of the axial and field rays, for example, we find with the aid of the boundary conditions (37.521) and (37.531) that

$$H \vartheta - h\theta = 1 .$$ (37.75)

We can use this result to show that it is possible to find the field ray and thus any other paraxial ray by a quadrature if the axial ray is known. The equations (37.72) applied to h, ϑ and H, θ give

$$\theta h - \vartheta \dot{H} = 0; \quad H\dot{\vartheta} - h\dot{\theta} = 0 .$$ (37.76)

From

$$H\dot{\vartheta} - h\dot{\theta} = 0 ,$$

$$H\vartheta - h\theta = 1 ,$$

it follows that

$$\theta \dot{\vartheta} - \vartheta \dot{\theta} = -\frac{\dot{\vartheta}}{h} ,$$

and hence

$$\left(\frac{\theta}{\vartheta}\right)' = \frac{\dot{\vartheta}}{h \vartheta^2} .$$ (37.77)

Since $\theta(z_0) = 0$ this yields

$$\theta = \vartheta \int_{z_0}^{z} \frac{\dot{\vartheta}}{h\vartheta^2} dz = - \vartheta \int_{z_0}^{z} \frac{1}{h} d\left(\frac{1}{\vartheta}\right) .$$ (37.78)

By introducing $\dot{\vartheta} = - Dh$ in the first integral, we find

$$\theta = - \vartheta (z) \int_{z_0}^{z} \frac{D}{\vartheta^2} \, dz \ . \tag{37.781}$$

The integrals (37.78) are improper since $h(z_0) = 0$. The convergence at $z = z_0$, however, is ensured by the form of (37.781) which is a proper integral. If $\theta(z)$ is found by these integrals we obtain $H(z)$ with the aid of the invariant (37.75).

We can easily generalize the result expressed by the formulae (37.78) and (37.781). Let us consider an arbitrary solution $x = x(z)$, $p = p(z)$. We have the relationships

$$H\dot{p} - x\dot{\theta} = 0 \ ,$$

$$Hp - x\theta = \Gamma \ ,$$

where Γ is a constant. By letting $z = z_0$ we find $\Gamma = p_0$. As above it follows that

$$\left(\frac{\theta}{p} \right)^{\cdot} = p_0 \frac{\dot{p}}{xp^2} \ , \tag{37.79}$$

and hence

$$\theta(z) = - p_0 p \int_{z_0}^{z} \frac{1}{x} \, d\left(\frac{1}{p} \right) = - p_0 p \int_{z_0}^{z} \frac{D(z)}{p^2} \, dz \ . \tag{37.791}$$

This demonstrates that the integrals on the right side define the same function $\theta(z)$ for all solutions $x(z)$, $p(z)$ of our differential equations, i.e., they are invariant integrals.

Let us assume that $p(z) \neq 0$ in the interval $z_0 \leq z \leq z_1$. Since $\theta(z_1) = - 1/f$ we find the interesting relationship

$$\frac{1}{f} = p_0 p_1 \int_{z_0}^{z_1} \frac{D(z)}{p^2} \, dz = p_0 p_1 \int_{z_0}^{z_1} \frac{1}{x} \, d\left(\frac{1}{p} \right) \ . \tag{37.792}$$

The focal length of the medium between z_0 and z_1 can be found by the integrals (37.792) from an arbitrary paraxial ray $x = x(z)$, $p = p(z)$, provided that $p(z) \neq 0$.

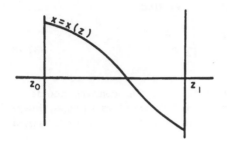

Figure 149

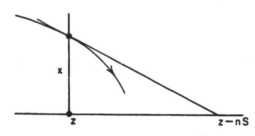

Figure 150

37.8 The problem of integrating the two linear equations

$$\dot{x} = \frac{1}{n}\, p; \quad \dot{p} = -\,Dx \qquad (37.81)$$

can be reduced to the problem of integrating one quadratic differential equation of <u>Riccati's</u> type. Let us consider the ratio

$$S(z) = \frac{x(z)}{p(z)} \qquad (37.82)$$

which determines by $z - nS(z)$ the point where the tangent to the ray intersects the z-axis. We find

$$\dot{S} = \frac{x}{p} - \frac{x}{p^2}\,\dot{p} = \frac{\dot{x}}{p} - S^2\,\frac{\dot{p}}{x} \;.$$
$$(37.83)$$

This gives because of (37.81) the Riccati equation

$$\dot{S} = \frac{1}{n} + DS^2 \;. \qquad (37.84)$$

Let us assume that we know the solution $S(z)$ which, for $z = z_0$ assumes the boundary value $S_0 = \frac{x_0}{p_0}$. From (37.81) it follows that

$$\frac{\dot{p}}{p} = -\,DS \;, \qquad (37.85$$

and hence

$$p = p_0 e^{\displaystyle -\int_{z_0}^{z_1} DS\;dz} \quad , \qquad x = p_0 S e^{\displaystyle -\int_{z_0}^{z_1} DS\;dz} \quad , \qquad (37.86)$$

i.e., the solutions $x(z)$ and $p(z)$ can be found by quadratures if we know the solution $S(z)$ of (37.84) which satisfies the boundary condition $S(z_0) = \frac{x_0}{p_0}$

§38. THE PATH OF ELECTRONS IN THE NEIGHBORHOOD OF THE AXIS OF AN INSTRUMENT OF REVOLUTION.

38.1 In the original definitions of the functions n(z) and D(z), the functions $n_0(z)$ and $n_1(z)$ are in general not related, but can be assigned arbitrarily. This, however, is not the case for an electron-optical instrument of revolution. We assume that only electrostatic fields are used in the instrument. The path of the electrons in this case can be found mathematically in the same manner as the light rays in a medium of rotational symmetry, provided that the index of refraction n(x,y,z) satisfies the differential equation

$$\Delta n^2 = \frac{\partial^2 n^2}{\partial x^2} + \frac{\partial^2 n^2}{\partial y^2} + \frac{\partial^2 n^2}{\partial z^2} = 0 . \tag{38.11}$$

Let us now assume that the electrostatic potential ϕ is known on the axis; then we know $n^2(0,0,z) = f(z)$ on the axis because of the relation $n^2 = 2(C - K\phi)$ which we have derived in §17.3. Our problem is to find the function $n^2 = U(z,\rho)$ which satisfies the equation $\Delta U = 0$ and assumes the values $U = f(z)$ for $\rho = 0$; ρ is the distance from the axis, i.e., $\rho = \sqrt{x^2 + y^2}$. We assume that f(z) is an analytic function of z. The problem then has a simple solution, namely

$$U(z,\rho) = \frac{1}{2\pi} \int_0^{2\pi} f(z + i \rho \cos \varphi) d\varphi . \tag{38.12}$$

It is clear that u(z,0) = f(z), and we have to show that $\Delta u = 0$. One easily verifies the equations

$$U_{zz} = \frac{1}{2\pi} \int_0^{2\pi} f''(z + i \rho \cos \varphi) d\varphi ,$$

$$u_{xx} + u_{yy} = u_{\rho\rho} + \frac{1}{\rho} u_\rho = \frac{1}{2\pi} \int_0^{2\pi} \left(\frac{i \cos \varphi}{\rho} f' - \cos^2\varphi \, f'' \right) d\varphi ,$$

and hence

$$\Delta U = \frac{1}{2\pi} \int_0^{2\pi} \left[\sin^2\varphi \, f'' (z + i \rho \cos \varphi) + \frac{i \cos \varphi}{\rho} f' (z + i \rho \cos \varphi) \right] d\varphi . \tag{38.13}$$

The integrand is the derivative with respect to φ of the expression

$$\frac{i \sin \varphi}{\rho} f' (z + i \rho \cos \varphi) . \tag{38.14}$$

Hence $\Delta U = 0$ since (38.14) is periodic in φ with period 2π.

We thus obtain the "index of refraction" $n(z,\rho)$ in the form

$$n^2 = \frac{1}{2\pi} \int_0^{2\pi} f \, (z + i \rho \cos \varphi) \, d\varphi \, . \tag{38.15}$$

38.2 By developing n^2 with respect to powers of ρ we find

$$n^2 = f(z) - \frac{1}{4} f''(z)\rho^2 + \ldots \, , \tag{38.21}$$

or, by introducing $\rho^2 = u_1$:

$$n^2 = f(z) - \frac{u}{4} f''(z) + \ldots \, .$$

This yields for $u = 0$:

$$\frac{\partial}{\partial u} \, n^2 = 2n_0 n_1 = - \frac{1}{4} f''(z)$$

and hence by (37.44) we obtain the following expression for <u>the refractive</u> <u>power of the equipotential</u> through the axial point z:

$$D = - \, 2n_1 = + \, \frac{1}{4} \frac{f''(z)}{\sqrt{f(z)}} \, . \tag{38.22}$$

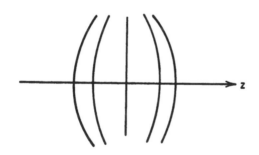

The refracting power D of an equipotential surface is thus completely determined by the values of the field on the axis of symmetry. By introducing (38.22) and $n = \sqrt{f}$ in the general equations (37.46) we obtain the paraxial paths of the electron as solutions of the differential equations

Figure 151

$$\sqrt{f} \, \dot{x} = p \, , \quad 4\sqrt{f} \, \dot{p} = - \, \ddot{f}(z)x \, ,$$
$$\tag{38.23}$$
$$\sqrt{f} \, \dot{y} = q \, , \quad 4\sqrt{f} \, \dot{q} = - \, \ddot{f}(z)y \, .$$

By eliminating p and q we get two second order equations for $x(z)$ and $y(z)$, namely

$$4f(z)\ddot{x} + 2\dot{f}(z)\dot{x} + \ddot{f}(z)x = 0 \, ,$$

$$4f(z)\ddot{y} + 2\dot{f}(z)\dot{y} + \ddot{f}(z)y = 0 \, . \tag{38.24}$$

38.3 The integration of these differential equations is equivalent to the same problem for the Hamilton–Jacobi equation (37.47) which in our case is

$$2\sqrt{f(z)}\ \psi_z + \psi_x{}^2 + \psi_y{}^2 + \frac{1}{4} f''(z)\ (x^2 + y^2) = 0\ . \tag{38.31}$$

We can reduce the number of variables in this equation immediately by the theorem: If $\psi(x,z;a)$ is an integral of the equation

$$2\sqrt{f(z)}\ \psi_z + \psi_x{}^2 + \frac{1}{4} x^2 f''(z) = 0 \tag{38.32}$$

which depends on the arbitrary parameter a, then

$$\psi(x,y,z;a,b) = \psi(x,z;a) + \psi(y,z,b)$$

is an integral of (37.591) which depends on two parameters a and b. The equivalent Riccati equation (37.84) for the quantity $S = \frac{x}{p}$ becomes in our case

$$\sqrt{f}\ \dot{S} = 1 + \frac{1}{4} \ddot{f} S^2\ . \tag{38.33}$$

§39. DIFFERENCE EQUATIONS FOR A CENTERED SYSTEM OF REFRACTING SURFACES OF REVOLUTION.

39.1 We consider next the case of an optical instrument which consists of a number of refracting surfaces of revolution. We assume that the system is centered, i.e., that the surfaces have the z-axis as a common axis of revolution. Instead of the canonical differential equations (37.46) we shall obtain a system of canonical difference equations. We will find that it is possible to repeat the above theory almost in every step.

Let us first introduce a suitable notation. The paraxial rays are continuous curves x(z) and y(z) which consist of sections of straight lines. The derivatives of these curves are discontinuous at points of the z-axis which are given by the vertices z_i of the refracting surfaces. It is now extremely convenient to use only even subscripts i in connection with data referring to the surfaces and odd subscripts for data which are connected with the media between the surfaces.

For example we denote by

$z_2, z_4, \ldots, z_k$ the position of the vertices of the surfaces

$R_2, R_4, \ldots, R_k$ the radii of curvature of the surfaces on the axis

$x_2, x_4, \ldots, x_k$

$y_2, y_4, \ldots, y_k$ the values of x(z) and y(z) at the points z_i .

and by

$t_3, t_5, \ldots, t_{k-1}$ the axial separations of the surfaces

$n_1, n_3, n_5, \ldots, n_{k-1}, n_{k+1}$ the indices of refraction

$p_1, p_3, p_5, \ldots, p_{k-1}, p_{k+1}$

                                                                the canonical variables which give

$q_1, q_3, q_5, \ldots, q_{k-1}, q_{k+1}$

the direction of the paraxial rays, i.e., the quantities

$$p_i = n_i \frac{a_i}{c_i} \, ,$$

$$q_i = n_i \frac{b_i}{c_i}$$

(39.11)

where a_i, b_i, c_i are the direction cosines of the ray.

We also extend the above notation to both the object and the image plane. Let z_0 and z_{k+2} be the position of these planes and t_1 and t_{k+1} the distances from the first and last surface respectively. We shall, however, take the freedom to denote the coordinates $x_0, y_0; p_1, q_1$ and $x_{k+2}, y_{k+2}, p_{k+1}, q_{k+1}$ in object and image space simply by

$$x, y, p, q$$

and

$$x', y', p', q' \, .$$

(39.12)

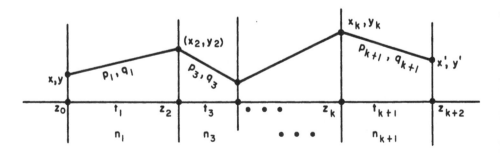

Figure 152

39.2 We now establish the difference equations which connect the quantities x_i, y_i, p_i, q_i. These equations are closely related to the canonical equations (37.46). In fact, from the first column in (37.46) it follows by integration — since p and n are constant between two surfaces i-1 and i+1 — that

$$x_{i+1} - x_{i-1} = \frac{t_i}{n_i} \, p_i \ ,$$

$$y_{i+1} - y_{i-1} = \frac{t_i}{n_i} \, y_i \ ,$$

(39.21)

These equations are geometrically evident from the interpretation (39.11) of the quantities p_i and q_i.

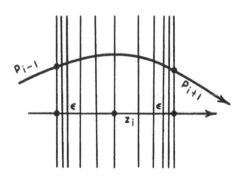

Figure 153

We can use the second column in (37.46) in order to derive the additional relations. Let us assume for a moment that n(z) is continuous in the neighborhood of the point z_i but increases rapidly from n_{i-1} to n_{i+1} if z goes from $z_i - \epsilon$ to $z_i + \epsilon$. Let R(z) be constant in this interval and equal to R_i, the radius of curvature of the surface. From (37.46) it follows that

$$p_{i+1} - p_{i-1} = - \int_{z_i - \epsilon}^{z_i + \epsilon} \frac{n'}{R_i} \, x(z) dz$$

$$= - \frac{x(z^*)}{R_i} (n_{i+1} - n_{i-1})$$

(39.22)

where z^* is a certain point in the interval of integration.

The curve $x = x(z)$ is assumed to approach the paraxial ray in case $\epsilon \to 0$, i.e., a continuous curve which consists of two linear sections joined together at z_i. Clearly $x(z^*) \to x_i$ and in the limit we obtain the relation

$$p_{i+1} - p_{i-1} = - D_i x_i$$

and similarly

$$q_{i+1} - q_{i-1} = - D_i y_i$$

(39.23)

where

$$D_i = \frac{n_{i+1} - n_{i-1}}{R_i} \tag{39.24}$$

is the power of the i-th surface.

We finally introduce the so-called "optical separations" of the surfaces

$$\delta_i = \frac{t_i}{n_i} \tag{39.25}$$

and obtain the following set of canonical difference equations

$$x_{i+1} - x_{i-1} = \delta_i p_i, \qquad p_{i+1} - p_{i-1} = - D_i x_i,$$

$$y_{i+1} - y_{i-1} = \delta_i q_i, \qquad q_{i+1} - q_{i-1} = - D_i y_i. \tag{39.26}$$

The coefficients

$$\delta_i = \frac{t_i}{n_i}; \qquad D_i = \frac{n_{i+1} - n_{i-1}}{R_i} \tag{39.27}$$

are given directly by the data of the instrument.

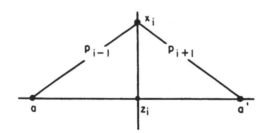

Figure 154

The physical meaning of the equations (39.26) can be found readily. The first column expresses, as we have seen, a simple geometrical fact. The second column is equivalent to the lens equation for one refracting surface. In fact let us consider the ratios

$$\frac{a}{n_{i-1}} = - \frac{x_i}{p_{i-1}} ;$$

$$\frac{a'}{n_{i+1}} = - \frac{x_i}{p_{i+1}} \tag{39.28}$$

for a ray in the x,z plane. Then a and a' are nothing but the distances of the points where the ray intersects the axis before and after refraction, from the vertex z_i. From (39.26) it follows

$$\frac{n_{i+1}}{a'} - \frac{n_{i-1}}{a} = D_i = \frac{n_{i+1} - n_{i-1}}{R_i} . \tag{39.29}$$

This, however, is the lens equation which we have derived in §21.39 for the case of a spherical surface.

39.3 The equations (39.26) are especially convenient for the problem of paraxial ray tracing, i.e., for the problem of determining a paraxial ray which has given initial data x_0, y_0, p_1, q_1. We remark again that the quantities x_i, y_i, p_i, q_i are not infinitely small but finite. The meaning of paraxial rays is that both paraxial rays and the light rays of the instrument have a first order manifold of rays in common in the neighborhood of the optical axis.

The problem of variation (37.31) becomes an ordinary extreme value problem for the quantities x_i, y_i, p_i, q_i, namely

$$V = \sum_{i=1,3,\ldots}^{k+1} (x_{i+1} - x_{i-1})p_i + (y_{i+1} - y_{i-1})q_i - \frac{1}{2}\delta_i (p_i^2 + q_i^2)$$

$$- \frac{1}{2} \sum_{i=2,4,\ldots}^{k} D_i (x_i^2 + y_i^2) = \text{Extr.} \tag{39.31}$$

The canonical equations (39.26) are the "Euler equations" of this problem. We can replace this "canonical" extreme value problem by a problem for the quantities x_i and y_i alone, by introducing

$$p_i = \frac{1}{\delta_i} (x_{i+1} - x_{i-1}) ,$$

$$q_i = \frac{1}{\delta_i} (y_{i+1} - y_{i-1}) .$$

It follows that

$$V = \sum_{i=1,3,\ldots}^{k+1} \frac{1}{2\delta_i} \left[(x_{i+1} - x_{i-1})^2 + (y_{i+1} - y_{i-1})^2 \right]$$

$$- \frac{1}{2} \sum_{i=2,4,\ldots}^{k} D_i (x_i^2 + y_i^2) = \text{Extr.} \tag{39.32}$$

39.4 The general solution of the equations (39.26) can be expressed as a linear combination of two particular linearly independent solutions, which we

call <u>Axial ray</u> and <u>Field ray</u>. The axial ray is $x_i = h_i$; $p_i = \vartheta_i$ defined by
the initial values

$$h_0 = 0 \ ,$$
$$\vartheta_1 = 1 \tag{39.41}$$

and the field ray $x_i = H_i$; $p_i = \theta_i$ by

$$H_0 = 1 \ ,$$
$$\theta_1 = 0 \ . \tag{39.411}$$

With the aid of these two rays we obtain every other paraxial ray in the form

$$x_i = x_0 H_i + p_0 h_i \ , \qquad y_i = y_0 H_i + q_0 h_i \ ,$$
$$p_i = x_0 \theta_i + p_0 \vartheta_i \ , \qquad q_i = y_0 \theta_i + q_0 \vartheta_i \ . \tag{39.42}$$

The image plane z_{k+2} is conjugate to the object plane z_0 if $h_{k+2} = h' = 0$. In
this case the equations (39.42) give

$$x' = x_0 H' \ , \qquad\qquad y' = y_0 H' \ ,$$
$$p' = x_0 \theta' + p_0 \vartheta' \ , \qquad q' = y_0 \theta' + q_0 \vartheta \ , \tag{39.43}$$

which shows, as before, that the focal length f of the instrument and the magnification, M, of the object plane are given by the quantities

$$M = H'; \qquad \theta' = -\frac{1}{f}; \qquad \vartheta' = \frac{1}{M} \ . \tag{39.44}$$

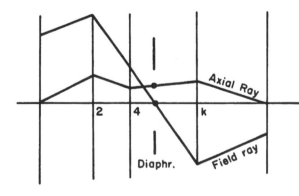

Figure 155

If the instrument contains a
diaphragm at a certain point
ζ of the axis it is preferable to
define axial ray and field ray
by the following conditions:

$$h_0 = 0$$
$$h(\zeta) = 1 \ , \tag{39.45}$$

$$H_0 = 1$$
$$H(\zeta) = 0 \tag{39.46}$$

so that the field ray goes through the center of the diaphragm. The general
solution of (38.26) then is given by

$$x_i = x_0 H_i + \xi h_i \ , \qquad y_i = y_0 H_i + \eta h_i \ ,$$
$$p_i = x_0 \theta_i + \xi \vartheta_i \ , \qquad q_i = y_0 \theta_i + \eta \vartheta_i \ ,$$
(39.47)

which represents a ray which intersects the object plane at the point x_0, y_0 and
the plane of the diaphragm at ξ, η. In this form one can make good use of
paraxial rays for approximately solving vignetting problems, i.e., the problem
of determining the manifold of rays which passes through the instrument.
Indeed, the quantities x_i and y_i approximate the points where the actual light
rays intersect the lens surfaces. In many cases this approximation is sufficient.

39.5 Lagrange's invariant.

Two solutions x_i, p_i and X_i, P_i of the difference equations

$$x_{i+1} - x_{i-1} = \delta_i p_i \ ,$$
$$p_{i+1} - p_{i-1} = - D_i x_i$$
(39.51)

are related to each other by an invariant similar to that of solutions of the
corresponding differential equations. We conclude from

$$\delta_i = \frac{x_{i+1} - x_{i-1}}{p_i} = \frac{X_{i+1} - X_{i-1}}{P_i} \ , i = 1,3,5, \ldots ,$$
$$- D_i = \frac{p_{i+1} - p_{i-1}}{x_i} = \frac{P_{i+1} - P_{i-1}}{X_i} \ , i = 2,4,6, \ldots$$
(39.52)

that

$$\begin{vmatrix} X_{i+1} & P_i \\ x_{i+1} & p_i \end{vmatrix} = \begin{vmatrix} X_{i-1} & P_i \\ x_{i-1} & p_i \end{vmatrix} \qquad i = 1,3, \ldots$$
(39.53)

and

$$\begin{vmatrix} X_i & P_{i+1} \\ x_i & p_{i+1} \end{vmatrix} = \begin{vmatrix} X_i & P_{i-1} \\ x_i & p_{i-1} \end{vmatrix} = \Gamma_i , i = 2,4, \ldots .$$
(39.54)

By introducing even subscripts i in (39.53) we obtain

$$\begin{vmatrix} X_i & P_{i-1} \\ x_i & P_{i-1} \end{vmatrix} = \begin{vmatrix} X_{i-2} & P_{i-1} \\ x_{i-2} & P_{i-1} \end{vmatrix} \quad , \ i = 2,4, \ldots \tag{39.55}$$

and hence by comparing with (39.54) we obtain the identity

$$\begin{vmatrix} X_i & P_{i+1} \\ x_i & P_{i+1} \end{vmatrix} = \begin{vmatrix} X_{i-2} & P_{i-1} \\ x_{i-2} & P_{i-1} \end{vmatrix} \quad , \ i = 2,4, \ldots \tag{39.56}$$

or $\Gamma_i = \Gamma_{i-2}$. We conclude from this that the Γ_i have the same value Γ for all i. Therefore our result is: <u>The determinants</u>

$$\Gamma = \begin{vmatrix} X_i & P_{i+1} \\ x_i & P_{i+1} \end{vmatrix} = \begin{vmatrix} X_i & P_{i-1} \\ x_i & P_{i-1} \end{vmatrix} \tag{39.57}$$

have the same value Γ for all surfaces, i.

In the case of the rays (39.41) and (39.411), for example, we have

$$\begin{vmatrix} H_i & \theta_{i+1} \\ h_i & \vartheta_{i+1} \end{vmatrix} = \begin{vmatrix} H_i & \theta_{i-1} \\ h_i & \vartheta_{i-1} \end{vmatrix} = 1 \ . \tag{39.58}$$

39.6 We can use the above invariant to express the data H_i, θ_i of the field ray by the data h_i, ϑ_i of the axial ray. We write (39.58) in the form

$$\frac{\theta_{i+1}}{\vartheta_{i+1}} - \frac{H_i}{h_i} = - \frac{1}{h_i \, \vartheta_{i+1}} \quad ,$$

$$\tag{39.61}$$

$$\frac{\theta_{i-1}}{\vartheta_{i-1}} - \frac{H_i}{h_i} = - \frac{1}{h_i \, \vartheta_{i-1}}$$

and obtain by subtraction

$$\frac{\theta_{i+1}}{\vartheta_{i+1}} - \frac{\theta_{i-1}}{\vartheta_{i-1}} = \frac{1}{h_i} \left(\frac{1}{\vartheta_{i-1}} - \frac{1}{\vartheta_{i+1}} \right) \ . \tag{39.62}$$

This gives

$$\frac{\theta_{i+1}}{\vartheta_{i+1}} = \sum_{\nu=2}^{i} \frac{1}{h_\nu}\left(\frac{1}{\vartheta_{\nu-1}} - \frac{1}{\vartheta_{\nu+1}}\right) , \tag{39.63}$$

i.e., θ_{i+1} is given by the data h_i, ϑ_i of the axial ray. By introducing (39.63) and (39.61) we obtain H_i expressed by the data h_i, ϑ_i.

Letting $i = k$ in (39.63) we find an expression for the total power $D = \frac{1}{f}$ of the k refracting surfaces. In fact, since $D = \frac{1}{f} = -\theta_{k+1}$ we obtain

$$D = \vartheta_{k+1} \sum_{\nu=2}^{k} \frac{1}{h_\nu}\left(\frac{1}{\vartheta_{\nu+1}} - \frac{1}{\vartheta_{\nu-1}}\right) . \tag{39.64}$$

We can carry out the above considerations for an <u>arbitrary paraxial ray</u>, x_i, p_i, instead of the axial ray. The result is:

$$D = p_1 p_{k+1} \sum_{\nu=2}^{k} \frac{1}{x_\nu}\left(\frac{1}{p_{\nu+1}} - \frac{1}{p_{\nu-1}}\right) . \tag{39.65}$$

By introducing $p_{\nu-1} - p_{\nu+1} = D_\nu x_\nu$ we can replace (39.65) by the equivalent formula:

$$\frac{D}{p_1 p_{k+1}} = \sum_{\nu=2}^{k} \frac{D_\nu}{p_{\nu-1} p_{\nu+1}} . \tag{39.66}$$

39.7 We can use the difference equations

$$x_{i+1} - x_{i-1} = \delta_i p_i ,$$
$$p_{i+1} - p_{i-1} = -D_i x_i \tag{39.71}$$

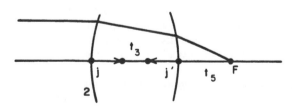

Figure 156

to derive expressions for the Gaussian constants of a lens system, i.e., for the focal length $f = 1/D$ and the location of the unit points. Let us illustrate this with the example of a single lens with two spherical surfaces. Through the lens we "trace" a paraxial ray from infinity with the initial data $p_1 = 0$, $x_2 = 1$.

From (39.71) it follows that

$$p_3 = - D_2 ,$$

$$x_4 = 1 + \delta_3 p_3 = 1 - \delta_3 D_2 , \qquad (39.72)$$

$$p_5 = p_3 - D_4 x_4 = - D_2 - D_4 (1 - \delta_3 D_2) .$$

We know by the initial conditions that $p_5 = - \dfrac{1}{f} = - D$. Hence

$$\frac{1}{f} = D = D_2 + D_4 - \delta_3 D_2 D_4 \qquad (39.73)$$

where

$$D_2 = \frac{n-1}{R_2} ; \quad D_4 = \frac{1-n}{R_4} ; \quad \delta_3 = \frac{t_3}{n} . \qquad (39.731)$$

The distance of the focal point F from the surface 4 is given by the ratio

$$t_5 = - \frac{x_4}{p_5} = x_4 f . \qquad (39.74)$$

Let j' be the abscissa of the unit point of the image space relative to the surface 4. We know that the focal point has the distance $F = n_5 f = f$ from this unit point. Hence $t_5 - j' = f$ or $j' = (x_4 - 1)f$. From (39.72) it follows that

$$j' = - \delta_3 D_2 f = - \delta_3 \frac{D_2}{D} . \qquad (39.75)$$

By tracing a paraxial ray backwards through the system so that $p_5 = 0$; $x_4 = 1$ we can obtain the unit point of the object space by similar considerations. The result is

$$j = \delta_3 \frac{D_4}{D} \qquad (39.76)$$

where j is the abscissa of the unit point relative to the surface 2. The above method can be applied, of course, to optical systems of any number of surfaces. However, the resulting formulae rapidly become complicated and unsuited for numerical computation. A direct numerical computation of the quantities x_i, p_i with the aid of the linear equations (39.71) is actually incomparably simpler.

39.8 <u>First Order Design of Optical Instruments</u>. The equations (39.71) provide us with a simple method of constructing optical instruments which satisfy given first order conditions. These first order conditions, as, for example, magnification, focal length, location of object and image plane, can be expressed by quite simple conditions for the data of paraxial rays. Very complicated formulae are obtained, however, if the geometrical data of the lenses are introduced. This fact suggests the use of the paraxial rays as intermediate parameters of the system in optical design. Let us consider, for example, the $k + 1$ parameters h_i, ϑ_i of the axial ray. We can assign these parameters and the refractive indices n_i arbitrarily and then determine an optical instrument which possesses the assigned axial ray. Indeed, from

$$h_{i+1} - h_{i-1} = \delta_i \vartheta_i \ ,$$

$$\vartheta_{i+1} - \vartheta_{i-1} = - D_i h_i$$

$$(39.81)$$

we obtain

$$\delta_i = \frac{h_{i+1} - h_{i-1}}{\vartheta_i} ; \quad D_i = \frac{\vartheta_{i-1} - \vartheta_{i+1}}{h_i} \ , \qquad (39.82)$$

i.e., the optical separations and the surface powers of the system. The geometric separations and the radii of the surfaces are given by

$$t_i = n_i \delta_i \ ,$$

$$R_i = \frac{n_{i+1} - n_{i-1}}{D_i} \ . \qquad (39.83)$$

Thus we find that an arbitrarily given paraxial ray represents an optical system which is uniquely determined when the indices of the media are chosen.

We illustrate this method with a simple example. We want a lens which forms a real magnified image of a given plane on another given plane. Let $M = -5$ and $L = 500$ mm. be the distance from the object to the image plane. We also give the "working distance", i.e., the distance t_1 of the surface 2 from the object plane and the thickness t_3 of the lens: $t_1 = 80$ mm.; $t_3 = 10$ mm., $t_5 = 410$ mm. . Let $n = 1.5$. The problem has a unique solution. We arbitrarily choose $\vartheta_1 = 5$ and thus have $\vartheta_5 = -1$ in order to obtain the magnification $M = -5$. It follows that $h_2 = 400$, $h_4 = 410$ and hence

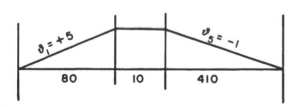

$$\vartheta_3 = \frac{1.5}{10}(410 - 400) = 1.5 \ .$$

The data of the axial ray are therefore known and we are in the position to determine

Figure 157

the geometrical data of the lens. The results of the different steps are as follows

$\vartheta_1 = 5$ $\delta_1 = 80$ $t_1 = 80$

$h_2 = 400$ $D_2 = 0.00815$ $R_2 = 57.14$

$\vartheta_3 = 1.5$ $\delta_3 = 6.6667$ $t_3 = 10$ $n_3 = 1.5$.

$h_4 = 410$ $D_4 = 0.00610$ $R_4 = -82.0$

$\vartheta_5 = -1$ $\delta_5 = 410$ $t_5 = 410$.

If the lens consists of more than two surfaces our method can still be used but

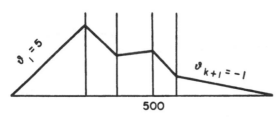

Figure 158

it leads to many solutions. Indeed, our conditions are satisfied as long as the axial ray has the values $\vartheta_1 = 5$ and $\vartheta_{k+1} = -1$ and intersects the axis at two points which are 500 mm. apart. Any broken line curve which satisfies these conditions represents an optical system which satisfies our conditions. We can impose other first order conditions, for example, that the optical system has a given focal length f. Then we can make use of the formula (39.64) which expresses $D = 1/f$ by the quantities h_1, ϑ_1 and take care to select only such sets h_1, ϑ_1 which satisfy this condition (39.44).

Any additional degree of freedom can be used by the designer to decrease the aberrations of the lens combination. The advantage of characterizing a lens system first by the intermediary parameters h_1, ϑ_1 becomes even more evident in connection with the aberration problem. We shall see in the next chapter that it is possible to express the 3rd order aberrations as functions of the data of paraxial rays and that these functions are relatively simple and are well suited for numerical calculations.

CHAPTER V

THE THIRD ORDER ABERRATIONS IN SYSTEMS OF ROTATIONAL SYMMETRY

The problem of this chapter is to develop the image functions

$$x_1 = x_1(x_0, y_0, p_0, q_0) \ ,$$

$$y_1 = y_1(x_0, y_0, p_0, q_0) \ ,$$

$$p_1 = p_1(x_0, y_0, p_0, q_0) \ ,$$

$$q_1 = q_1(x_0, y_0, p_0, q_0) \ ,$$

one step beyond the first order development in the neighborhood of the principal ray. In general this leads to certain algebraic functions which are of second order in the initial ray coordinates x_0, y_0, p_0, q_0. If we call the departure from the first order functions the _aberrations of the rays_, then the second order aberrations $\Delta x_1, \Delta y_1; \Delta p_1, \Delta q_1$ are given by homogeneous quadratic polynominals of x_0, y_0, p_0, q_0.

In the case that the principal ray is the axis of an optical system of revolution we will obtain no second order aberrations. This follows from the fact that the above image functions are _odd_ functions of x_0, y_0, p_0, q_0. The development of the image functions one step beyond the first order gives algebraic functions of third order. The departures $\Delta x_1, \Delta y_1, \Delta p_1, \Delta q_1$ from the Gaussian approximations are determined by certain homogeneous third order polynomials of x_0, y_0, p_0, q_0. These polynomials represent the _third order aberrations_ of the instrument. In this chapter we shall be concerned with the theory of these aberrations.

§40. GENERAL TYPES OF THIRD ORDER ABERRATIONS.

40.1 We first derive the possible types of aberrations which we must expect in the third order approximation. For this purpose we make use of Hamilton's mixed characteristic $W(x_0, y_0, p_1, q_1)$ from which we obtain the image functions in the form

$$x_1 = -\frac{\partial W}{\partial p_1} \ , \qquad p_0 = -\frac{\partial W}{\partial x_0} \ ,$$

$$\tag{40.11}$$

$$y_1 = -\frac{\partial W}{\partial q_1} \ , \qquad q_0 = -\frac{\partial W}{\partial y_0} \ .$$

269

We know that W is a function, $W(u,v,w)$ of the invariants of rotation,

$$u = x_0^2 + y_0^2 ,$$

$$v = p_1^2 + q_1^2 , \tag{40.12}$$

$$w = 2(x_0 p_1 + y_0 q_1) .$$

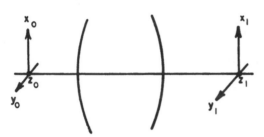

Figure 159

Let us assume that the reference planes $z = z_0$ and $z = z_1$ are conjugate in the Gaussian optics of the instrument.

This means, according to our results in §34, that the image functions of first order have the form

$$x_1 = Mx_0 , \quad p_0 = \frac{M}{f} x_0 + Mp_1 ,$$

$$\tag{40.13}$$

$$y_1 = My_0 , \quad q_0 = \frac{M}{f} y_0 + Mq_1$$

and hence, that $W(u,v,w)$ can be developed in the form

$$W = - \frac{1}{2} \left(\frac{M}{f} u + Mw \right) + W_2(u,v,w) + \dots . \tag{40.14}$$

The expression $W_2(u,v,w)$ represents a homogeneous polynomial of second order in u,v,w. The third order aberrations of our system, i.e., the departures

$$\Delta x_1 = x_1 - Mx_0 , \qquad \Delta p_0 = p_0 - \frac{M}{f} x_0 - Mp_1 ,$$

$$\tag{40.15}$$

$$\Delta y_1 = y_1 - My_0 , \qquad \Delta q_0 = q_0 - \frac{M}{f} y_0 - Mq_1 ,$$

from the functions (40.13) are determined by the polynomial $W_2(u,v,w)$. We obtain

$$\Delta x_1 = - 2 \left(\frac{\partial W_2}{\partial w} x_0 + \frac{\partial W_2}{\partial v} p_1 \right) , \quad \Delta p_0 = - 2 \left(\frac{\partial W_2}{\partial u} x_0 + \frac{\partial W_2}{\partial w} p_1 \right) ,$$

$$\tag{40.16}$$

$$\Delta y_1 = - 2 \left(\frac{\partial W_2}{\partial w} y_0 + \frac{\partial W_2}{\partial v} q_1 \right) , \quad \Delta q_0 = - 2 \left(\frac{\partial W_2}{\partial u} y_0 + \frac{\partial W_2}{\partial w} q_1 \right) .$$

40.2 We assume that W_2 (u,v,w) has the form

$$W_2 = - \left[\frac{1}{4}Fu^2 + \frac{1}{4}Av^2 + \frac{C - D}{8}w^2 + \frac{1}{2}Duv + \frac{1}{2}Euw + \frac{1}{6}Buw \right] , \qquad (40.21)$$

where A,B,C,D,E,F are certain constants depending on the instrument and on the choice of object and image planes. By applying (40.16) we obtain the following expressions for the third order aberrations:

$$\Delta x_1 = \left[Eu + \frac{1}{3}Bv + \frac{1}{2}(C - D)w \right] x_0 + \left[Du + Av + \frac{1}{3}Bw \right] p_1 ,$$

$$\Delta y_1 = \left[Eu + \frac{1}{3}Bv + \frac{1}{2}(C - D)w \right] y_0 + \left[Du + Av + \frac{1}{3}Bw \right] q_1 ,$$

$$\qquad\qquad (40.22)$$

$$\Delta p_0 = \left[Fu + Dv + Ew \right] x_0 + \left[Eu + \frac{1}{3}Bv + \frac{1}{2}(C - D)w \right] p_1 ,$$

$$\Delta q_0 = \left[Fu + Dv + Ew \right] y_0 + \left[Eu + \frac{1}{3}Bv + \frac{1}{2}(C - D)w \right] q_1 .$$

The quality of the image is determined by the two functions Δx_1 and Δy_1. We see that only the five coefficients A,B,C,D,E enter in these functions. This shows that, in the realm of third order optics, five different types of aberrations have to be considered. The sixth coefficient F, however, is necessary in order to characterize the complete manifold of rays into which the rays of the object space are transformed by the optical instrument. By knowing F in addition to A,B,C,D,E we are, for example, in a position to determine the quality of images for any object plane, not merely for the object plane which was chosen originally.

For the physical interpretation of the five coefficients A,B,C,D,E we may assume $y_0 = 0$ without loss of generality. Thus we introduce in (40.22):

$$y_0 = 0 ,$$

$$u = x_0^2 ,$$

$$v = p_1^2 + q_1^2 ,$$

$$w = 2x_0 p_1 ,$$

and obtain, by arranging with respect to powers of x_0:

$$\Delta x_1 = A(p_1^2 + q_1^2)p_1 + \frac{1}{3}Bx_0(3p_1^2 + q_1^2) + Cx_0^2p_1 + Ex_0^3,$$

$$\Delta y_1 = A(p_1^2 + q_1^2)q_1 + \frac{2}{3}Bx_0p_1q_1 + Dx_0^2q_1,$$

$$\Delta p_0 = \frac{1}{3}B(p_1^2 + q_1^2)p_1 + (Cp_1^2 + Dq_1^2)x_0 + 3Ex_0^2p_1 + Fx_0^3,$$
(40.23)

$$\Delta q_0 = \frac{1}{3}B(p_1^2 + q_1^2)q_1 + (C - D)x_0p_1q_1 + Ex_0^2q_1.$$

Let us first consider the special set of light rays which lie in the "primary plane", i.e., in the meridional plane through the point $(x_0, 0)$. Since $y_1 = q_1 = q_0 = 0$, we obtain from (40.23):

$$\Delta x_1 = Ap_1^3 + Bx_0p_1^2 + Cx_0^2p_1 + Ex_0^3,$$

$$\Delta p_0 = \frac{1}{3}Bp_1^3 + Cx_0p_1^2 + 3Ex_0^2p_1 + Fx_0^3.$$
(40.24)

This equation demonstrates that five of the 3rd order aberrations, namely those connected with the coefficients A,B,C,E,F, can be observed by investigating meridional rays alone. Only one aberration is reserved for skew rays alone, namely that determined by the coefficient D. For an optical system which is perfectly corrected in 3rd order for meridional rays we have A = B = C = E = F = 0. In this case the image functions (40.23) become

$$\Delta x_1 = 0, \qquad\qquad \Delta y_1 = Dx_0^2q_1,$$

$$\Delta p_0 = Dx_0q_1^2, \qquad \Delta q_0 = -Dx_0p_1q_1.$$
(40.25)

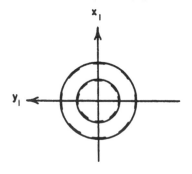

Figure 160

The points of the object plane are imaged as sections of lines which are normal to the optical axis. In many problems connected with third order aberrations we can limit ourselves to considering only the functions (40.24) for meridional rays. This is justified since we shall find that C and D are related to each other by a simple equation, the so called Petzval equation (§42.) which allows us to determine D as soon as C is known.

40.3 For the interpretation of the quantities A,B,C,D,E let us consider the following optical instrument.

A small diaphragm is placed at the first focal point F_0 of the objective. This diaphragm selects a narrow manifold of rays which are nearly parallel to the axis after refraction. If, in (40.23) we let $p_1 = q_1 = 0$, we obtain a bundle of rays which are parallel to the axis after refraction. Obviously this can be interpreted as physically closing our diaphragm down to the limit so that only a one parameter manifold of rays passes

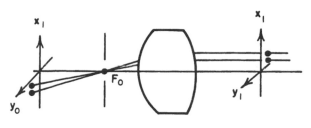

Figure 161

the objective. We have called this bundle the bundle of field rays selected by the diaphragm.

The aberrations Δx_1 and Δy_1 in (40.23) can be considered as a superposition of different types of aberrations. Let us study these types separately.

<u>Spherical Aberration</u>: The image of the axial point $x_0 = y_0 = 0$ is given by the expression

$$\Delta x_1 = A\rho^3 \cos \varphi ,$$
$$\Delta y_1 = A\rho^3 \sin \varphi ,$$

(40.31)

if we introduce polar coordinates ρ,φ by

$$p_1 = \rho \cos \varphi$$
$$q_1 = \rho \sin \varphi$$

(40.32)

We have represented the spherical aberration of the axial bundle §31.14 by the expressions

$$\Delta x_1 = - \ell (\rho) \cos \varphi ,$$
$$\Delta y_1 = - \ell (\rho) \sin \varphi ,$$

(40.33)

and introduced ℓ (p) as the lateral spherical aberration. By comparing (40.33) and (40.31) we obtain

$$\ell (\rho) = - A\rho^3$$

(40.34)

which represents the exact function $\ell(\rho)$ for small values of ρ. The longitudinal spherical aberration $L(\rho)$ is related to $\ell(\rho)$ by (31.19). We find, therefore,

$$L(\rho) = -n_1 A\rho^2 \tag{40.341}$$

in the neighborhood of $\rho = 0$.

Coma: We consider next the functions

$$x_1 = \frac{1}{3}Bx_0 (3p_1^2 + q_1^2) ,$$

$$\tag{40.35}$$

$$y_1 = \frac{1}{3}Bx_0 2p_1q_1$$

which represent the aberrations of our instrument for small quantities x_0 if the system is free of spherical aberration. In polar coordinates we have

$$\Delta x_1 = \frac{B}{3}x_0\rho^2(2 + \cos 2\varphi) ,$$

$$\tag{40.351}$$

$$\Delta y_1 = \frac{B}{3}x_0\rho^2\sin 2\varphi .$$

Let us consider the manifold of rays from $(x_0,0)$ for which ρ has a constant value. Physically we may obtain these rays by placing a ring diaphragm at the focal plane F_0, i.e., a diaphragm which stops all rays except those through a narrow circular ring. The intersection figure with the image plane of these rays is then given by a circle which, according to (40.351), has its center at

$$m = \frac{2}{3}Bx_0\rho^2$$

and the radius   (40.352)

$$R = \frac{B}{3}x_0\rho^2 = \frac{m}{2} .$$

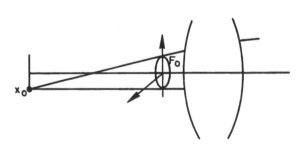

Figure 162

The superposition of the circles which belong to different values of ρ yields the characteristic coma figure which we have already studied in §31. By comparing the above expressions for m and R with those obtained in §31 for wide apertures, ρ, we find

$$\psi(\rho) = \frac{1}{M_0} \left(M(\rho) - M_0\right) = \frac{B}{3M_0} \rho^2 , \tag{40.353}$$

which is the development of $\psi(\rho)$ in the neighborhood of $\rho = 0$.

Astigmatism and Curvature of Field. If the optical system is corrected for spherical aberration and coma, i.e., if $A = B = 0$, then we obtain from

$$\Delta x_1 = Cx_0^2\rho \cos \varphi$$

$$\Delta y_1 = Dx_0^2\rho \sin \varphi$$

(40.36)

the aberration which appears first if x_0 increases. The intersection figure of the rays through our ring diaphragm is an ellipse with axes $Cx_0^2\rho$ and $Dx_0^2\rho$. This aberration is caused by the fact that first order bundles of rays taken about a given field ray as a principal ray are astigmatic after refraction as we have seen in §35. In order to find the position of the primary and secondary focal lines on the field ray from $(x_0,0)$, let us determine the intersection figure with a reference plane at a small distance z_1 from the image plane. We have

$$\bar{x}_1 = x_1 + \frac{z_1}{n_1}p_1 = M_0x_0 + \left(Cx_0^2 + \frac{z_1}{n_1}\right)\rho \cos \varphi,$$

(40.361)

$$\bar{y}_1 = y_1 + \frac{z_1}{n_1}q_1 = \left(Dx_0^2 + \frac{z_1}{n_1}\right)\rho \sin \varphi,$$

i.e., again an elliptic cross-section.

The positions of the focal lines are determined by those values z_1 for which one of the axes of these ellipses becomes zero. We find

$$z_1 = - n_1Cx_0^2 : \text{Primary focus}$$

(40.362)

$$z_2 = - n_1Dx_0^2 : \text{Secondary focus} .$$

The two sets of focal lines on the field rays determine the field curves of the optical instrument. We obtain the equation of these curves in the neighborhood of $x_1 = y_1 = 0$ by introducing $x_0 = \frac{1}{M} x_1$ in (40.362). This gives

$$z_1 = - \frac{n_1}{M^2}Cx_1^2 : \text{Primary field curve}$$

(40.363)

$$z_2 = - \frac{n_1}{M^2}Dx_1^2 : \text{Secondary field curve} .$$

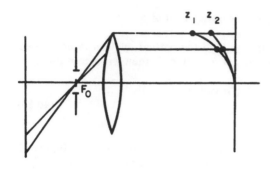

Figure 163

The two surfaces of revolution which are given by these field curves are the curved image surfaces of the object plane. The distance

$$z_2 - z_1 = -\frac{n_1}{M^2}(D-C)x_1^2 \text{ meas-}$$

ures the <u>astigmatism</u> of the instrument; the mean value

$$\frac{1}{2}(z_2 + z_1) = -\frac{1}{2}\frac{n_1}{M^2}(C+D)x_1^2,$$

the <u>curvature of field</u>.

<u>Distortion</u>. By closing the diaphragm down to its limit so that only the field rays pass the objective, we obtain a certain optical projection of the object plane. This projection, however, does not give a true image in general. From (40.23) we find for $p_1 = q_1 = 0$ that

$$\Delta x_1 = Ex_0^3$$

$$\Delta y_1 = 0 \tag{40.37}$$

or

$$x_1 = Mx_0 + Ex_0^3$$

$$y_1 = 0 . \tag{40.371}$$

The departure of the projected point x_1 from the point Mx_0 of an undistorted projection increases as Ex_0^3. From

$$\frac{\Delta x_1}{x_1} = \frac{E}{M}x_0^2 \tag{40.372}$$

it follows that the image point is moved away from the axis if $\frac{E}{M} > 0$ and towards the axis if $\frac{E}{M} < 0$. The two cases are known as pin cushion and barrel distortion respectively, as is illustrated by the distorted images of a square in Fig. 164.

Figure 164

40.4 <u>The Combined Effect of Third Order Aberrations</u>. Let us finally consider the case where all third order aberrations are present. We introduce polar coordinates in (40.23). This yields

$$\Delta x_1 = \frac{2}{3} B\rho^2 x_0 + E x_0^3 + (A\rho^3 + C\rho x_0^2) \cos \varphi - \frac{B}{3} \rho^2 x_0 \cos 2\varphi$$

$$\tag{40.41}$$

$$\Delta y_1 = \qquad\qquad (A\rho^3 + D\rho x_0^2) \sin \varphi + \frac{B}{3} \rho^2 x_0 \sin 2\varphi .$$

For a constant ρ these equations represent a certain algebraic curve which can be constructed as follows. We consider an ellipse which has its center at the point

$$\Delta x_1 = \frac{2}{3} B\rho^2 x_0 + E x_0^3$$

$$\tag{40.42}$$

$$\Delta y_1 = 0$$

and axes given by the coefficients of $\cos \varphi$ and $\sin \varphi$ in (40.41). The center m

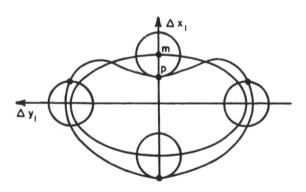

Figure 165

of a circle of radius $R = \frac{B}{3} \rho^2 x_0$ moves once around this ellipse while a point on the circle rotates twice about m. A point P on the circumference of the circle then describes the curve (40.41). The shape of the curve depends on the relative size of the axes of the ellipse and the radius R. Two curves which are obtained if the ellipse degenerates into a straight line are shown in Fig. 166. It is easy to observe these curves with ordinary lenses by placing a ring diaphragm in front of the lens and using a small light source as an object.

40.5 In order to give a direct physical interpretation of the coefficients A,B,C,D,E in (40.23) we have considered an optical instrument with a diaphragm placed at the first focal point of the instrument. The coefficients B,C,D,E are then directly connected with the aberrations of coma, curvature or field, astigmatism, and distortion, for the instrument with the diaphragm. The position of the diaphragm is, however, not without influence on the aberrations of

Figure 166

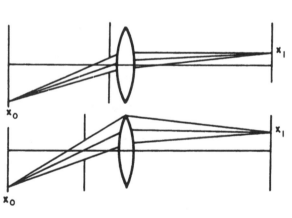

Figure 167

the instrument, especially if small diaphragms are used. The action of a small diaphragm is to select a certain manifold of rays from the total manifold of object rays. These selected rays are then used for the formation of images. By moving the diaphragm to a new position we obtain from the same object point a selected manifold of rays which is different from the original one (Fig. 167). This new manifold passes through different parts of the instrument and thus is differently refracted. Consequently, we have to expect a considerable influence on the quality of images of off-the-axis object points if we shift the diaphragm. In fact, it is possible to remove certain types of aberrations simply by shifting the diaphragm along the z-axis.

Mathematically we can express this situation by the fact that in (40.23) we can introduce new coordinates ξ_1, η_1, instead of p_1, q_1 by certain linear transformations which leave the general form of these equations unchanged. Let us first consider a diaphragm in the real part of the image space. A ray in this space is completely determined by its intersections x_1, y_1 and ξ_1, η_1

with these two planes. Let ζ_1 by the z coordinate of the ξ_1, η_1 plane of the diaphragm relative to the image plane $z_1 = 0$. Then we have the relations

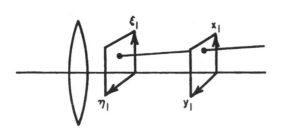

Figure 168

$$\xi_1 = x_1 + \zeta_1 \frac{p_1}{r_1}$$

$$\eta_1 = y_1 + \zeta_1 \frac{q_1}{r_1}$$

(40.51)

which we can use to introduce ξ_1 and η_1 instead of p_1 and q_1 in (40.23). Since we are interested only in third order approximations we can use, instead of (40.51), the linear relations

$$\xi_1 = x_1 + \frac{\zeta_1}{n_1} p_1 = Mx_0 + \frac{\zeta_1}{n_1} p_1 \ ,$$

$$\eta_1 = y_1 + \frac{\zeta_1}{n_1} q_1 = My_0 + \frac{\zeta_1}{n_1} q_1 \ ,$$

(40.52)

which are first order approximations of (40.51). In other words, if we introduce in the first two equations of (40.23) the quantities ξ_1, η_1 by means of the relationships

$$p_1 = \frac{n_1}{\zeta_1} (\xi_1 - Mx_0) \ ,$$

$$q_1 = \frac{n_1}{\zeta_1} (\eta_1 - My_0) \ ,$$

(40.53)

we obtain the third order development of Δx_1 and Δy_1 as functions of x_0, y_0 and ξ_1, η_1.

We can verify without difficulty that Δx_1 and Δy_1 assume the same form as they had in (40.23), namely,

$$\Delta x_1 = \overline{A} (\xi_1^2 + \eta_1^2)\xi_1 + \frac{1}{3}\overline{B}x_0 (3\xi_1^2 + \eta_1^2) + \overline{C}x_0^2\xi_1 + \overline{E}x_0^3 \ ,$$

(40.54)

$$\Delta y_1 = \overline{A}(\xi_1^2 + \eta_1^2)\eta_1 + \frac{2}{3}\overline{B}x_0\xi_1\eta_1 + \overline{D}x_0^2\eta_1 \ ,$$

where $\overline{A}, \overline{B}, \overline{C}, \overline{D}, \overline{E}$ are new coefficients which are linear combinations of the original coefficients.

This can be seen without calculation as follows. We know that $dW = -x_1 dp_1 - y_1 dq_1 - p_0 dx_0 - y_0 dy_0$ is a total differential. By introducing

$$p_1 = \frac{n_1}{\xi_1} (\xi_1 - x_1)$$

$$q_1 = \frac{n_1}{\xi_1} (\eta_1 - y_1)$$

we obtain

$$dW = -\frac{n_1}{\xi_1} (x_1 d\xi_1 + y_1 d\eta_1) + \frac{n_1}{\xi_1} (x_1 dx_1 + y_1 dy_1) - p_0 dx_0 - q_0 dy_0 \ ,$$

or

$$d\left[\frac{\xi_1}{n_1} W - \frac{1}{2}(x_1{}^2 + y_1{}^2)\right] = -(x_1 d\xi_1 + y_1 d\eta_1) - \frac{\xi_1}{n_1}(p_0 dx_0 + q_0 dy_0) \ . \quad (40.55)$$

It follows that the right side is a total differential of a function $\overline{W}$. This function depends only on the invariants of rotation $u = x_0{}^2 + y_0{}^2$, $v = \xi_1{}^2 + \eta_1{}^2$, $w = 2(x_0\xi_1 + y_0\eta_1)$ and hence we are in a position to repeat the derivation of the formulae (40.22) and (40.23) directly for the coordinates x_0, y_0, ξ_1, η_1, with the same general result.

In order to obtain the new coefficients $\overline{A}, \overline{B}, \overline{C}, \overline{E}$, we can make use of the observation that these coefficients appear in the formula (40.24) for the aberrations of meridional rays. Consequently, letting $\tau = \frac{n_1}{\xi_1}$, we have the identity

$$\overline{A}\xi_1{}^3 + \overline{B}\xi_1{}^2 x_0 + \overline{C}\xi_1 x_0{}^2 + \overline{E}x_0{}^3 = A\tau^3(\xi_1 - Mx_0)^3 + B\tau^2(\xi_1 - Mx_0)^2 x_0$$

$$+ C\tau(\xi_1 - Mx_0)x_0{}^2 + Ex_0{}^3$$

(40.56)

from which follows

$$\overline{A} = A\tau^3 \ ,$$

$$\overline{B} = \tau^2(B - 3M\tau A) \ ,$$

$$\overline{C} = \tau(C - 2M\tau B + 3M^2\tau^2 A) \ ,$$

(40.57)

$$\overline{E} = E - M\tau C + M^2\tau^2 B - M^3\tau^3 A \ .$$

For the coefficients $\overline{A}, \overline{B}, \overline{C}, \overline{D}, \overline{E}$ we can give the same physical explanation as above. The only difference is that the bundle of field rays is now determined

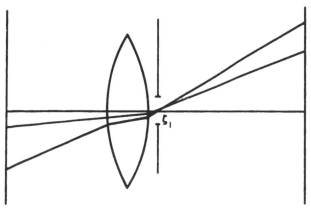

by the center of a diaphragm at a position different from the former one. The formulae (40.57) show that there always exists one position for which $\overline{B} = 0$, when the image is coma free. There are possibly two real values of τ with $\overline{C} = 0$, and at least one position where the instrument has no distortion.

The above results are valid also if $z = \zeta_1$ lies in the virtual part of the image plane. This corresponds physically to the case where the diaphragm is placed so that the instrument produces a virtual Gaussian image at $z = \zeta_1$.

Figure 169

§41. THE THIRD ORDER COEFFICIENTS AS FUNCTIONS OF THE POSITION OF OBJECT AND PUPIL PLANE.

41.1 The coefficients A,B,C,D,E which determine the image formation of a given instrument with respect to a particular object plane and pupil plane vary if we change the positions of these planes. If the object plane is unchanged, and only the pupil plane is displaced, the variation of the third order coefficients is given by the formulae (40.57). In the following we shall derive the algebraic functions which determine the 3rd order aberrations of a given optical instrument for any choice of object or pupil plane.

We consider only the coefficients A,B,C,E, and use the Petzval equation (§42.62) to find D for a known C. Let U_0 and U_1 be the unit points of the instrument and z_0, z_1 the positions of the object and image planes which are assumed to be conjugate. The object and image spaces shall be homogeneous with constant indices of refraction n_0 and n_1 respectively. We can characterize the positions z_0 and z_1 by the magnification, M, of the object plane. If f is the focal length of the instrument we have

$$\frac{z_0}{n_0} = \frac{1-M}{M} f \; ; \; \frac{z_1}{n_1} = (1-M) f \; . \tag{41.11}$$

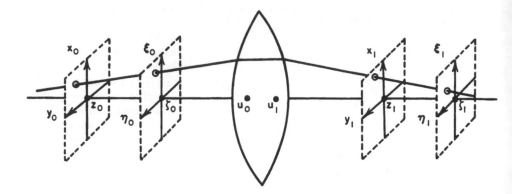

Figure 170

Next we consider two other pairs of conjugate planes at the points ζ_0 and ζ_1. These planes are called the entrance pupil plane and the exit pupil plane. Physically these planes are determined by the position of a diaphragm in the instrument. If the diaphragm is placed in the real part of the object space, then it coincides with the entrance pupil plane. Its Gaussian image gives the exit pupil plane. If the diaphragm is in the real part of the image space, then it coincides with the exit pupil plane; its conjugate plane in the object space is the entrance pupil plane. If, finally, the diaphragm is inside the instrument entrance and exit pupil planes are given by the Gaussian images of the diaphragm in the object and image spaces respectively. Let us assume that m is the magnification of the entrance pupil plane. Then we have

$$\frac{\zeta_0}{n_0} = \frac{1-m}{m}f \; ; \; \frac{\zeta_1}{n_1} = (1-m)f \; . \tag{41.12}$$

41.2 Since the aberrations A,B,C,E are determined by the meridional rays of the instrument, in this section we consider only light rays which lie in the x,z plane. Instead of characterizing these rays by the canonical variables x_0,p_0; x_1,p_1 we shall use the intersections x_0,ξ_0 and x_1,ξ_1 with the above conjugate planes. To a first order approximation these different variables are related by the formulae

$$p_0 = n_0\frac{\xi_0 - x_0}{\zeta_0 - z_0} \; , \; p_1 = n_1\frac{\xi_1 - x_1}{\zeta_1 - z_1} \; , \tag{41.21}$$

or on account of (41.11) and (41.12) by

$$p_0 = \frac{Mm}{M-m}\frac{\xi_0 - x_0}{f} \; , \; p_1 = \frac{1}{M-m}\frac{\xi_1 - x_1}{f} \; . \tag{41.22}$$

Our aim in this section is to determine the third order aberration polynomials $\Delta x_1 = x_1 - Mx_0$ and $\Delta \xi_1 = \xi_1 - m\xi_0$ which we assume are in the form

$$\Delta x_1 = A\xi_0^3 + B\xi_1^2 x_0 + C\xi_1 x_0^2 + Ex_0^3$$

$$\Delta \xi_1 = e\xi_0^3 + c\xi_0^2 x_0 + b\xi_1 x_0^2 + ax_0^3 \ .$$

(41.23)

The first polynomial gives the aberrations of the object plane, the second polynomial the aberrations of the entrance pupil plane.

41.3 We base our investigation upon Hamilton's angular characteristic.

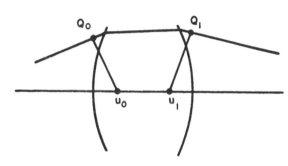

Figure 171

Let T_0 (p_0, p_1) be the angular characteristic of the instrument with the unit points as reference points. It determines the optical distance of the base points Q_0, Q_1 of the perpendiculars dropped from the unit points onto a ray in the x,z plane. Since the unit planes are conjugate with the magnification $M = 1$ we obtain from (34.34), the first order relations

$$x = f(p_0 - p_1) \ ,$$

$$-x' = -f(p_0 - p_1) \ ,$$

(41.31)

if x and x' are the intersections of the ray with the unit planes. This implies that the function T_0 (p_0, p_1) must have a development of the form:

$$T_0 \ (p_0 p_1) = f\left[\frac{1}{2} (p_0 - p_1)^2 - P (p_0, p_1)\right] \ + \ldots \ ,$$

(41.32)

where $P(p_0, p_1)$ is a homogeneous polynomial of 4th order. We assume that P is given by

$$P = \frac{1}{4!} (P_0 p_0^4 + 4P_1 p_0^3 p_1 + 6P_2 p_0^2 p_1^2 + 4P_3 p_0 p_1^3 + P_4 p_1^4) \ .$$

(41.33)

The coefficients, P_i, are determined by the optical instrument.

The angular characteristic $T(p_0, p_1; z_0, z_1)$ which belongs to the object and image planes is given by the general formula

$$T(p_0, p_1; z_0, z_1) = T_0 (p_0, p_1) + z_1 \sqrt{n_1^2 - p_1^2} - z_0 \sqrt{n_0^2 - p_0^2} \ .$$

(41.34)

By developing the right side with respect to p_0 and p_1 and disregarding terms of an order greater than the fourth we obtain, with the aid of (41.11) and (41.12),

$$T = \frac{f}{2M} (p_0 - Mp_1)^2 - fP + \frac{f}{8} \left[\frac{1 - M}{M} \frac{p_0^4}{n_0^2} - (1 - M)\frac{p_1^4}{n_1^2} \right] . \qquad (41.35)$$

The intersections x_0 and x_1 of the ray with the object and image planes are obtained by

$$x_1 = -\frac{\partial T}{\partial p_1} \; ; \; x_0 = \frac{\partial T}{\partial p_0}$$

which gives

$$\Delta x_1 = x_1 - Mx_0 = -\left(\frac{\partial T}{\partial p_1} + M\frac{\partial T}{\partial p_0} \right) . \qquad (41.36)$$

By introducing (41.35) in this equation we find

$$\frac{1}{f}\Delta x_1 = \frac{\partial P}{\partial p_1} + M\frac{\partial P}{\partial p_0} + \frac{1}{2}(1 - M)\left(\frac{p_1^3}{n_1^2} - \frac{p_0^3}{n_0^2}\right) . \qquad (41.37)$$

41.4 The right side in (41.37) is a homogeneous polynomial of third order in p_0 and p_1. We obtain the aberration Δx_1 as a function of the variables x_0, ξ_0 of the ray by replacing p_0 and p_1 in (41.37) by the expressions (41.22). We are allowed to use these first order approximations of the functions $p_0 (x_0, \xi_0)$ and $p_1 (x_0, \xi_0)$ because higher order terms in them would contribute only terms of higher order than 3 in (41.37). We therefore have the result: The third order polynomial

$$\Delta x_1 = A\xi_0^3 + B\xi_0^2 x_0 + C\xi_0 x_0^2 + Ex_0^3 \qquad (41.41)$$

for an object plane of magnification M and an entrance pupil plane of magnification m can be found from (41.37) by replacing p_0 and p_1 by the expressions

$$p_0 = \frac{Mm}{M - m} \frac{\xi_0 - x_0}{f} \; ; \; p_1 = \frac{1}{M - m} \frac{m\xi_0 - Mx_0}{f} \qquad (41.42)$$

and arranging the result in the form (41.41).

By repeating the same procedure for the pupil planes we obtain the additional result: The polynomial

$$\Delta\xi_1 = e\xi_0^3 + c\xi_0^2 x_0 + b\xi_0 x_0^2 + ax_0^3$$

is given by the polynomial

$$\frac{1}{f}\Delta\xi_1 = \frac{\partial P}{\partial p_1} + m\frac{\partial P}{\partial p_0} + \frac{1}{2}(1 - m)\left(\frac{p_1^3}{n_1^2} - \frac{p_0^3}{n_0^2}\right) \tag{41.43}$$

if the expressions (41.42) are introduced on the right side.

41.5 One verfies readily that

$$\frac{1}{m(M - m)^3 f^3}\ \frac{\partial}{\partial\xi_0}P\left(Mm(\xi_0 - x_0), m\xi_0 - Mx_0\right) = \frac{\partial P}{\partial p_1} + M\frac{\partial P}{\partial p_0}$$

and   (41.51)

$$-\frac{1}{M(M - m)^3 f^3}\ \frac{\partial}{\partial x_0}P\left(Mm(\xi_0 - x_0), m\xi_0 - Mx_0\right) = \frac{\partial P}{\partial p_1} + m\frac{\partial P}{\partial p_0} .$$

Hence we can summarize our results in the following form: The aberrations Δx_1 and $\Delta\xi_1$ are given by the expressions

$$\frac{1}{f}\Delta x_1 = \frac{1}{(M - m)^3 f^3}\left[\frac{1}{m}\frac{\partial Q}{\partial\xi_0} + \frac{1}{2}(1 - M)\left(M^3 m^3(\xi_0 - x_0)^3 - (m\xi_0 - Mx_0)^3\right)\right]$$

$$\tag{41.52}$$

$$\frac{1}{f}\Delta\xi_1 = \frac{1}{(M - m)^3 f^3}\left[-\frac{1}{M}\frac{\partial Q}{\partial x_0} + \frac{1}{2}(1 - m)\left(M^3 m^3(\xi_0 - x_0)^3 - (m\xi_0 - Mx_0)^3\right)\right]$$

where Q is the polynomial

$$Q = P\left(Mm(\xi_0 - x_0), m\xi_0 - Mx_0\right) . \tag{41.53}$$

By arranging the right sides in (41.52) in the form of (41.23) we obtain the coefficients A,B,C,E and a,b,c,e as algebraic functions of the two magnifications M and m. These functions are known explicitly as soon as the five coefficients P_0, P_1, P_2, P_3, P_4 of the expression (41.33) have been found. They allow us to determine the aberrations of a given optical instrument for any object plane and for any entrance pupil plane.

41.6 We can use the formula (41.37) to investigate the third order aberrations A,B,C,E for the case that the entrance pupil plane is at the first focal point of the instrument. The function Δx_1 then assumes the form

$$\Delta x_1 = Ap_1^3 + Bp_1^2 x_0 + Cp_1 x_0^2 + Ex_0^3 \tag{41.61}$$

which we originally obtained in §40. This polynomial is gotten from (41.37) by replacing p_0 by the expression

$$p_0 = M\left(\frac{x_0}{f} + p_1\right) . \qquad (41.62)$$

Letting $f = 1$ and $n_0 = n_1 = 1$ we find

$$\Delta x_1 = \frac{\partial}{\partial p_1} P\left[M(x_0 + p_1), p_1\right] + \frac{1}{2}(1 - M)\left[p_1^3 - M^3(x_0 + p_1)^3\right] , \qquad (41.63)$$

and hence

$$A = \frac{1}{3!}\left[P_0 M^4 + 4P_1 M^3 + 6P_2 M^2 + 4P_3 M + P_4\right] + \frac{1}{2}(1 - M)(1 - M^3) ,$$

$$B = \frac{M}{2!}\left[P_0 M^3 + 3P_1 M^2 + 3P_2 M + P_3\right] \qquad\qquad - \frac{3}{2}(1 - M)M^3 ,$$

$$\qquad\qquad\qquad\qquad\qquad\qquad\qquad\qquad\qquad\qquad\qquad\qquad\qquad (41.64)$$

$$C = \frac{M^2}{2!}\left[P_0 M^2 + 2P_1 M + P_2\right] \qquad\qquad\qquad - \frac{3}{2}(1 - M)M^3 ,$$

$$E = \frac{M^3}{3!}\left[P_0 M + P_1\right] \qquad\qquad\qquad\qquad\qquad - \frac{1}{2}(1 - M)M^3 .$$

We can demonstrate with the aid of these formulae that it is impossible to correct an optical instrument for all pairs of conjugate planes simultaneously. Let us assume that the unit planes are aberration free. It follows that $P_0 + P_1 = 0$; $P_1 + P_2 = 0$; $P_2 + P_3 = 0$; $P_4 + P_3 = 0$ and hence

$$A = \frac{1}{6}P_0(M - 1)^4 + \frac{1}{2}(1 - M)(1 - M^3) ,$$

$$B = \frac{1}{2}P_0 M(M - 1)^3 - \frac{3}{2}(1 - M)M^3 ,$$

$$\qquad\qquad\qquad\qquad\qquad\qquad\qquad\qquad\qquad\qquad\qquad\qquad\qquad (41.65)$$

$$C = \frac{1}{2}P_0 M^2(M - 1)^2 - \frac{3}{2}(1 - M)M^3 ,$$

$$E = \frac{1}{6}P_0 M^3(M - 1) - \frac{1}{2}(1 - M)M^3 .$$

It obviously is impossible to choose P_0 so that the polynomials (41.65) are identically zero which would be necessary for an instrument which is corrected for all pairs of conjugate planes.

§42. INTEGRAL EXPRESSIONS FOR THE THIRD ORDER COEFFICIENTS.

42.1 In this section the problem is to find explicit expressions for the third order coefficients A,B,C,D,E which determine the image quality for a given pair of conjugate planes. We shall solve this problem for the case of a continuous medium, but the resulting integral formulae can be applied directly to discontinuous media and lead to summation formulae in that case.

Let us assume that two planes $z = z_0$ and $z = z_1$ in an optical medium are conjugate in the realm of first order optics. The paraxial rays in this medium are the solutions of the linear differential equations

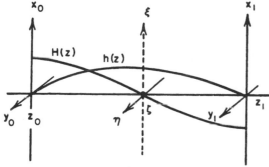

Figure 172

$$\dot{x} = \frac{1}{n}\, p, \quad \dot{p} = -Dx \ ,$$

$$\dot{y} = \frac{1}{n}\, q, \quad \dot{q} = -Dy \ . \tag{42.11}$$

We represent solutions of these equations by two particular solutions h, ϑ and H, θ which satisfy the boundary conditions

$$h(z_0) = 0 \ , \quad H(z_0) = 1 \ ,$$

$$h(\zeta) = 1 \ , \quad H(\zeta) = 0 \ . \tag{42.12}$$

The plane $z = \zeta$ is the pupil plane of the medium which may be determined physically by the position of a diaphragm. For our purpose we can consider it simply as a third reference plane. A paraxial ray which intersects the plane $z = z_0$ at the point x_0, y_0 and the plane $z = \zeta$ at ξ, η is then given by the expressions

$$x = x_0 H(z) + \xi h(z) \ , \qquad p = x_0 \theta(z) + \xi\, \vartheta(z) \ ,$$

$$y = y_0 H(z) + \eta h(z) \ , \qquad q = y_0 \theta(z) + \eta\, \vartheta(z) \ . \tag{42.13}$$

Let us now consider the actual light rays of the medium. We shall denote them by $X(z), Y(z), P(z), Q(z)$ to distinguish them from the paraxial rays. The functions X, Y, P, Q, are solutions of the canonical equations

$$\dot{X} = 2H_V P \ , \qquad \dot{P} = -2H_U X \ ,$$

$$\dot{Y} = 2H_V Q \ , \qquad \dot{Q} = -2H_U Y \ , \tag{42.14}$$

where $U = X^2 + Y^2$ and $V = P^2 + Q^2$.

A special light ray shall be characterized by the points of intersection x_0, y_0 and ξ, η with the planes $z = z_0$ and $z = \zeta$ respectively, so that X, Y, P, Q can be considered as functions of z and the parameters x_0, y_0, ξ, η. We develop these functions in a Taylor series with respect to x_0, y_0, ξ, η and write

$$X = X_1 + X_3 + X_5 + \ldots$$

$$Y = Y_1 + Y_3 + Y_5 + \ldots$$

$$\text{(42.15)}$$

$$P = P_1 + P_3 + P_5 + \ldots$$

$$Q = Q_1 + Q_3 + Q_5 + \ldots$$

where X_i, Y_i, P_i, Q_i are homogeneous polynomials of degree i in the parameters x_0, y_0, ξ, η. The coefficients of these polynomials are functions of z.

42.2 We now develop the Hamiltonian $H(U, V, z)$ with respect to U and V. We write

$$H(U, V, z) = H_0 + H_1 U + H_2 V + \frac{1}{2}(H_{11}U^2 + 2H_{12}UV + H_{22}V^2) + \ldots \quad \text{(42.21)}$$

where the coefficients H_1 and H_{1k} are certain functions of z which we shall determine later. We introduce both this expression and the series (42.15), in the canonical equations (42.14). Let us consider the first equation for example. We have

$$\dot{X}_1 + \dot{X}_3 + \ldots = 2(H_2 + H_{12}U + H_{22}V + \ldots)(P_1 + P_3 + \ldots) \quad \text{(42.22)}$$

or

$$\dot{X}_1 + \dot{X}_3 + \ldots = 2H_2 P_1 + 2H_2 P_3 + 2(H_{12}U_1 + H_{22}V_1)P_1 + \ldots \quad \text{(42.23)}$$

where $U_1 = X_1^2 + Y_1^2$ and $V_1 = P_1^2 + Q_1^2$.

The omitted terms on both sides of this equation represent polynomials in x_0, y_0, ξ, η of a degree higher than three. An identity between homogeneous polynomials can only be true if the polynomials of the same degree on both sides are identical. Thus it follows that

$$\dot{X}_1 = 2H_2 P_1 ,$$

$$\text{(42.24)}$$

$$\dot{X}_3 = 2H_2 P_3 + 2(H_{12}U_1 + H_{22}V_1)P_1 .$$

Similar considerations can be carried out for the other three equations (42.14). We obtain the first order equations

$$\dot{X}_1 = 2H_2P_1 , \qquad \dot{P}_1 = -2H_1X_1 ,$$

$$\dot{Y}_1 = 2H_2Q_1 , \qquad \dot{Q}_1 = -2H_1Y_1 \qquad\qquad (42.25)$$

and the third order equations

$$\dot{X}_3 = 2H_2P_3 + 2(H_{12}U_1 + H_{22}V_1)P_1$$

$$\dot{Y}_3 = 2H_2Q_3 + 2(H_{12}U_1 + H_{22}V_1)Q_1 \qquad\qquad (42.26)$$

$$\dot{P}_3 = -2H_1X_3 - 2(H_{11}U_1 + H_{12}V_1)X_1$$

$$\dot{Q}_3 = -2H_1Y_3 = -2(H_{11}U_1 + H_{12}V_1)Y_1 . \qquad\qquad (42.261)$$

The first order equations (42.25) are of course identical with the paraxial equations (42.11), i.e., we have $2H_1 = D(z)$ and $2II_2 = \dfrac{1}{n(z)}$. Hence the general solution of (42.25) is given by the paraxial rays $x(z)$, $y(z)$, $p(z)$, $q(z)$ and can be expressed by the formulae (42.13).

The third order equations (42.26) can be written in the form

$$\dot{X}_3 - \frac{1}{n}P_3 = 2(H_{12}U + H_{22}V)p$$

$$\dot{Y}_3 - \frac{1}{n}Q_3 = 2(H_{12}U + H_{22}V)q \qquad\qquad (42.27)$$

$$\dot{P}_3 + DX_3 = -2(H_{11}U + H_{12}V)x$$

$$\dot{Q}_3 + DY_3 = -2(H_{11}U + H_{12}V)y \qquad\qquad (42.271)$$

where $U = x^2 + y^2$ and $V = p^2 + q^2$.

The third order aberrations X_3, Y_3; P_3, Q_3 are the solutions of a system of nonhomogeneous equations. The nonhomogeneous terms are known functions of z once the solutions $x(z)$, $y(z)$, $p(z)$, $q(z)$ of the paraxial equations are known. The differential operators on the left side are the same operators which determine the homogeneous canonical equations for the paraxial rays.

Our original problem was to find a solution X,Y,P,Q of the canonical equations (42.14) which satisfies the boundary conditions

$$X(z_0) = x_0 , \qquad X(\zeta) = \xi ,$$

$$Y(z_0) = y_0 , \qquad Y(\zeta) = \eta . \qquad\qquad (42.28)$$

If we determine the first order polynomials $X_1 = x$, $Y_1 = y$, $P_1 = p$, $Q_1 = q$ by the equations (42.13), then these boundary conditions are satisfied by the functions (42.15) when the higher order polynomials X_1, Y_1, P_1, Q_1 satisfy the boundary conditions

$$X_1(z_0) = Y_1(z_0) = 0$$

$$X_1(\zeta) = Y_1(\zeta) = 0 \; .$$

(42.29)

This leads us to the following problem for the third order polynomials X_3, Y_3, P_3, Q_3: To find a solution of the nonhomogeneous canonical equations (42.27) and (42.271) which satisfies the boundary conditions

$$X_3(z_0) = Y_3(z_0) = X_3(\zeta) = Y_3(\zeta) = 0 \; .$$

(42.291)

42.3 We solve this problem as follows. We consider the two equations

$$\dot{X}_3 - \frac{1}{n}P_3 = 2(H_{12}U + H_{22}V)p$$

$$\dot{h} - \frac{1}{n}\vartheta = 0 \; .$$

(42.31)

$h(z)$ and $\vartheta(z)$ represent the paraxial ray defined in (42.12). From these equations it follows that

$$\dot{X}_3\,\vartheta - \dot{h}P_3 = 2(H_{12}U + H_{22}V)p\,\vartheta \; .$$

(42.32)

In a similar manner we conclude from

$$\dot{P}_3 + DX_3 = -2(H_{11}U + H_{12}V)x \; ,$$

$$\dot{\vartheta} + Dh = 0$$

(42.33)

that

$$X_3\,\dot{\vartheta} - h\dot{P}_3 = 2(H_{11}U + H_{12}V)xh \; .$$

(42.34)

Hence by adding (42.32) and (42.34) we obtain

$$\frac{d}{dz}(X_3\,\vartheta - hP_3) = 2(H_{12}U + H_{22}V)p\,\vartheta + 2(H_{11}U + H_{12}V)xh \; .$$

(42.35)

We integrate this equation from z_0 to z_1 and make use of the condition $X_3(z_0) = 0$; $h(z_0) = h(z_1) = 0$. We obtain

$$X_3(z_1)\,\vartheta(z_1) = 2\int_{z_0}^{z_1} \left[(H_{12}U + H_{22}V)p\,\vartheta + (H_{11}U + H_{12}V)xh \right] dz$$

(42.351)

and similarly,

$$Y_3(z_1) \vartheta(z_1) = 2 \int_{z_0}^{z_1} \left[(H_{12}U + H_{22}V)q\,\vartheta + (H_{11}U + H_{12}V)yh \right] dz \quad . \quad (42.352)$$

These integrals represent the desired third order aberration in the image plane $z = z_1$. Let us introduce

$$\Delta x_1 = X_3(z_1)$$

$$\Delta y_1 = Y_3(z_1)$$
(42.36)

and the Lagrangian invariant $\Gamma = \begin{vmatrix} H\theta \\ h\,\vartheta \end{vmatrix}$ of the two paraxial rays H,θ and h, ϑ.
For $z = z_1$ we have $\Gamma = H(z_1)\,\vartheta(z_1) = M\,\vartheta(z_1)$. Hence it follows from the above integrals that

$$\Delta x_1 = \frac{2M}{\Gamma} \int_{z_0}^{z_1} \left[(H_{12}U + H_{22}V)p\,\vartheta + (H_{11}U + H_{12}V)xh \right] dz$$

(42.37)

$$\Delta y_1 = \frac{2M}{\Gamma} \int_{z_0}^{z_1} \left[(H_{12}U + H_{22}V)q\,\vartheta + (H_{11}U + H_{12}V)yh \right] dz \quad .$$

The right sides become third order polynomials of x_0, y_0, ξ, η if instead of x, y, p, q we introduce the expressions

$$x = x_0 H + \xi h , \qquad p = x_0\theta + \xi\,\vartheta ,$$

$$y = y_0 H + \eta h , \qquad q = y_0\theta + \eta\,\vartheta ,$$
(42.38)

and hence, for $U = x^2 + y^2$ and $V = p^2 + q^2$, we have the expressions

$$U = H^2 u + h^2 v + Hhw ,$$

$$V = \theta^2 u + \vartheta^2 v + \theta\,\vartheta w \quad .$$
(42.39)

The quantities u, v, w are defined by

$$u = x_0^2 + y_0^2$$

$$v = \xi^2 + \eta^2$$
(42.391)

$$w = 2(x_0\xi + y_0\eta) \quad .$$

42.4 In order to determine the third order coefficients explicitly we use the result of §40 which is an expression for Δx_1 in the form of the following polynomial

$$\frac{1}{M}\Delta x_1 = \left[Eu + \frac{1}{3}Bv + \frac{1}{2}(C - D)w\right]x_0 + \left[Du + Av + \frac{1}{3}Bw\right]\xi \ . \tag{42.41}$$

By introducing $p = x_0\theta + \xi\vartheta$, $x = x_0 H + \xi h$ in the first equation (42.37) we obtain

$$\frac{\Delta x_1}{M} = \frac{2}{\Gamma}x_0 \int_{z_0}^{z_1} \left[(H_{11}Hh + H_{12}\theta\,\vartheta)U + (H_{12}Hh + H_{22}\theta\,\vartheta)V\right] dz$$

$$\tag{42.42}$$

$$+ \frac{2}{\Gamma}\xi \int_{z_0}^{z_1} \left[(H_{11}h^2 + H_{12}\vartheta^2)U + (H_{12}h^2 + H_{22}\vartheta^2)V\right] dz \ ,$$

and hence, by comparison with (42.41), we obtain the formulae

$$Eu + \frac{1}{3}Bv + \frac{1}{2}(C - D)w = \frac{2}{\Gamma}\int_{z_0}^{z_1} \left[(H_{11}Hh + H_{12}\theta\,\vartheta\,)U + (H_{12}Hh + H_{22}\theta\,\vartheta)V\right] dz \ ,$$

$$\tag{42.43}$$

$$Du + Av + \frac{1}{3}Bw = \frac{2}{\Gamma}\int_{z_0}^{z_1} \left[(H_{11}h^2 + H_{12}\vartheta^2)U + (H_{12}h^2 + H_{22}\vartheta^2)V\right] dz \ .$$

These equations give the coefficients A,. . . ,E directly if we replace U and V by the expressions (42.39). The result is

$$A = \frac{2}{\Gamma}\int_{z_0}^{z_1} \left[H_{11}h^4 + 2H_{12}h^2\vartheta^2 + H_{22}\vartheta^4\right] dz \ ,$$

$$B = \frac{6}{\Gamma}\int_{z_0}^{z_1} \left[H_{11}h^3H + H_{12}h\vartheta\,(H\vartheta + h\theta) + H_{22}\theta\,\vartheta^3\right] dz \ ,$$

$$\tag{42.44}$$

$$C = \frac{6}{\Gamma}\int_{z_0}^{z_1} \left[H_{11}h^2H^2 + 2H_{12}Hh\theta\,\vartheta + H_{22}\theta^2\vartheta^2\right] dz + 2\Gamma\int_{z_0}^{z_1} H_{12}dz \ ,$$

$$E = \frac{2}{\Gamma} \int_{z_0}^{z_1} \left[H_{11} h H^3 + H_{12} H \theta (H \vartheta + h \theta) + H_{22} \theta^3 \vartheta \right] dz \ ,$$

$$D = \frac{2}{\Gamma} \int_{z_0}^{z_1} \left[H_{11} H^2 h^2 + H_{12} (H^2 \vartheta^2 + h^2 \theta^2) + H_{22} \theta^2 \vartheta^2 \right] dz \ .$$

The third order coefficients are thus given by certain integral forms of 4th order of the functions $H(z)$, $\theta(z)$; $h(z)$, $\vartheta(z)$, i.e., of the paraxial rays which we have defined above.

42.5 The quantities H_{ik} are given by the second derivatives of the Hamiltonian $H = - \sqrt{n^2(u,z) - v}$ with respect to u and v for $u = v = 0$. Letting

$$n(0,z) = n(z) \ ,$$

$$n_u(0,z) = n_1(z) \ , \tag{42.51}$$

$$n_{uu}(0,z) = n_2(z) \ ,$$

we obtain

$$H_{11} = - n_2(z) \ ,$$

$$H_{12} = - \frac{n_1(z)}{2n^2} \ , \tag{42.52}$$

$$H_{22} = \frac{1}{4n^3} \ .$$

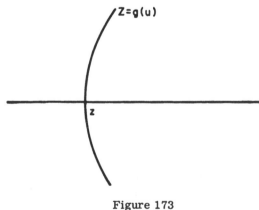

Figure 173

We can express n_1 and n_2 by the geometric characteristics of the surfaces $n(u,z) = $ const., the refracting surfaces of the medium. Let $Z = g(u)$ be the refracting surface which passes through the point z of the z axis. Then we have the identity

$$n\big(u, g(u)\big) \equiv n(0,z) \ . \tag{42.53}$$

By differentiating and then letting $u = 0$ it follows that

$$n_1 + \dot{n} g'(0) = 0 \ ,$$

$$n_2 + 2\dot{n}_1 g'(0) + \ddot{n}\big(g'(0)\big)^2 + \dot{n} g''(0) = 0 \ . \tag{42.54}$$

We represent $Z = g(u)$ in the neighborhood of $u = 0$ by

$$g(u) = z + \frac{u}{2R} + \frac{a}{2}u^2 + \dots \tag{42.55}$$

where $R(z)$ is the radius of curvature of the surface at its vertex, and $a(z) = g''(0)$ is the coefficient of u^2 in the development (42.55). We then obtain

$$2n_1 = -\frac{\dot{n}}{R} \; ,$$

$$n_2 = \frac{1}{4}\left(\frac{\dot{n}}{R^2}\right)' - \dot{n}a \; , \tag{42.56}$$

and hence

$$H_{11} = a\dot{n} - \frac{1}{4}\left(\frac{\dot{n}}{R^2}\right)' \; ,$$

$$H_{12} = -\frac{1}{4}\frac{1}{R}\left(\frac{1}{n}\right)' \; , \tag{42.57}$$

$$H_{22} = \frac{1}{4n^3} \; .$$

42.6 <u>Petzval's theorem</u>. By (42.44) we obtain the equation

$$3D - C = \frac{6}{\Gamma} \int_{z_0}^{z_1} (H\vartheta - h\theta)^2 H_{12}dz - 2 \Gamma\int_{z_0}^{z_1} H_{12}dz \; .$$

Since $H\vartheta - h\theta = \Gamma$ this yields

$$3D - C = 4\Gamma \int_{z_0}^{z_1} H_{12}dz \; , \tag{42.61}$$

and hence by (42.57)

$$\frac{1}{\Gamma}(C - 3D) = \int_{z_0}^{z_1} \frac{1}{R} d\left(\frac{1}{n}\right) \; . \tag{42.62}$$

This relation between C and D is known as the <u>equation of Petzval</u>. It is of great importance for optical design. In fact, we conclude that both C and D can be zero only if the <u>Petzval sum</u>

$$\int_{z_0}^{z_1} \frac{1}{R} d\frac{1}{n} = 0 \; . \tag{42.63}$$

An optical instrument which is corrected for Astigmatism and curvature of field is called an <u>anastigmat</u>. The condition (42.63) is a necessary condition for such instruments.

Let us assume that object and image space are homogeneous so that $n_0 = n_1 = 1$. One verifies by a consideration similar to that in §40.36 that the two field curves of the optical instrument are given by the parabolas

$$z_1 = - \frac{C}{\Gamma} x_1^2 \ ,$$

$$z_1 = - \frac{D}{\Gamma} x_1^2 \ ,$$ (42.64)

in the neighborhood of $x_1 = 0$. The curvatures of these parabolae are equal to

$$\frac{1}{\rho_1} = - \frac{2C}{\Gamma} \ ; \ \frac{1}{\rho_2} = - \frac{2D}{\Gamma}$$ (42.65)

respectively. By introducing these radii in (42.62) we find Petzval's theorem in geometrical form:

$$\frac{3}{\rho_2} - \frac{1}{\rho_1} = 2 \int_{z_0}^{z_1} \frac{1}{R} \, d \frac{1}{n} \ .$$ (42.66)

42.7 The quantities H_{ik} in the case of an electrostatic field. The integral expressions for the third order coefficients can be applied directly to the problem of finding the aberrations of an electron optical instrument, provided that only electrostatic lenses are used. The "index of refraction" $n(u,z)$ satisfies the equation $\Delta n^2 = 0$ and is given by the integral

$$n^2(u,z) = \frac{1}{2\pi} \int_0^{2\pi} f \left(z + i\sqrt{u} \, \cos \varphi \right) d\varphi$$ (42.71)

as we have seen in §38. As before we develop this function one step further so that we may determine the quantities H_{ik} by (42.52). It follows that

$$n^2(u,z) = f(z) - \frac{1}{4} u f''(z) + \frac{1}{64} u^2 f''''(z) + \ldots \ ,$$ (42.72)

and this gives

$$2n_1 = - \frac{1}{4} \frac{f''(z)}{\sqrt{f}}$$

$$2n_2 = - \frac{1}{32} \frac{(f'')^2 - ff''''}{\sqrt{f^3}}$$ (42.73)

hence we obtain by (42.52):

$$2H_{11} = \frac{1}{32} \frac{(f'')^2 - ff''''}{\sqrt{f^3}} \quad ,$$

$$2H_{12} = \frac{1}{8} \frac{f''}{\sqrt{f^3}} \quad , \tag{42.74}$$

$$2H_{22} = \frac{1}{2} \frac{1}{\sqrt{f^3}} \quad .$$

42.8 <u>Spherical surfaces</u>. Let us assume that the refracting surfaces in our medium are spherical. We represent these surfaces in the form

$$Z = g(u) = z + R\left(1 - \sqrt{1 - \frac{u^2}{R^2}}\right) , \tag{42.81}$$

and determine the coefficients of u and u^2 in the power development of the right side. It follows that

$$g(u) = z + \frac{1}{2R} u + \frac{1}{8R^3} u^2 + \ldots , \tag{42.82}$$

and hence $a = \frac{1}{4R^3}$. With the aid of the general formulae (42.57) we obtain

$$2H_{11} = \frac{1}{2}\left(\frac{\dot{n}}{R^3} - \left(\frac{\dot{n}}{R^2}\right)'\right) \quad ,$$

$$2H_{12} = \frac{1}{2} \frac{\dot{n}}{n^2 R} \quad , \tag{42.83}$$

$$2H_{22} = \frac{1}{2} \frac{1}{n^3} \quad .$$

It is possible to simplify the integral formulae for the Seidel coefficients A,B,C,E considerably in this case, so that they become extremely well suited for numerical computation. This simplification is obtained by replacing the quantity R in (42.83) by the data of the paraxial ray h,ϑ . We have the equation

$$\dot{\vartheta} = - \frac{\dot{n}}{R} h ,$$

and hence

$$\frac{1}{R} = - \frac{\dot{\vartheta}}{nh} \quad . \tag{42.84}$$

By introducing this in (42.83) and subsequently in the integral formulae (42.44) we obtain the coefficients A,B,C,E expressed by the two paraxial rays h, ϑ and H, θ and by the index of refraction $n(z)$ on the axis. We give the result without the details of the derivation.

Let S, P and ω be defined by the formulae

$$S = \frac{h \left[\left(\frac{\vartheta}{n}\right)\right]^2 \left(\frac{\vartheta}{n^2}\right)'}{\left(\frac{1}{n}\right)'}; \quad P = \frac{1}{R}\left(\frac{1}{n}\right)' \quad ; \quad \omega = \frac{\left(\frac{\theta}{n}\right)'}{\left(\frac{\vartheta}{n}\right)'} \quad . \tag{42.85}$$

Then we obtain the coefficients A,B,C,E of the polynomial

$$\frac{\Delta x_1}{M} = \frac{x_1}{M} - x_0 = A\xi^3 + B\xi^2 x_0 + C\xi x_0^2 + E x_0^3 \ ,$$

which determine the aberrations of meridional rays, by the formulae:

$$A = -\frac{1}{2\Gamma} \int_{z_0}^{z_1} S \, dz \ ,$$

$$B = -\frac{3}{2\Gamma} \int_{z_0}^{z_1} S\omega \, dz$$

$$\tag{42.86}$$

$$C = -\frac{3}{2\Gamma} \int_{z_0}^{z_1} S\omega^2 dz - \frac{1}{2}\Gamma \int_{z_0}^{z_1} P \, dz$$

$$E = -\frac{1}{2\Gamma} \int_{z_0}^{z_1} S\omega^3 dz - \frac{1}{2}\Gamma \int_{z_0}^{z_1} P\omega \, dz \ .$$

If we have C we get the coefficient D by Petzval's equation (42.62).

42.9 <u>A finite number of spherical surfaces</u>. In the case of an optical instrument which consists of a number of spherical surfaces, i.e., of ordinary lenses, the above integral formulae when rewritten in the form of Stieltjes integrals lead directly to summation formulae. These formulae may also be derived independently by the method of difference equations. The result is completely analogous to (42.85) and (42.86).

We define, for every surface i of the system, the quantities

$$S_i = h_i \frac{\left(\Delta_i \frac{\vartheta}{n}\right)^2 \left(\Delta_i \frac{\vartheta}{n^2}\right)}{\left(\Delta_i \frac{1}{n}\right)^2} \quad ; P_i = \frac{1}{R_i}\left(\Delta_i \frac{1}{n}\right); \omega = \frac{\Delta_i \frac{\theta}{n}}{\Delta_i \frac{\vartheta}{n}} . \tag{42.91}$$

The coefficients A,B,C,E of the polynomial

$$\frac{1}{M}\Delta x_1 = \frac{1}{M}x_1 - x_0 = A\xi^3 + B\xi^2 x_0 + C\xi x_0^2 + E x_0^3 \tag{42.92}$$

are then given by the sums:

$$A = -\frac{1}{2\Gamma}\sum S_i ,$$

$$B = -\frac{3}{2\Gamma}\sum S_i \omega_i ,$$

$$\tag{42.93}$$

$$C = -\frac{3}{2\Gamma}\sum S_i \omega_i^2 - \frac{1}{2}\Gamma\sum P_i ,$$

$$E = -\frac{1}{2\Gamma}\sum S_i \omega_i^3 - \frac{1}{2}\Gamma\sum P_i \omega_i .$$

For the numerical computation of these coefficients it is only necessary to compute the data h_1, ϑ_1 and H_1, θ_1 of two paraxial rays with the aid of the recursion formulae

$$h_{i+1} = h_{i-1} + \delta_i \vartheta_i , \qquad H_{i+1} = H_{i-1} + \delta_i \theta_i ,$$

$$\tag{42.94}$$

$$\vartheta_{i+1} = \vartheta_{i-1} - D_i h_i , \qquad \theta_{i+1} = \theta_{i-1} - D_i H_i .$$

These rays must satisfy the boundary conditions

$$h_0 = 0 , \qquad H_0 = 1 ,$$

$$\tag{42.95}$$

$$h(\zeta) = 1 , \qquad H(\zeta) = 0 .$$

We are then in a position to determine the quantities S_i, P_i and ω_i and finally, by summation, the coefficients of the polynomial (42.92).

We mention, however, that the only essential conditions in (42.95) are $h_0 = 0$ and $H(\zeta) = 0$ which require that the axial ray h_1, ϑ_1, passes through the object point and the field ray H_1, θ_1 passes through the pupil point. If $h(\zeta) \neq 1$ and $H(z_0) \neq 1$ then by (42.93) and (42.91) we may obtain the coefficients of the polynomial

$$\frac{1}{M} \Delta x_1' = A\xi'^3 + B\xi'^2 x_0' + C\xi' x_0'^2 + E x_0'^3 \,, \tag{42.96}$$

where $x_1' = \dfrac{x_1}{H_0}$; $x_0' = \dfrac{x_0}{H_0}$, $\xi' = \dfrac{\xi}{h(\zeta)}$.

§43. CHROMATIC ABERRATIONS.

43.1 The index of refraction of an optical medium is a function of the wave length of light. The aberrations caused by this dependence on wave length are called chromatic aberrations. It is customary to consider the index of refraction which corresponds to the wave length

$$\lambda = 5896 \text{ Å}$$

of the sodium D line as the normal index. The correction of the instrument is carried out on the basis of this index for the monochromatic aberrations which we have discussed above. By combinations of Flint and Crown glasses it is possible to eliminate the chromatic aberrations to a certain extent. In general it is sufficient to investigate the instrument for two additional wave lengths, the C and F lines of the solar spectrum, for which $\lambda = 6563$ Å and $\lambda = 4861$ Å respectively.

The chromatic aberrations affect even the Gaussian optics of an instrument. The elimination of this part of the chromatic aberration thus is the first necessary step in optical design. In the following we investigate how the Gaussian optics of a medium is affected if the function $n(z)$ is replaced by a function $n + \Delta n$ in the canonical equations. We assume that Δn is sufficiently small so that we can neglect terms in which $(\Delta n)^2, (\Delta n)^3$, etc. appear.

43.2 From the canonical equations

$$\dot{x} = \frac{1}{n} p$$

$$\dot{p} = - Dx \tag{43.21}$$

it follows that

$$\Delta \dot{x} = \frac{1}{n} \Delta p + p\Delta \, \frac{1}{n}$$

$$\Delta \dot{p} = - D\Delta x - x\Delta D \tag{43.22}$$

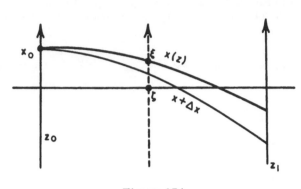

where Δx and Δp give the departure of a paraxial ray in the medium $n + \Delta n$ from the corresponding ray in the medium n. We mean by corresponding ray the ray which has the same boundary values x_0 and p_0 at the plane $z = z_0$. Thus the functions Δx and Δp are the solutions of the

Figure 174

nonhomogeneous canonical equations (43.22) which satisfy the boundary conditions

$$\Delta x = 0 \text{ at } z = z_0 \ ,$$

$$\Delta p = 0 \text{ at } z = z_0 \ .$$

(43.23)

We solve this problem in the same way as the similar problem in §42.3 We consider the first equation (43.22) and the equation for the axial ray

$$\dot{h} = \frac{1}{n} \ \vartheta \ .$$

(43.24)

As in §42.3 it follows that

$$\vartheta \Delta \dot{x} - \dot{h} \Delta p = p \vartheta \Delta \frac{1}{n} \ .$$

(43.25)

From the second equation (43.22) and the relation

$$\dot{\vartheta} = - Dx$$

(43.26)

we find similarly that

$$\dot{\vartheta} \Delta x - h \Delta \dot{p} = xh \Delta D \ ,$$

(43.27)

and hence

$$\frac{d}{dz} \left(\vartheta \Delta x - h \Delta p \right) = p \vartheta \Delta \frac{1}{n} + xh \Delta D \ .$$

(43.28)

With the aid of the boundary conditions $\Delta x = h = 0$ at $z = z_0$ and $h(z_1) = 0$ this yields by integration

$$\vartheta(z_1)\Delta x_1 = \int_{z_0}^{z_1} \left(p\vartheta \Delta\frac{1}{n} + xh\Delta D \right) dz \ . \tag{43.29}$$

By introducing the invariant $\Gamma = H\vartheta - h\theta = H(z_1)\vartheta(z_1) = M\vartheta(z_1)$ we can write this in the form

$$\frac{1}{M}\Delta x_1 = \frac{1}{\Gamma} \int_{z_0}^{z_1} \left(p\vartheta\Delta\frac{1}{n} + xh\Delta D \right) dz \ . \tag{43.291}$$

43.3 The original paraxial ray $x(z), p(z)$ is given by

$$x = x_0 H(z) + \xi h(z); \quad p = x_0\theta(z) + \xi\vartheta(z)$$

as a linear combination of the two paraxial rays $H(z),\theta(z)$ and $h(z), \vartheta(z)$ defined by the conditions (42.12). By introducing these expressions in (43.291) we obtain a linear polynomial in x_0 and ξ, namely

$$\frac{1}{M}\Delta x_1 = K\xi + Lx_0 \ , \tag{43.31}$$

where K and L are given by the integrals

$$K = \frac{1}{\Gamma} \int_{z_0}^{z_1} \left(\vartheta^2\Delta\frac{1}{n} + h^2\Delta D \right) dz \ ,$$

$$\tag{43.32}$$

$$L = \frac{1}{\Gamma} \int_{z_0}^{z_1} \left(\theta\vartheta\Delta\frac{1}{n} + Hh\Delta D \right) dz \ .$$

The quantity K is called the <u>Axial Color</u> of the medium since it determines the chromatic aberrations of the axial bundle from point $x_0 = 0$. The coefficient L measures the <u>Lateral Color</u> of the medium, namely the chromatic aberrations of the bundle of field rays $\xi = 0$.

43.4 It is possible to simplify the expressions for K and L. We have

$$\Delta\frac{1}{n} = -\frac{\Delta n}{n^2} \quad \text{and} \quad \Delta D = \frac{\Delta\dot{n}}{R} = D\frac{\Delta\dot{n}}{\dot{n}} \ , \tag{43.41}$$

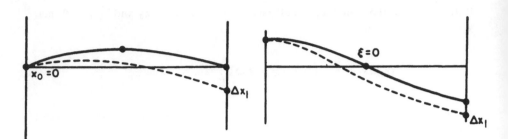

Figure 175

and hence

$$K = \frac{1}{\Gamma} \int_{z_0}^{z_1} \left(Dh^2 \frac{\Delta \dot{n}}{\dot{n}} - \frac{\vartheta^2}{n^2} \Delta n \right) dz \ .$$

(43.42)

We introduce $D = -\frac{\dot{\vartheta}}{h}$ and $\frac{\dot{\vartheta}}{n} = \dot{h}$. It follows that

$$K = -\frac{1}{\Gamma} \int_{z_0}^{z_1} \left(h \dot{\vartheta} \frac{\Delta \dot{n}}{\dot{n}} + \dot{h} \frac{\vartheta}{n} \Delta n \right) dz$$

(43.43)

By partial integration of the second term, using $h(z_0) = h(z_1) = 0$, we obtain

$$K = \frac{1}{\Gamma} \int_{z_0}^{z_1} h \left(\left(\frac{\vartheta}{n} \Delta n \right)^{\cdot} - \dot{\vartheta} \frac{\Delta \dot{n}}{n} \right) dz \ .$$

(43.44)

This however can be written as follows

$$K = \frac{1}{\Gamma} \int_{z_0}^{z_1} h \left(\frac{\vartheta}{n} \right)^{\cdot} \frac{\left(\frac{\Delta n}{n} \right)^{\cdot}}{\left(\frac{1}{n} \right)^{\cdot}} \ dz \ .$$

(43.45)

In a similar manner we obtain

$$L = \frac{1}{\Gamma} \int_{z_0}^{z_1} h \left(\frac{\theta}{n} \right)^{\cdot} \frac{\left(\frac{\Delta n}{n} \right)^{\cdot}}{\left(\frac{1}{n} \right)^{\cdot}} \ dz \ .$$

(43.46)

The function

$$\mu = \frac{\Delta n}{n} = \frac{n\lambda\,(z) - n\lambda_0\,(z)}{n\lambda_0\,(z)} \tag{43.47}$$

measures the dispersion of the optical medium. It is closely related to Abbe's ν-value

$$\nu = \frac{n_D - 1}{n_F - n_C} \tag{43.48}$$

which is used in practice to characterize the dispersion of optical media. Indeed, letting $\lambda_0 = \lambda_D$ and $\lambda = \lambda_F$ and then $\lambda = \lambda_C$ we obtain

$$\mu_F - \mu_C = \frac{n_F - n_C}{n_D} = \frac{1}{\nu}\,\frac{n_D - 1}{n_D} \quad . \tag{43.481}$$

By introducing $\mu\,(z)$ in (43.46) we find

$$K = \frac{1}{\Gamma} \int_{z_0}^{z_1} h\left(\frac{\vartheta}{n}\right)^{\cdot} \frac{\dot{\mu}}{\left(\frac{1}{n}\right)^{\cdot}}\, dz \ ,$$

$$\tag{43.49}$$

$$L = \frac{1}{\Gamma} \int_{z_0}^{z_1} h\left(\frac{\theta}{n}\right)^{\cdot} \frac{\dot{\mu}}{\left(\frac{1}{n}\right)^{\cdot}}\, dz \ .$$

If the optical instrument consists of a finite number of refracting surfaces with homogeneous media between them, we obtain, instead of (43.49), the summation formulae

$$K = \frac{1}{\Gamma} \sum h_i \left(\Delta_i\,\frac{\vartheta}{n}\right) \frac{\Delta_i \mu}{\Delta_i \frac{1}{n}} \ ,$$

$$\tag{43.491}$$

$$L = \frac{1}{\Gamma} \sum h_i \left(\Delta_i\,\frac{\theta}{n}\right) \frac{\Delta_i \mu}{\Delta_i \frac{1}{n}} \quad .$$

These formulae can be used with advantage to determine the chromatic aber-rations of an optical system numerically. We mention explicitly that $h(z)$, $\vartheta\,(z)$, $\theta(z)$ in (43.49) and h_i, ϑ_i, θ_i in (43.491) are the paraxial rays for the normal index of refraction for D-light. By considering h_i, ϑ_i as parameters which determine an optical system (refer to §39.8) we can interpret the relations (43.491) for given values of K and L as conditions which must be satisfied by these parameters. On the basis of these conditions and other conditions derived from the equations (42.93) for Seidel coefficients it is possible to solve many problems in optical design by direct algebraic methods. For details we refer to the extended literature of technical optics.

We only mention that it seems to us that by using the parameters h_i, ϑ_i in any application one obtains these conditions in the simplest mathematical form. To introduce the geometric parameters R_i, t_i of the system in the above formulae invariably complicates the algebraic equations.

43.5 "Chromatic" Aberrations in electron optics. The same general method of correcting chromatic aberration can be applied to electron optics. If only electrostatic fields are present the index of refraction is defined by

$$n^2 = 2(C - \kappa\theta) \tag{43.51}$$

where C is proportional to the original kinetic energy of the electrons. When the electrons have different original speeds we will have an analogue to chromatic aberration. We assume that the departures from a constant C are small. Unlike the optical case, the curvature $1/R$ of the "refracting surfaces" depends on C and the derivation of (43.49) does not apply. Instead, from (43.51) we obtain $n\Delta n = \Delta C$, and from (43.51) and (38.22) together, $n\Delta D + D\Delta n = 0$. Proceeding as before, we obtain from the general formula (43.32)

$$K = -\frac{\Delta C}{\Gamma} \int_{z_0}^{z_1} \frac{1}{n^2} \left[\dot{h}\vartheta - h\dot{\vartheta} \right] dz \,,$$

$$\tag{43.52}$$

$$L = -\frac{\Delta C}{\Gamma} \int_{z_0}^{z_1} \frac{1}{n^2} \left[\dot{h}\theta - H\dot{\vartheta} \right] dz \,.$$

The electron-optical instrument thus is "chromatically" corrected if the integrals in (43.52) are zero. In this case it is corrected for all values of C in the neighborhood of C_0. The electron optical instrument differs in this respect from an achromatic optical instrument. If an optical instrument is corrected for two wave lengths it is in general not corrected for other wave lengths. It requires special glass combinations such as Flint, Crown, and Fluorite to obtain correction for even three colors simultaneously. The so-called Apochromatic microscope objectives are optical systems with this latter type of chromatic correction.

CHAPTER VI

DIFFRACTION THEORY OF OPTICAL INSTRUMENTS

§44 FORMULATION OF THE PROBLEM.

44.1 The propagation of light waves in a nonhomogeneous optical medi--
um is called the diffraction theory of optical instruments. Let us assume that
there is a periodic oscillator at a certain point in the medium. We expect that
the electromagnetic field which is finally established by the oscillator will be
periodic in time and will have the form

$$E = u e^{-i\omega t} ,$$

$$H = v e^{-i\omega t} .$$

(44.11)

u and v are complex vectors which satisfy the equations

$$\text{curl } v + i k \epsilon u = 0 ,$$

$$\text{curl } u - i k \mu v = 0 .$$

(44.12)

The frequency of the oscillator is given by

$$\frac{\omega}{2\pi} = \frac{c}{\lambda} ,$$

(44.13)

k is the quantity

$$k = \frac{\omega}{c} = \frac{2\pi}{\lambda} .$$

(44.14)

44.2 We have derived in §16 the simplest type of progressing wave which
can be interpreted as the electromagnetic radiation from a dipole in a homogeneous medium $\epsilon = \mu = 1$. The result was

$$u = \left(\frac{1}{r} + \frac{i}{kr^2} \right) (m \times \rho) e^{ikr}$$

$$v = \rho \times u - \left[\frac{m}{k^2 r^3} - \left(\frac{3}{k^2 r^3} - \frac{2i}{kr^2} \right) (m \cdot \rho) \rho \right] e^{ikr}$$

(44.21)

where $m = a + ia^*$ is a given complex vector and ρ is the unit vector $\frac{1}{r} (x,y,z)$.

For small values of λ, i.e., large values of k, the principal part of the wave is given by

$$u_0 = \frac{1}{r}(m \times \rho)e^{ikr}$$

(44.22)

$$v_0 = \rho \times u_0 \ .$$

This electromagnetic field (44.22) thus represents the approximate solution which belongs to geometrical optics. We observe, however, another fact which is of importance in the following: The vectors (44.22) give the principal term of the wave (44.21) not only for large values of k but also for large r. We can express this as follows: <u>The wave (44.21) and the solution (44.22) of geometrical optics have the same boundary values at infinity</u> in the sense that

$$\lim_{r\to\infty} r(u - u_0) = 0 \ ,$$

(44.23)

$$\lim_{r\to\infty} r(v - v_0) = 0 \ .$$

44.3 Let us next consider a medium which is homogeneous in the two half spaces $z < \ell_0$ and $z > \ell_1$. The index of refraction in these half spaces shall be n_0 and n_1 respectively. In the domain $\ell_0 \leq z \leq \ell_1$ we assume $n(x,y,z)$ to be sectionally continuous. The problem is to find a solution of the equations (44.12) which represents the radiation from a periodic oscillator placed at the point (x_0,y_0,z_0) of the object space $z < \ell_0$. Of special interest is the

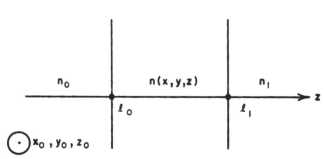

Figure 176

electromagnetic field in the image space $z > \ell_1$. A complete solution of this problem is very difficult. Even in comparatively simple geometrical configurations, for example a single lens, we obtain the vectors $u(x,y,z)$ and $v(x,y,z)$ as the superposition of infinitely many particular solutions which correspond to the internal reflections inside the lens. A similar complication is found if the medium is continuous. It is true, as we have seen in Chapter I, that the field of geometrical optics does not show any external or internal reflections. An equivalent result was obtained for the propagation of sudden discontinuities of the field, light signals. However this does not preclude the existence of scattered reflected waves as part of the complete solution of our problem. In

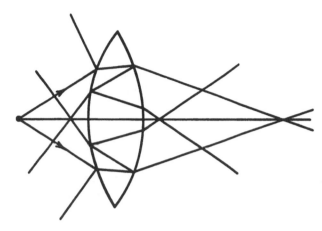

Figure 177

the case of a stratified medium the existence of these scattered reflected waves can be demonstrated directly by an explicit solution of the problem. In fact, these waves are used today to eliminate the reflections from a single air-glass surface by interference with the aid of film layers evaporated on the surface.

In general, we are not interested in the complete solution u(x,y, z) and v(x,y,z) of the above problem. The wave trains which reach the image space after one or more internal reflections or by scattering do not contribute to the image formation of the instrument. For this reason one carefully tries to eliminate these waves either by absorption or today by coating the surfaces with thin films. Of prime interest, however, is the wave which is transmitted directly without any internal reflections. Our problem is to determine only those parts of the wave from an oscillator which represent transmitted waves of this type.

44.4 The solution (44.21) for a homogeneous medium is a wave of the above type. We can write this solution in the form

$$u = Ue^{ikn_0 r} = Ue^{ik\psi}$$

$$v = Ve^{ikn_0 r} = Ve^{ik\psi}$$

(44.41)

where $\psi = n_0 r = n_0\sqrt{(x-x_0)^2 + (y-y_0)^2 + (z-z_0)^2}$ is the solution of the equation

$$\sqrt{\psi_x^2 + \psi_y^2 + \psi_z^2} = n_0$$

(44.42)

which determines the spherical wave fronts about the point x_0, y_0, z_0. We expect therefore that we can represent the transmitted waves in a nonhomogeneous medium by the expressions

$$u = Ue^{ik\psi}$$

$$v = Ve^{ik\psi}$$

(44.43)

where $\psi(x_0, y_0, z_0; x, y, z)$ satisfies the equation

$$\sqrt{\psi_x{}^2 + \psi_y{}^2 + \psi_z{}^2} = n(x, y, z) \tag{44.44}$$

and determines the "spherical" wave fronts in the medium about the point x_0, y_0, z_0. By assuming (44.43) in §16 we have derived the electromagnetic field

$$u_0 = U_0 e^{ik\psi}$$

$$v_0 = V_0 e^{ik\psi} \tag{44.45}$$

which is associated with geometrical optics. We have found that the complex vectors $U_0(x, y, z)$ and $V_0(x, y, z)$ satisfy a system of linear differential equations along a light ray, namely

$$\frac{dU_0}{d\tau} + \frac{1}{2}\Delta_\mu\psi U_0 + \left(U_0 \cdot \frac{\text{grad } n}{n}\right)\text{grad }\psi = 0$$

$$\frac{dV_0}{d\tau} + \frac{1}{2}\Delta_\epsilon\psi V_0 + \left(V_0 \cdot \frac{\text{grad } n}{n}\right)\text{grad }\psi = 0 \tag{44.46}$$

where the parameter τ is defined by $n d\tau = ds$ where s is the geometrical length of the rays. If the light ray passes a surface of discontinuity in $n(x, y, z)$ then these differential equations have to be replaced by difference equations namely the two of Fresnel's formulae (16.55) which give the transmitted part of the vectors U_0 and V_0. With the aid of these formulae we are in a position to determine the field $U_0(x, y, z)$ and $V_0(x, y, z)$ in the image space $z > \ell_1$ if the

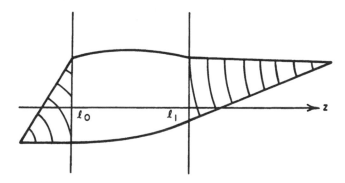

Figure 178

field U_0, V_0 in the object space $z < \ell_0$ is known. We then have from

$$u_0 = U_0 e^{ik} \qquad\qquad v_0 = V_0 e^{ik} \tag{44.47}$$

an approximation of the exact solution which is the better the greater the frequency $\dfrac{\omega}{2\pi} = \dfrac{kc}{2\pi}$ of the oscillator.

44.5 The exact solutions $u(x,y,z)$ and $v(x,y,z)$ in the image space $\epsilon = n_1^2$, $\mu = 1$) must satisfy the equations

$$\text{curl } v + ikn_1^2 u = 0$$
$$\text{curl } u - ikv \quad = 0 \tag{44.51}$$

where n_1 is a constant. It follows that u is a solution of the second order equation

$$\Delta u + k^2 n_1^2 u = 0 \tag{44.52}$$

such that

$$\text{div } u = 0 \ . \tag{44.53}$$

We consider now not only the real part $z > \ell_1$ of the image space but also its virtual part $z \leqq \ell_1$. We also extend the electromagnetic wave

$$E = U(x,y,z)e^{ik(\psi - ct)}$$
$$H = V(x,y,z)e^{ik(\psi - ct)} \tag{44.54}$$

backwards into the virtual part of the image space. In other words we regard (44.54) as part of a progressing wave which is defined in the image space as a whole.

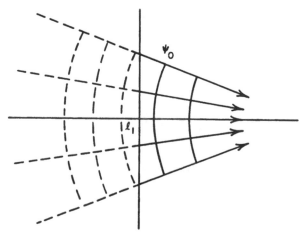

Figure 179

In a similar way we can supplement the electromagnetic field (44.47) by a virtual extension. Let us consider for example a wave front $\psi = \psi_0$ in the neighborhood of the plane $z = \ell_1$. The light rays are given by the normals of this surface. We first construct the virtual extension of these rays and then the surfaces parallel to $\psi = \psi_0$ which are normal to these ray extensions. With the aid of the differential relations

$$\frac{dU_0}{d\tau} + \frac{1}{2}\Delta\psi U_0 = 0$$

$$\frac{dV_0}{d\tau} + \frac{1}{2}\Delta\psi V_0 = 0$$

(44.55)

we finally obtain the vectors U_0 and V_0 on the fictitious wave fronts in the virtual image space.

44.6   We are now in a position to formulate the hypothesis which is the basis of all the following considerations.

The light wave which is directly transmitted into the image space can be expressed in the form

$$E = ue^{-ikct}$$

$$H = ve^{-ikct}$$

(44.61)

where u and v are solutions of the equations (44.51) which satisfy the following conditions

a) $u(x,y,z)$ and $v(x,y,z)$ are regular at any finite point of the x,y,z space.

b)   The boundary values of u and v at infinity are given by the functions

$$u_0 = U_0 e^{ik\psi}$$

$$v_0 = V_0 e^{ik\psi}$$

(44.62)

which represent the electromagnetic field of geometrical optics. The boundary values are attained in the sense that

$$\lim_{\psi \to \pm\infty} \psi(u - u_0) = 0$$

(44.63)

$$\lim_{\psi \to \pm\infty} \psi(v - v_0) = 0 \quad .$$

Both hypotheses can be made physically plausible. The assumption b) is supported by the evidence in the case of the spherical wave (44.21) in a homogeneous medium.

The condition a) however may be considered with an element of doubt. Let us, for example, assume that our optical instrument has two perfect

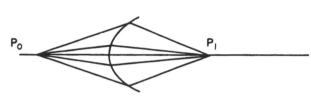

Figure 180

conjugate points at P_0 and P_1. We consider P_0 as a point source which emits a light wave given by the expressions (44.21) for u and v. Both u and v tend to ∞ if the point P_0 is approached. After refraction however we exclude singularities by the condition a). In view of the fact that P_1 is a perfect conjugate point to P_0 it does not seem to be a condition which is self evident. On the other hand, this condition has considerable optical consequences. Indeed, it excludes a complete reconcentration of radiated energy even at points which are perfectly conjugate in the sense of geometrical optics. The impossibility of perfect definition i.e., a limit for the resolution of any instrument therefore is introduced in our theory from the very beginning.

We shall assume the two hypotheses in the following to be true and determine the solution of (44.51) which satisfies the conditions a) and b). We remark, however, that it must be possible either to prove or to disprove both conditions by mathematical considerations namely by a construction of the transmitted wave inside the optical medium.

§45 THE BOUNDARY VALUE PROBLEM OF THE EQUATION $\Delta u + k^2 u = 0$ FOR A PLANE BOUNDARY.

45.1 In order to solve the problem formulated in §44 we derive the following boundary value problem: To find a function u(x,y,z) which satisfies the equation

$$\Delta u + k^2 u = 0 \qquad\qquad (45.11)$$

in the half space z > 0 and which attains given boundary values

$$u(x,y,0) = f(x,y) \ . \qquad\qquad (45.12)$$

We assume that f(x,y) is sectionally continuous in the xy plane. Outside a certain circle of radius R_0 f(x,y) shall be continuous and shall have continuous derivatives such that

$$\left| f(x,y) \right| < \frac{B}{R} \ ; \ \ \left| f_x \right| < \frac{B}{R} \ ; \ \ \left| f_y \right| < \frac{B}{R} \qquad\qquad (45.13)$$

where $R = \sqrt{x^2 + y^2}$ and B is a constant independent of x and y. We require moreover that the solution u(x,y,z) is regular for z > 0 and satisfies the following conditions

a) In the domain $R = \sqrt{x^2 + y^2 + z^2} > R_0$ of the half space z > 0 there exists a constant C such that u and the derivative $\frac{\partial u}{\partial R}$ satisfy the inequalities

$$|u| < \frac{C}{R} \qquad \left|\frac{\partial u}{\partial R}\right| < \frac{C}{R} \qquad\qquad (45.14)$$

b) In any solid sector $-\frac{\pi}{2} + \delta < \theta < \frac{\pi}{2} - \delta$ of the domain $R > R_0$; z > 0

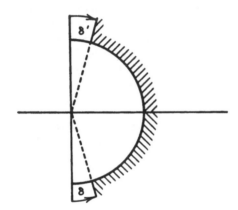

there exists a constant $D(\delta)$ such that for all points (x,y,z) of the sector we have

$$\left|\frac{\partial u}{\partial R} - iku\right| < \frac{D}{R^2} \qquad\qquad (45.141)$$

where $R = \sqrt{x^2 + y^2 + z^2}$.

These last conditions are essential for the uniqueness of the solution of the boundary value problem. In fact, we can easily find solutions u which are zero on the plane z = 0, for example

Figure 181 $u = \sin kz$.   (45.15)

Hence, if $u_0(x,y,z)$ is a solution with the boundary values f(x,y) then all functions $u = u_0 + a \sin kz$ are solutions with the same boundary values. Functions of the type (45.15) are however excluded by the conditions (45.14). In other words we shall see that the only solution of (45.11) which satisfies the conditions (45.14) and (45.141) and has the boundary values f ≡ 0 is the function $u(x,y,z) \equiv 0$.

The condition (45.141) moreover insures that u represents, as z→∞, a wave of the type of an outgoing spherical wave. Indeed, in the case of

$$u = \frac{1}{r} e^{ikr}$$

we have

$$\frac{\partial u}{\partial r} - iku = -\frac{1}{r^2} e^{ikr}$$

i.e.,

$$\left| \frac{\partial u}{\partial r} - iku \right| \leq \frac{1}{r^2} \ . \tag{45.16}$$

It is physically plausible that we will have to deal with waves of this type in our problems. Let us for example assume that $f(x,y) = 0$ outside the circle $\sqrt{x^2 + y^2} = a$. The light wave which is sent out from such a circular surface will assume more and more the form of a spherical wave the greater the distance z from the plane $z = 0$.

45.2 Let us assume that $u(x,y,z)$ is a solution of the above problem. We

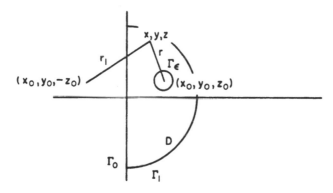

Figure 182

consider a point x_0,y_0,z_0 of the half space $z > 0$. Let r be the distance of a point (x,y,z) from (x_0,y_0,z_0) and r_1 the distance from the mirror image $(x_0,y_0, -z_0)$ of (x_0,y_0,z_0). The function

$$v(x,y,z) = \frac{1}{r} e^{ikr} - \frac{1}{r_1} e^{ikr_1} \tag{45.21}$$

is then a solution of the equation (45.11) which is zero on the plane $z = 0$ and regular in the half space $z > 0$ with the exception of the point x_0,y_0,z_0. Now let D be the domain which is obtained by removing a small sphere of radius ϵ about the point (x_0,y_0,z_0) from the interior of a hemisphere of radius R about the origin. We assume of course that R is so great that this point lies inside the hemisphere. We denote the plane section of the boundary Γ of D by Γ_0, the hemisphere by Γ_1 and the sphere of radius ϵ by Γ_ϵ . We apply Green's theorem to the domain D:

$$\iiint_D (u\Delta v - v\Delta u) \ dxdydz = - \iint_\Gamma \left(u\frac{\partial v}{\partial \nu} - v\frac{\partial u}{\partial \nu} \right) do \tag{45.22}$$

where $\dfrac{\partial}{\partial \nu}$ means differentiation in the direction of the interior normal of Γ. Since $u\Delta v - v\Delta u = 0$ and $u = f$, $v = 0$ on Γ_0 it follows that

$$\iint_{\Gamma_1} \left(u\frac{\partial v}{\partial \nu} - v\frac{\partial u}{\partial \nu}\right) do + \iint_{\Gamma_\epsilon} \left(u\frac{\partial v}{\partial \nu} - v\frac{\partial u}{\partial \nu}\right) do = - \iint_{\Gamma_0} f\frac{\partial v}{\partial z} \, dxdy \ . \qquad (45.23)$$

This equation holds for all values $R > 0$ and $\epsilon > 0$ and remains true in case $R \to \infty$ and $\epsilon \to 0$ provided that the three different integrals converge to definite limits.

First we consider the integral over Γ_ϵ. Let $v_0 = \dfrac{1}{r}e^{ikr}$ and $v_1 = \dfrac{1}{r_1}e^{ikr_1}$ and hence $v = v_0 - v_1$. Since v_1 is regular at (x_0, y_0, z_0) we have

$$\lim_{\epsilon \to 0} \iint_{\Gamma_\epsilon} \left(u\frac{\partial v_1}{\partial \nu} - v_1\frac{\partial u}{\partial \nu}\right) do = 0 \ . \qquad (45.24)$$

Furthermore

$$\iint_{\Gamma_\epsilon} \left(u\frac{\partial v_0}{\partial \nu} - v_0\frac{\partial u}{\partial \nu}\right) do = \epsilon^2 e^{ik\epsilon} \iint \left[u\left(-\frac{1}{\epsilon^2} + \frac{ik}{\epsilon}\right) - \frac{1}{\epsilon}\frac{\partial u}{\partial \nu}\right] d\omega$$

where $d\omega$ is the surface element of the unit sphere. It follows that this integral has the limit $- 4\pi u(x_0, y_0, z_0)$. We thus obtain the equation

$$4\pi u = \iint_{\Gamma_0} f\frac{\partial v}{\partial z} \, dxdy + \iint_{\Gamma_1} \left(u\frac{\partial v}{\partial \nu} - v\frac{\partial u}{\partial \nu}\right) do \qquad (45.25)$$

which is valid for all hemispheres which include the point x_0, y_0, z_0.

One verifies readily that the derivative $\dfrac{\partial v}{\partial z}$ on Γ_0 has the value

$$\frac{\partial v}{\partial z} = - \frac{2z_0}{r} \frac{\partial}{\partial r}\left(\frac{1}{r}e^{ikr}\right) \qquad (45.26)$$

where $r = \sqrt{(x-x_0)^2 + (y-y_0)^2 + z_0^2}$

For $r > 1$ this yields the inequality

$$\left|\frac{\partial v}{\partial z}\right| < \frac{2(1 + |k|)}{r^2} z_0 \qquad (45.261)$$

and hence it follows from (45.13) that the first integral in (45.25) converges if $R \to \infty$. We therefore conclude that the integral over Γ_1 must also tend to a

finite limit if $R \to \infty$. We show next that this limit is zero as a consequence
of the conditions (45.14).

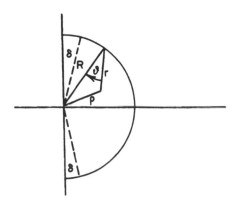

Figure 183

Let us first consider the
integral

$$I_0 = \iint_{\Gamma_1} \left(u \frac{\partial v_0}{\partial \nu} - v_0 \frac{\partial u}{\partial \nu} \right) do ,$$

where $v_0 = \frac{1}{r} e^{ikr}$. Since

$$\frac{\partial}{\partial \nu} = - \frac{\partial}{\partial R} = - \cos \vartheta \frac{\partial}{\partial r} \text{ on } \Gamma_1$$

we have

$$I_0 = R^2 \iint_{\omega} \left[u \left(\frac{1}{r^2} - \frac{ik}{r} \right) \cos \vartheta \right.$$

$$\left. + \frac{1}{r} \frac{\partial u}{\partial R} \right] e^{ikr} d\omega \qquad (45.27)$$

where $d\omega$ is the surface element of the unit sphere ω. We write this in the form

$$I_0 = R^2 \iint_{\omega} \left[u \frac{\cos \vartheta}{r^2} + ik \frac{(1 - \cos \vartheta)}{r} u + \frac{1}{r} \left(\frac{\partial u}{\partial R} - iku \right) \right] e^{ikr} d\omega \qquad (45.271)$$

and remark that $\cos \vartheta > 0$ and thus $1 - \cos \vartheta \leqq 1 - \cos^2 \vartheta \leqq \frac{\rho^2}{R^2}$. We divide the
domain of integration into two parts, ω_1 and ω_2, separated by the conical sur-
face of angular opening $\frac{\pi}{2} - \delta$ (refer to Fig. 183). We assume that $R > R_0$ so
that on ω_1 the conditions (45.14) are satisfied and on ω_2 both conditions (45.14)
and (45.141) are satisfied. Since $\frac{R}{r} \to 1$ if $R \to \infty$ we conclude from (45.271) and
the inequalities (45.14), (45.141) that I_0 can be estimated as follows:

$$|I_0| < A_0 \delta + \frac{A_1 (\delta)}{R} . \qquad (45.28)$$

The constant A_0 is independent of δ and R and $A_1 (\delta)$ is independent of R. Thus
if we choose δ such that $A_0 \delta < \frac{\epsilon}{2}$ and then R such that $\frac{1}{R} A_1 (\delta) < \frac{\epsilon}{2}$ we find
$|I_0| < \epsilon$ where $\epsilon > 0$ can be arbitrarily small. This means that $I_0 \to 0$ if $R \to \infty$.

The same estimates can be carried out in the case of the integral

$$I_1 = \iint\limits_{\Gamma_1} \left(u\frac{\partial v_1}{\partial \nu} - v_1\frac{\partial u}{\partial \nu} \right) d\omega, \text{ where } v_1 = \frac{1}{r_1}e^{ikr_1}$$

with the result that $I_1 \to 0$ if $R \to \infty$.

The formula (45.25) thus yields the following representation of $u(x_0, y_0, z_0)$:

$$4\pi u = \iint f\frac{\partial v}{\partial z} \, dxdy$$

or with the aid of (45.26):

$$2\pi u(x_0, y_0, z_0) = - z_0 \iint\limits_{-\infty}^{\infty} f(x,y)\left(\frac{ik}{r^2} - \frac{1}{r^3} \right) e^{ikr} \, dxdy \tag{45.29}$$

where $r = \sqrt{(x-x_0)^2 + (y-y_0)^2 + z_0^2}$.

This result demonstrates the <u>uniqueness of the solution</u> of our boundary value problem. Let us assume that $u(x,y,z)$ is a solution of (45.11) which satisfies the conditions (45.14) and (45.141) and has the boundary values $f \equiv 0$. Since u must satisfy the relation (25.29) we find that $u \equiv 0$ for $z > 0$.

45.3 It is not difficult to prove that the integral (45.29) represents the solution of our problem for any function $f(x,y)$ which satisfies the conditions (45.13). Let us write the integral in the form

$$2\pi u(x,y,z) = - z \iint\limits_{-\infty}^{\infty} f(\xi,\eta)\left(\frac{ik}{r^2} - \frac{1}{r^3} \right) e^{ikr} \, d\xi \, d\eta \tag{45.31}$$

where $r = \sqrt{(\xi-x)^2 + (\eta-y)^2 + z^2}$.

We verify directly that

$$K(\xi,\eta;x,y,z) = - \frac{1}{2\pi}z\left(\frac{ik}{r^2} - \frac{1}{r^3} \right) e^{ikr} = - \frac{1}{2\pi}\frac{z}{r}\frac{\partial}{\partial r}\left(\frac{e^{ikr}}{r} \right)$$

$$= - \frac{1}{2\pi}\frac{\partial}{\partial z}\left(\frac{e^{ikr}}{r} \right) \tag{45.32}$$

is a solution of $\Delta K + k^2 K = 0$ for any values of ξ, η. By differentiating with respect to x, y, z under the integral in (45.31) we obtain integrals which converge uniformly because of the first condition (45.13). This insures that the derivatives of u(x,y,z) in (45.31) can be obtained by differentiating under the integral signs. Hence

$$\Delta u + k^2 u = \int\int f\ (\xi,\eta)\ (\Delta K + k^2 K) d\xi\, d\eta = 0 \tag{45.33}$$

i.e., u is a solution of (45.11).

45.4 We show next that u(x,y,z) assumes the boundary values f(x,y) if z→0 at any point (x,y) where f(x,y) is continuous. We write (45.31) in the form

$$2\pi u(x,y,z) = -z \int\int_{-\infty}^{\infty} f\ (x + \xi,\ y + \eta)\ \frac{1}{r}\ \frac{\partial}{\partial r}\left(\frac{e^{ikr}}{r}\right)\, d\xi\, d\eta \tag{45.41}$$

where $r = \sqrt{\xi^2 + \eta^2 + z^2}$ and introduce the mean values

$$Q(x,y;\rho) = \frac{1}{2\pi} \int_0^{2\pi} f(x + \rho \cos\varphi,\ y + \rho \sin\varphi)\ d\varphi \tag{45.42}$$

of the function f on a circle of radius ρ about the point (x,y). It follows that

$$u(x,y,z) = -z \int_0^{\infty} Q(x,y;\rho) \frac{\partial}{\partial r}\left(\frac{e^{ikr}}{r}\right)\frac{\rho\, d\rho}{r} \tag{45.43}$$

in which now $r = \sqrt{\rho^2 + z^2}$.

Instead of ρ we can introduce r as the variable of integration. This yields

$$u(x,y,z) = -z \int_z^{\infty} Q\left(x,y,\sqrt{r^2 - z^2}\right) \frac{\partial}{\partial r}\left(\frac{e^{ikr}}{r}\right)\, dr \ . \tag{45.44}$$

Finally, letting r = zs we may write

$$u(x,y,z) = -\int_1^{\infty} Q\left(x,y,z\sqrt{s^2 - 1}\right) \frac{\partial}{\partial s}\left(\frac{e^{ikzs}}{s}\right)\, ds \ . \tag{45.45}$$

This last expression allows us to prove formally that u(x,y,0) = f(x,y). Indeed, letting z = 0 in (45.45) we obtain

$$u(x,y,z) = -Q(x,y,0) \int_1^{\infty} \frac{\partial}{\partial s}\left(\frac{1}{s}\right)\, ds = Q(x,y,0) \ . \tag{45.46}$$

However, Q(x,y,0) = f(x,y) if f(x,y) is continuous at the point x,y.

45.5 The above formal proof that $u(x,y,0) = f(x,y)$ can easily be completed with a rigorous consideration. First we consider only a finite part of the interval of integration in (45.41) namely a circle of radius $z\sqrt{T^2 - 1}$ where T is a number which can be chosen arbitrarily large. By transformations similar to the above we can write this part of the integral in the form

$$- \int_1^T Q\left(x,y,z\sqrt{s^2 - 1}\right) \frac{\partial}{\partial s}\left(\frac{e^{ikzs}}{s}\right) ds \ . \tag{45.51}$$

The remaining part of the integral is estimated as follows. From the first condition (45.13) we conclude that a constant B exists such that

$$\left| f(x + \xi, y + \eta) \right| < \frac{B}{\rho} \ ; \ \rho = \sqrt{\xi^2 + \eta^2} \ . \tag{45.52}$$

On the other hand for $r = \sqrt{\rho^2 + z^2} > 1$ we have

$$\left| \frac{1}{r} \frac{\partial}{\partial r}\left(\frac{e^{ikr}}{r}\right) \right| < \frac{1 + |k|}{r^2} \leq \frac{1 + |k|}{\rho^2}$$

and we thus conclude that the remaining part of the integral is not greater than

$$B_1 z \int_{z\sqrt{T^2-1}}^{\infty} \frac{d\rho}{\rho^2} = \frac{B_1}{\sqrt{T^2-1}} \tag{45.53}$$

where B_1 is another constant independent of z and T. Hence we obtain the inequality

$$\left| u + \int_1^T Q\left(x,y,z\sqrt{s^2 - 1}\right) \frac{\partial}{\partial s}\left(\frac{e^{ikzs}}{s}\right) ds \right| < \frac{B_1}{\sqrt{T^2-1}} \tag{45.54}$$

which is valid for all values of z and T. We choose a sequence $z \to 0$ such that $u(x,y,z)$ converges to a certain limit $u^*(x,y,0)$. For any sequence $z \to 0$ the integral in (45.54) converges towards the value $- Q(x,y,0)\left(1 - \frac{1}{T}\right)$. Hence we obtain the inequality

$$\left| u^* - \left(1 - \frac{1}{T}\right) Q(x,y,0) \right| < \frac{B_1}{\sqrt{T^2-1}} \tag{45.55}$$

which holds for all values of T. This is only possible if $u^* = Q(x,y,0)$, i.e., if $u(x,y,z) \to Q(x,y,0)$ for any approach $z \to 0$.

45.6 Thus we have demonstrated that the function $u(x,y,z)$ defined by the integral (45.31) is a solution of (45.11) which assumes the boundary values $f(x,y)$ at any point where f is continuous. It remains to be shown that u also satisfies the conditions (45.14) and (45.141). If the function $f(x,y)$ is zero out-side a certain finite domain these conditions follow directly from the fact that the kernel

$$K = - \frac{z}{2\pi r} \frac{\partial}{\partial r} \left(\frac{1}{r} e^{ikr} \right)$$

(45.61)

satisfies these conditions. For functions $f(x,y)$ which satisfy only the condi-tions (45.13) one has to proceed in a manner similar to the above by consid-ering first a finite domain of integration and then estimating the rest.

45.7 The corresponding boundary value problem for Maxwell's equa-tions.

By a similar consideration to the above we can solve the problem of finding a function $u(x,y,z)$ which satisfies the differential equation $\Delta u + k^2 u = 0$ for $z > 0$ such that on $z = 0$ the normal derivative $\frac{\partial u}{\partial z}$ $(x,y,0)$ assumes given boundary values $g(x,y)$. Instead of (45.21) we now consider the function

$$v = \frac{1}{r} e^{ikr} + \frac{1}{r_1} e^{ikr_1}$$

(45.71)

where $r = \sqrt{(x-x_0)^2 + (y-y_0)^2 + (z-z_0)^2}$ and $r_1 = \sqrt{(x-x_0)^2 + (y-y_0)^2 + (z+z_0)^2}$

We now have $\frac{\partial v}{\partial z} = 0$ on $z = 0$. With the aid of Green's theorem we obtain the solution in the form

$$2\pi u(x,y,z) = - \iint\limits_{-\infty}^{\infty} g(\xi,\eta) \frac{e^{ikr}}{r} \, d\xi \, d\eta$$

(45.72)

where $r = \sqrt{(x-\xi)^2 + (y-\eta)^2 + z^2}$. In this case we have to assume that $g(x,y)$ satisfies the condition

$$|g(x,y)| < \frac{B}{R^2} \; ; \; R^2 = x^2 + y^2$$

(45.721)

in order to insure the convergence of the integral (45.72). Otherwise the same requirements for $u(x,y,z)$ are made as before, namely the conditions (45.14) and (45.141). We can use the expression (45.72) to find the solution in the half space $z > 0$ of Maxwell's equations

$$\text{curl } v + iku = 0$$

$$\text{curl } u - ikv = 0$$

(45.73)

which assumes given boundary values on z = 0. These boundary values are not arbitrary. However, we can give the components

$$u_1(x,y,0) = f(x,y)$$

$$u_2(x,y,0) = g(x,y)$$

(45.731)

provided that f and g have continuous derivatives and satisfy the conditions for $R > R_0$:

$$|f| < \frac{B}{R} \qquad |f_x| < \frac{B}{R^2}$$

(45.74)

$$|g| < \frac{B}{R} \qquad |g_y| < \frac{B}{R^2} \ .$$

Since div u = 0 we have $\dfrac{\partial u_3(x,y,0)}{\partial z} = - (f_x + g_y)$. With the aid of (45.31) and (45.72) we obtain the solution of our problem by

$$u_1 = - \frac{1}{2\pi} \int\int f(\xi,\eta) \frac{\partial}{\partial z} \frac{e^{ikr}}{r} \, d\xi \, d\eta$$

$$u_2 = - \frac{1}{2\pi} \int\int g(\xi,\eta) \frac{\partial}{\partial z} \frac{e^{ikr}}{r} \, d\xi \, d\eta$$

(45.75)

$$u_3 = + \frac{1}{2\pi} \int\int (f\xi + g\eta) \frac{e^{ikr}}{r} \, d\xi \, d\eta$$

where $r = \sqrt{(x-\xi)^2 + (y-\eta)^2 + z^2}$. It is easily verified that div u = 0 for all x,y,z. Thus if the vector u has been found we obtain v from ikv = curl u. If f(x,y) and g(x,y) are only assumed to be sectionally continuous with sectionally continuous derivatives then we obtain the vector u by the integrals

$$u_1 = - \frac{1}{2\pi} \int\int f(\xi,\eta) \frac{\partial}{\partial z} \left(\frac{e^{ikr}}{r} \right) d\xi \, d\eta$$

$$u_2 = - \frac{1}{2\pi} \int\int g(\xi,\eta) \frac{\partial}{\partial z} \left(\frac{e^{ikr}}{r} \right) d\xi \, d\eta$$

(45.76)

$$u_3 = \frac{1}{2\pi} \int\int \left[f(\xi,\eta) \frac{\partial}{\partial x} \left(\frac{e^{ikr}}{r} \right) + g(\xi,\eta) \frac{\partial}{\partial y} \left(\frac{e^{ikr}}{r} \right) \right] d\xi \, d\eta \ .$$

These formulae are identical with (45.75) if $f(\xi,\eta)$ and $g(\xi,\eta)$ have continuous derivatives. We remark that in (45.76) we can loosen the restrictions (45.74) and assume only that

$$|f| < \frac{B}{R}, |f_x| < \frac{B}{R}, |f_y| < \frac{B}{R} \qquad |g| < \frac{B}{R}, |g_x| < \frac{B}{R}, |g_y| < \frac{B}{R} \ . \quad (45.77)$$

§46. DIFFRACTION OF CONVERGING SPHERICAL WAVES.

46.1 Let us assume that P_0 and P_1 are perfect conjugate points on the

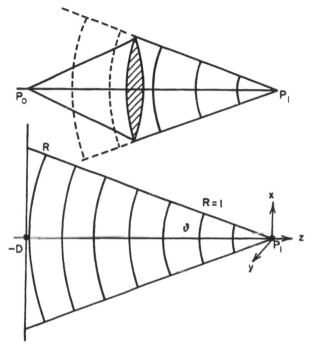

Figure 184

axis of rotation of an optical instrument. We consider the spherical wave fronts which converge after refraction to the point P_1. The equation of these wave fronts is given by Hamilton's point characteristic

$$\psi(x,y,z) = V(0,0,z_0; x,y,z) = C$$

and must have the form

$$\psi(x,y,z) = C - Rn_1 \tag{46.11}$$

where $R = \sqrt{x^2 + y^2 + z^2}$ is the distance from P_1 and the constant C is the optical distance of the points P_0 and P_1. Let us assume that we know the vectors U_0, V_0 on one of these wave fronts, for example on $\psi(x,y,z) = C - n_1$ of radius R = 1. The vectors U_0 and V_0 are electromagnetic vectors of the geometric optical approximation. On any other of the spherical wave fronts after refraction and also on those which are obtained by virtual extension we then obtain the corresponding vectors U_0^* and V_0^* simply by the relations

$$U_0^* = \frac{1}{R} U_0 (\vartheta, \varphi)$$

$$\text{(46.12)}$$

$$V_0^* = \frac{1}{R} V_0 (\vartheta, \varphi)$$

where ϑ, φ are polar coordinates of the unit sphere. We choose the point P_1 as origin of a cartesian coordinate system so that

$$x = R \sin \vartheta \cos\varphi$$

$$y = R \sin \vartheta \sin\varphi \qquad \text{(46.13)}$$

$$z = - R \cos \vartheta$$

Our problem is to find a solution u,v of Maxwell's equations (44.12) which has the same boundary values at infinity as the solution of geometrical optics

$$u_0 = \frac{1}{R} U_0 (\vartheta, \varphi) e^{ik(C - n_1 R)}$$

$$\text{(46.14)}$$

$$v_0 = \frac{1}{R} V_0 (\vartheta, \varphi) e^{ik(C - n_1 R)}$$

which means that

$$\lim_{R \to \infty} R(u - u_0) = 0$$

$$\text{(46.15)}$$

$$\lim_{R \to \infty} R(v - v_0) = 0 \ .$$

46.2   We solve this problem as follows. We consider a plane in the image space normal to the z axis at z = - D. The geometric optical field on this plane has the vectors

$$u_0 (\xi, \eta) = \frac{1}{R} U_0 (\vartheta, \varphi) e^{ik(C - n_1 R)}$$

$$\text{(46.21)}$$

$$v_0 (\xi, \eta) \ \frac{1}{R} V_0 (\vartheta, \varphi) e^{ik(C - n_1 R)}$$

where $R = \sqrt{\xi^2 + \eta^2 + D^2}$.

The vector u must be a solution of the equation

$$\Delta u + k^2 n_1^2 u = 0 \ . \qquad \text{(46.22)}$$

We determine first the solution which has the boundary values (46.21) on the plane z = - D. If a limit of these solutions u_D (x,y,z) is obtained as D→∞ then we can expect that this limit represents the desired wave.

From (45.31) it follows that

$$2\pi u_D (x,y,z) = - (D + z)e^{ikC} \int\int U_0 e^{-ikn_1(R-r)} \left(\frac{ikn_1}{r^2} - \frac{1}{r^3}\right) \frac{d\xi d\eta}{R} \qquad (46.221)$$

where $R = \sqrt{\xi^2 + \eta^2 + D^2}$ and $r = \sqrt{(\xi-x)^2 + (\eta-y)^2 + (z+D)^2}$.

We introduce the polar coordinates ϑ ,φ by

$$\xi = D \tan\vartheta \cos\varphi$$
$$\eta = D \tan\vartheta \sin\varphi . \qquad (46.23)$$

Thus we obtain

$$d\xi d\eta = D^2 \frac{\sin\vartheta d\vartheta d\varphi}{\cos^3 \vartheta}$$

$$R = \frac{D}{\cos\vartheta} \qquad (46.24)$$

$$r = \sqrt{(z + D)^2 + D^2\tan^2 \vartheta - 2D\tan\vartheta (x \cos\varphi + y \sin\varphi)}$$

and one verifies readily that

$$\frac{D}{R} = \cos\vartheta$$

$$\frac{D}{r} \to \cos\vartheta \qquad (46.25)$$

$$r - R \to z \cos\vartheta - \sin\vartheta (x \cos\varphi + y \sin\varphi)$$

in case D→∞. The integral (46.22) converges as D→∞; the result is

$$u(x,y,z) = - \frac{ikn_1}{2\pi} e^{ikC} \int\int U_0(\vartheta , \varphi)e^{-ikn_1 \left[\sin\vartheta (x \cos\varphi + y \sin\varphi) - z\cos\vartheta\right]} \sin\vartheta d\vartheta d\varphi .$$

$$(46.26)$$

This integral is known as Debye's solution.

By introducing cartesian coordinates

$$\xi = \sin \vartheta \cos \varphi$$

$$\eta = \sin \vartheta \sin \varphi \qquad (46.27)$$

$$\zeta = - \cos \vartheta$$

on the unit sphere ω we can write our result in the form

$$u(x,y,z) = - \frac{ikn_1}{2\pi} e^{ikC} \iint_\omega U_0 e^{-ikn_1(x\xi + y\eta + z\zeta)} d\omega . \qquad (46.28)$$

We notice immediately that div u = 0; indeed we have

$$\text{div } u = - \frac{(ikn_1)^2}{2\pi} e^{ikC} \iint_\omega (\xi U_0^1 + \eta U_0^2 + \zeta U_0^3) e^{-ikn_1(x\xi + y\eta + z\zeta)} d\omega$$

$$(46.281)$$

where U_0^1, U_0^2, U_0^3 are the components of the vector U_0. We know that U_0 is tangential to the unit sphere, hence $\xi U_0^1 + \eta U_0^2 + \zeta U_0^3 = 0$ and therefore div u = 0.

In a similar way we can construct the vector $v(x,y,z)$ as a solution of the equation $\Delta v + k^2 n_1^2 v = 0$. It follows that

$$v = - \frac{ikn_1}{2\pi} e^{ikC} \iint_\omega V_0 e^{-ikn_1(x\xi + y\eta + z\zeta)} d\omega . \qquad (46.29)$$

We now find that the vectors u and v satisfy Maxwell's equations (44.12) in consequence of the relations (16.28) for U_0 and V_0, namely

$$\text{grad } \psi \times V_0 + n_1^2 U_0 = 0$$

$$(46.291)$$

$$\text{grad } \psi \times U_0 - \quad V_0 = 0 .$$

§47. DIFFRACTION OF IMPERFECT SPHERICAL WAVES.

47.1 The method of §46 can be applied almost directly to wave fronts which are not perfectly spherical. Let us therefore consider an optical instrument which transforms a set of diverging spherical wave fronts into a converging set of wave fronts which are not aberration free. The wave fronts are given by Hamilton's point characteristic

$$\psi(x_0, y_0, z_0; x,y,z) = C .$$

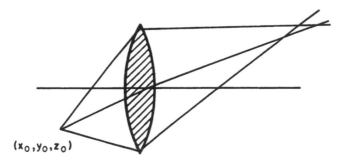

(x_0,y_0,z_0)

Figure 185

The object point (x_0,y_0,z_0) is not restricted to an axial location. The wave fronts in the image space are a set of parallel surfaces with common normals. These normals are the light rays of the bundle. As before we construct the virtual extension of this set of wave fronts by first extending the light rays and then drawing the orthogonal surfaces ψ = const. in the virtual image space. Let us consider the surface elements $d\sigma$ which are determined by a narrow tube of light rays. We call such elements corresponding surface elements of the wave fronts. The ratio $\dfrac{d\sigma}{d\sigma_0}$ of two such surface elements is equal to the ratio $\left|\dfrac{K_0}{K}\right|$ where K_0 and K are the Gaussian curvatures of the wave fronts at the points in which these surfaces are intersected by the rays of the tube.

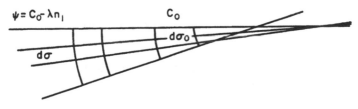

$\psi = C_0 - \lambda n_1$ C_0 $d\sigma_0$ $d\sigma$

Figure 186

We now choose a certain wave front $\psi = C_0$ as a reference surface and assume that this surface is so far away from the region of convergence that it has no singularities. Let R_1 and R_2 be the principal radii of curvature on $\psi = C_0$, i.e.,

$$K_0 = \frac{1}{R_1 R_2}$$

(47.11)

The Gaussian curvature K of a virtual wave front $\psi = C_0 - \lambda n_1$ which has a distance λ from C_0 is then given by the function

$$K = \frac{1}{(R_1 + \lambda)(R_2 + \lambda)} . \qquad (47.12)$$

Let us now assume that the vectors U_0 and V_0 are known on the surface C_0. With the aid of the result in §16.5 we obtain the corresponding vectors U_0^* and V_0^* on any other wave front, namely

$$U_0^* = \sqrt{\left|\frac{K}{K_0}\right|} \, U_0$$

$$(47.13)$$

$$V_0^* = \sqrt{\left|\frac{K}{K_0}\right|} \, V_0$$

where

$$\frac{K}{K_0} = \frac{R_1 R_2}{(R_1 + \lambda)(R_2 + \lambda)} . \qquad (47.14)$$

In other words: The vectors u_0 and v_0 which represent the electromagnetic field of geometric optics on the wave front $\psi = C_0 - \lambda n_1$ are given by the expressions

$$u_0 = \sqrt{\left|\frac{K}{K_0}\right|} \, U_0 e^{ik(C_0 - \lambda n_1)}.$$

$$(47.15)$$

$$v_0 = \sqrt{\left|\frac{K}{K_0}\right|} \, V_0 e^{ik(C_0 - \lambda n_1)} .$$

The problem is to find a solution u, v of Maxwell's equations (44.12) which has the same boundary values at infinity as (47.15), i.e.,

$$\lim_{\lambda \to \infty} \lambda(u - u_0) = 0$$

$$(47.16)$$

$$\lim_{\lambda \to \infty} \lambda(v - v_0) = 0 .$$

47.2 We solve this problem as before by first determining the solution $u_D(x,y,z)$ which has the boundary values (47.15) on the plane $z = -D$. This solution is given by the integral

$$2\pi u_D(x,y,z) = -(z + D)e^{ikC_0} \iint \sqrt{\left|\frac{K}{K_0}\right|} \, U_0 e^{-ikn_1(\lambda - r)} \left(\frac{ikn_1}{r^2} - \frac{1}{r^3}\right) d\xi \, d\eta \qquad (47.21)$$

where $r = \sqrt{(\xi - x)^2 + (\eta - y)^2 + (z + D)^2}$ and λ the distance of the point ξ, η from its corresponding point ξ_0, η_0 on $\psi = C_0$. The point ξ, η on $z = -D$ and its corresponding point ξ_0, η_0 on the surface C_0 are related by the equations

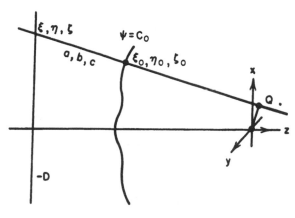

$$\xi = \xi_0 + \lambda a$$

$$\eta = \eta_0 + \lambda b \qquad (47.22)$$

$$-D = \zeta_0 + \lambda c$$

where a, b, c are the direction cosines of the surface normal at ξ_0, η_0, ζ_0, i.e.,

$$a = -\frac{1}{n_1} p_1$$

$$b = -\frac{1}{n_1} q_1 \qquad (47.23)$$

$$c = -\frac{1}{n_1} \sqrt{n_1^2 - p_1^2 - q_1^2} \quad .$$

Figure 187

We transform the integral (47.21) into an integral over the surface C_0 by introducing the expressions

$$d\xi \, d\eta = -\frac{1}{c} \left| \frac{K_0}{K} \right| d\sigma_0$$

$$\qquad (47.24)$$

$$r = \sqrt{(\xi_0 - x + \lambda a)^2 + (\eta_0 - y + \lambda b)^2 + (\zeta_0 - z + \lambda c)^2} \quad .$$

This yields

$$2\pi u_D = (z + D) e^{ikC_0} \int\!\!\int \sqrt{\left| \frac{K_0}{K} \right|} \, U_0 e^{-ikn_1(\lambda - r)} \left(\frac{ikn_1}{r^2} - \frac{1}{r^3} \right) \frac{d\sigma_0}{c} \quad . \qquad (47.25)$$

It remains to determine the limit of this integral as $D \rightarrow \infty$, i.e., $\lambda \rightarrow \infty$.

We conclude from (47.14) that

$$\frac{1}{\lambda} \sqrt{\left| \frac{K_0}{K} \right|} \rightarrow \sqrt{|K_0|} \qquad (47.250)$$

and from (47.22) we find that

$$\frac{D}{\lambda} \rightarrow -c \qquad (47.251)$$

and from (47.24) that

$$\frac{r}{\lambda} \to 1 \; . \tag{47.252}$$

Furthermore

$$r - \lambda = \sqrt{(\xi_0-x)^2 + (\eta_0-y)^2 + (\zeta_0-z)^2 + 2\lambda \left(a(\xi_0-x) + b(\eta_0-y) + c(\zeta_0-z) \right) + \lambda^2} - \lambda$$

from which it follows that

$$r - \lambda \to a(\xi_0-x) + b(\eta_0-y) + c(\zeta_0-z) \; . \tag{47.26}$$

With the aid of these relations we readily verify that the integral (47.25) tends to a limit if $\lambda \to \infty$. The result is

$$u = -\frac{ikn_1}{2\pi} e^{ikC_0} \iint \sqrt{|K_0|} \; U_0 e^{ikn_1 \left[a(\xi_0-x)+b(\eta_0-y)+c(\zeta_0-z) \right]} \, d\sigma_0 \; . \tag{47.27}$$

We can simplify this integral by introducing the optical direction cosines p,q,r of the rays. Then we recognize that the expression

$$C_0 - (p\xi_0 + q\eta_0 + r\zeta_0) = W(x_0,y_0,z_0;p,q) \tag{47.28}$$

is nothing but <u>Hamilton's mixed characteristic W</u> which is the optical length between the object point P_0 and the base point Q_1 of the perpendicular dropped

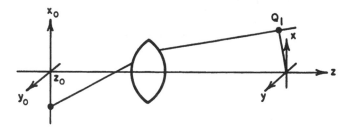

Figure 188

from the point x=y=z=0 onto the ray. By introducing this function in (47.27) we obtain the solution

$$u = -\frac{ikn_1}{2\pi} \iint \sqrt{|K_0|} \; U_0 e^{ik[W+px+qy+rz]} d\sigma_0 \; . \tag{47.29}$$

We verify as before that div $u = 0$ as a consequence of the fact that the vector U_0 is normal to the ray, i.e., $pU_0{}^1 + qU_0{}^2 + rU_0{}^3 = 0$. If we define the magnetic vector v by

$$v = -\frac{ikn_1}{2\pi} \iint \sqrt{|K_0|} \; V_0 e^{ik[W + px + qy + rz]} d\sigma_0 \qquad (47.291)$$

then u and v are solutions of Maxwell's equations

$$\text{curl } v + ikn_1{}^2 u = 0$$

$$\text{curl } u + ikv \quad = 0 \; .$$

This follows, because, on the wave front C_0, U_0 and V_0 satisfy the relations

$$\text{grad } \psi \times V_0 + n_1{}^2 U_0 = 0$$

$$\text{grad } \psi \times U_0 - V_0 = 0 \; .$$

47.3   The integrals (47.29) and (47.291) are independent of the particular wave front $\psi = C_0$ over which the integration is carried out. Indeed we transform to another wave front by introducing

$$d\sigma_0 = \left|\frac{K}{K_0}\right| d\sigma$$

and $\qquad\qquad\qquad\qquad\qquad\qquad\qquad\qquad\qquad\qquad\qquad$ (47.31)

$$U_0 = \sqrt{\left|\frac{K_0}{K}\right|} \; U_0{}^* \; .$$

It follows that

$$u = -\frac{ikn_1}{2\pi} \iint \sqrt{|K|} \; U_0{}^* e^{ik[W + px + qy + rz]} d\sigma \qquad (47.32)$$

which is the same integral form as before. We can summarize our results as follows: <u>The solution of Maxwell's equations which corresponds to a geometric optical wave in the sense that both waves have the same boundary values at infinity is given by integrals over an arbitrary wave front</u>; namely

$$u = -\frac{ikn_1}{2\pi} \iint \sqrt{|K|} \; U_0 e^{ik[W + px + qy + rz]} d\sigma$$

$$\qquad\qquad\qquad\qquad\qquad\qquad\qquad\qquad\qquad\qquad\qquad (47.33)$$

$$v = -\frac{ikn_1}{2\pi} \int \iint \sqrt{|K|} \; V_0 e^{ik[W + px + qy + rz]} d\sigma \; .$$

U_0 and V_0 represent the electromagnetic vectors of geometric optics on the wave front, $d\sigma$ is the surface element and K the Gaussian curvature of the wave front. Finally $W(x_0, y_0, z_0; p, q)$ is Hamilton's mixed characteristic.

The integrals (47.33) are the basis of the diffraction theory of optical instruments.

47.4 Parametric representation of the wave fronts.

For the evaluation of the integrals (47.33) we have to determine one of the wave fronts $\psi = C$. These surfaces are given by the characteristic function $W(x_0,y_0,z_0; p,q)$. Let us consider a light ray of direction (p,q,r). Its intersection with the plane $z_1 = 0$ is given by the functions

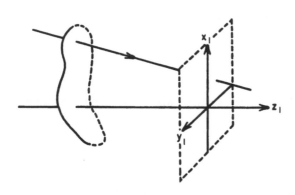

Figure 189

$$x_1 = - W_p$$

$$y_1 = - W_q \qquad (47.41)$$

and hence we obtain the intersection x,y,z with a wave front by

$$x = - W_p + \lambda_p$$

$$y = - W_q + \lambda_q \qquad (47.42)$$

$$z = \qquad \lambda_r$$

where $\lambda = \lambda(p,q)$ is a function of p and q. Since the wave fronts are normal to the light rays it follows that the two vectors (x_p,y_p,z_p) and (x_q,y_q,z_q) where

$$x_p = - W_{pp} + \lambda + \lambda_p p \; ; \; x_q = - W_{pq} + \qquad \lambda_q p$$

$$y_p = - W_{pq} \qquad + \lambda_p q \; ; \; y_q = - W_{qq} + \lambda + \lambda_q q \qquad (47.43)$$

$$z_p = \qquad \lambda r_p + \lambda_p r \; ; \; z_q = \qquad \lambda r_q + \lambda_q r$$

are normal to the vector (p,q,r). This yields the relations

$$n^2 \lambda_p = p W_{pp} + q W_{pq}$$

$$\qquad (47.44)$$

$$n^2 \lambda_q = p W_{pq} + q W_{qq}$$

and hence

$$n^2(\lambda - \lambda_0) = pW_p + qW_q - W \qquad (47.45)$$

where λ_0 is an arbitrary constant.

Let us consider the special wave front which belongs to $\lambda_0 = 0$. By introducing λ from (47.45) into (47.42) we obtain the parametric representation

$$n^2x = (p^2 - n^2)W_p + pqW_q - pW$$

$$n^2y = pqW_p + (q^2 - n^2)W_q - qW \qquad (47.46)$$

$$n^2z = rpW_p + rqW_q - rW$$

in which $r = \sqrt{n^2 - p^2 - q^2}$.

47.5 We can use the above result to transform the integrals (47.33) into a simpler form. Let $X(p,q) = (x,y,z)$ be the parametric representation of the wave front (47.46). The normals to this surface are given by the vector

$$\xi(p,q) = \frac{1}{n}(p,q,r) . \qquad (47.51)$$

We consider the two quadratic differential forms

$$dX^2 = Edp^2 + 2Fdpdq + Gdq^2$$

$$- (dX \cdot d\xi) = Ldp^2 + 2Mdpdq + Ndq^2 . \qquad (47.52)$$

The Gaussian quantities E,F,G and L,M,N are given by the scalar products

$$E = X_p^2 ; \quad F = X_p \cdot X_q ; \quad G = X_q^2 \qquad (47.53)$$

$$- L = X_p \cdot \xi_p ; - 2M = X_p \cdot \xi_q + X_q \cdot \xi_p ; - N = X_q \cdot \xi_q$$

and thus can be found by (47.46) if $W(p,q)$ is known.

With the aid of these quantities we obtain the Gaussian curvature K by

$$K = \frac{LN - M^2}{EG - F^2} \tag{47.54}$$

and the surface element $d\sigma$ of the wave front by

$$d\sigma = \sqrt{EG - F^2} \, dpdq \quad . \tag{47.55}$$

It follows that

$$\sqrt{|K|} \, d\sigma = \sqrt{|LN - M^2|} \, dpdq$$

and hence by introducing this in (47.33) we find:

$$u = -\frac{ik}{2\pi} \int\int \sqrt{n^2 \, |LN - M^2|} \, U e^{ik(W + xp + yq + zr)} \, dpdq$$

$$\tag{47.56}$$

$$v = -\frac{ik}{2\pi} \int\int \sqrt{n^2 \, |LN - M^2|} \, V e^{ik(W + xp + yq + zr)} \, dpdq \quad .$$

It is not difficult to obtain explicit expressions for L,M,N. The result is:

$$nL = W_{pp} - \frac{p^2 + r^2}{n^2 r^2} \, (pW_p + qW_q - W)$$

$$nN = W_{qq} - \frac{q^2 + r^2}{n^2 r^2} \, (pW_p + qW_q - W) \tag{47.57}$$

$$nM = W_{pq} - \frac{pq}{n^2 r^2} \, (pW_p + qW_q - W) \quad .$$

We introduce the two vectors

$$P^*(p,q) = -\sqrt{n^2 \, |LN - M^2|} \, U$$

$$\tag{47.58}$$

$$Q^*(p,q) = -\sqrt{n^2 \, |LN - M^2|} \, V$$

which are constant along a given light ray (p,q) in the image space. Our integrals assume the form

$$u = \frac{ik}{2\pi} \int\int P^*(p,q) e^{ik(W + xp + yq + zr)} \, dpdq$$

$$\tag{47.59}$$

$$v = \frac{ik}{2\pi} \int\int Q^*(p,q) e^{ik(W + xp + yq + zr)} \, dpdq$$

over a certain domain D in the p,q plane which lies inside the circle $p^2 + q^2 = n^2$.

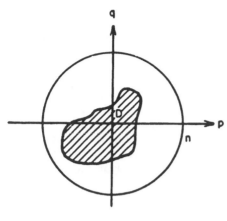

The vectors P* and Q* are related to the unit vectors P and Q which we have introduced in §16.41; the difference is that, in general P* and Q* do not have the absolute value 1 as do P and Q. One can show with the aid of the relations 16.5 that $|P*|^2$ and $|Q*|^2$ determine the <u>intensity</u> of the refracted wave, i.e., the flux $\frac{d\psi}{d\omega}$ per unit of solid angle ω. We find

Figure 190

$$I = \frac{d\psi}{1\omega} = \frac{nc}{8\pi}r^2\,|P*|^2 = \frac{c}{8\pi n}r^2\,|Q*|^2$$

(47.591)

where $r^2 = n^2 - p^2 - q^2$.

§48. DIFFRACTION OF UNPOLARIZED LIGHT.

48.1 We can assume that the vectors P* and Q* in (47.59) have the form

$$P* = (m \ x \ \rho)\phi(p,q)$$

$$Q* = \rho \ x \ P* = \left(m\rho^2 - (m \cdot \rho)\rho\right)\,\phi(p,q)$$

(48.11)

where ρ is the vector $\rho = (p,q,r)$ and where $m = (m_1, m_2, m_3)$ is an arbitrary unit vector. The components $m_1(p,q)$, $m_2(p,q)$, $m_3(p,q)$ of m are functions of p and q such that

$$m \cdot \overline{m} = m_1\overline{m}_1 + m_2\overline{m}_2 + m_3\overline{m}_3 = 1$$

(48.12)

The vector m determines the polarization of the electromagnetic field of geometrical optics which corresponds to the field given by (47.59). Consequently it also determines the polarization of (47.59) at infinity. Finally, the scalar function $\phi(p,q)$ is a measure of the intensity of the wave at infinity.

The wave u,v is linearly polarized at infinity if m is a real unit vector. In fact, we have linear polarization at infinity according to (2.69) if U satisfies the condition $U \ x \ \overline{U} = 0$. This implies

$$(m \ x \ \rho) \ x \ (\overline{m} \ x \ \rho) = 0$$

or

$$(m \ x \ \rho) \cdot \overline{m} = (\overline{m} \ x \ m) \cdot \rho = 0 \ .$$

(48.13)

This condition, however, is satisfied, if m is a real vector or proportional to a real vector.

48.2 In what is to follow we investigate the special types of waves which are obtained if a constant real unit vector m is introduced in (48.11). For physical reasons these waves can be considered to represent qualitatively the converging waves which leave an optical instrument after refraction. As we have seen in §44, the magnetic field of a dipole attains the boundary values m x ρ, where m is constant, at infinity. For a linearly polarized dipole the vector m is real. We can expect that the structure of this wave is not changed essentially if the wave passes through an optical instrument. We should how-ever consider the following results with the reservation that they refer only to idealized waves which represent the actual situation approximately.

48.3 We introduce the expression

$$P* = (m \times \rho)\phi(p,q)$$

$$Q* = \left(m\rho^2 - (m \cdot \rho)\rho\right) \phi(p,q)$$

(48.31)

in (47.59) and we assume m is constant. It is easy to verify that the result can be written in the following form

$$u = m \times \text{grad } F$$

$$v = \frac{1}{ik} \left[m\Delta F - \text{grad } (m \cdot \text{grad } F)\right]$$

(48.32)

where F is the scalar function

$$F(x,y,z) = \frac{1}{2\pi} \int\int \phi(p,q)e^{ik(W + xp + yq + zr)} \, dpdq .$$

(48.33)

The equations (48.32) assume an interesting form if we introduce the differ-ential operator

$$\nabla = \frac{1}{ik} \text{ grad} = \frac{\lambda}{2\pi i} \text{ grad}$$

and instead of F we write the function

$$F* = \frac{2\pi i}{\lambda} F .$$

(48.34)

It follows that

$$u = m \times \nabla F*$$

$$v = m\nabla^2 F* - \nabla(m \cdot \nabla F*) .$$

(48.35)

The wave which corresponds to an electromagnetic field in geometric optics of the form

$$P^* = (m \times \rho)\phi$$

$$Q^* = \left(m\rho^2 - (m \cdot \rho)\rho\right)\phi$$

(48.36)

thus can be found by replacing the vector ρ in (48.36) by the operator

$$\nabla = \frac{\lambda}{2\pi i}\left(\frac{\partial}{\partial x}, \frac{\partial}{\partial y}, \frac{\partial}{\partial z}\right)$$

(48.37)

and the function ϕ by the transformed function

$$F^* = \frac{i}{\lambda}\int\int \phi\,(p,q)e^{ik(W + xp + yq + zr)}\,dpdq \ .$$

(48.38)

48.4 We consider an unpolarized light wave as an assembly of linearly polarized waves of the above type (48.32) in which the unit vectors m are distributed at random. Let us determine the average energy and the average flux of this assembly of waves. The electric energy of an individual polarized wave is given by the expression

$$\frac{n^2}{16\pi}|u|^2 = \frac{n^2}{16\pi}\left(|m_2 F_z - m_3 F_y|^2 + |m_3 F_x - m_1 F_z|^2 + |m_1 F_y - m_2 F_x|^2\right) \ .$$

(48.41)

We take the mean value of this function of m_1, m_2, m_3 over the unit sphere $m_1^2 + m_2^2 + m_3^2 = 1$. Since

$$\frac{1}{4\pi}\int\int m_i\,m_k\,d\omega = \begin{cases} 0, \ i \neq k \\[2mm] 1/3, \ i = k \end{cases}$$

(48.42)

it follows that the average density of the electric energy is given by

$$w_E = \frac{n^2}{24\pi}\left(|F_x|^2 + |F_y|^2 + |F_z|^2\right) \ .$$

(48.43)

Now let us consider the energy which is contained in a domain D of the x,y,z space. From Green's theorem we have

$$\int\int\int w_E\,dxdydz = -\frac{n^2}{24\pi}\int\int\int_D \bar{F}\Delta F\,dxdydz + \frac{n^2}{24\pi}\int\int_\Gamma \bar{F}\frac{\partial F}{\partial \nu}\,d\sigma$$

(48.44)

where Γ is the boundary surface of D. Since F satisfies the equation $\Delta F = - k^2 n^2 F$ we obtain

$$\iiint_D \mathcal{W}_E dxdydz = \frac{k^2 n^4}{24\pi} \iiint_D |F|^2 dxdydz + \frac{n^2}{24\pi} \iint_\Gamma \overline{F} \frac{\partial F}{\partial \nu} d\sigma . \qquad (48.45)$$

From the definition (48.33) of F it follows that the derivative $\frac{\partial F}{\partial \nu}$ on Γ is of the order of magnitude of $k = \frac{2\pi}{\lambda}$. This implies that the surface integral in (48.45) is of the same order of magnitude and is consequently small compared with the first integral in (48.45) if the wave length is small. Since this is the case in optical problems we can consider the electric energy of an unpolarized light wave to be given to a sufficient approximation by the expression

$$\mathcal{W}_E = \frac{k^2 n^4}{24\pi} |F|^2 . \qquad (48.46)$$

48.5 The <u>average magnetic energy</u> of the unpolarized wave can be obtained in a similar manner. In (48.32) we replace ΔF by $- k^2 n^2 F$ and determine the expression, $\frac{1}{16\pi} |v|^2$, which gives the magnetic energy of an individual polarized wave as a quadratic function of m_1, m_2, m_3. We then take the mean value of this function over the unit sphere and obtain the average energy in the form

$$\mathcal{W}_H = \frac{1}{24\pi k^2} \left[n^4 k^4 |F|^2 + |F_{xx}|^2 + |F_{yy}|^2 + |F_{zz}|^2 + 2|F_{xy}|^2 + 2|F_{xz}|^2 + 2|F_{yz}|^2 \right] .$$

$$(48.51)$$

By Green's formula we find

$$\iiint_D \left(|F_{xx}|^2 + |F_{xy}|^2 + |F_{xz}|^2 \right) dxdydz = n^2 k^2 \iiint_D |F_x|^2 dxdydz + \iint_\Gamma \overline{F}_x \frac{\partial F_x}{\partial \nu} d\sigma$$

$$\iiint_D \left(|F_{yx}|^2 + |F_{yy}|^2 + |F_{yz}|^2 \right) dxdydz = n^2 k^2 \iiint_D |F_y|^2 dxdydz + \iint_\Gamma \overline{F}_y \frac{\partial F_y}{\partial \nu} d\sigma$$

$$\iiint_D \left(|F_{zx}|^2 + |F_{zy}|^2 + |F_{zz}|^2 \right) dxdydz = n^2 k^2 \iiint_D |F_z|^2 dxdydz + \iint_\Gamma F_z \frac{\partial F_z}{\partial \nu} d\sigma$$

$$(48.52)$$

and conclude for reasons similar to the above that $\mathcal{W}_H$ is given in sufficient approximation by

$$\mathcal{W}_H = \frac{1}{24\pi k^2} \left[n^4 k^4 |F|^2 + n^2 k^2 \left(|F_x|^2 + |F_y|^2 + |F_z|^2 \right) \right] . \tag{48.53}$$

To the expression $|F_x|^2 + |F_y|^2 + |F_z|^2$ we apply the result of the preceeding section 48.4. This yields

$$\mathcal{W}_H = \frac{n^4 k^2}{24\pi} |F|^2 \tag{48.54}$$

which demonstrates that we can consider the average electric and magnetic energies of an unpolarized light wave to be equal. The electromagnetic energy $\mathcal{W} = \mathcal{W}_E + \mathcal{W}_H$ thus is given by the formula

$$\mathcal{W} = \frac{n^4 k^2}{12\pi} |F|^2 . \tag{48.55}$$

48.6 We finally consider the flux vector $\frac{c}{16\pi} (u \times \overline{v} + \overline{u} \times v)$. The components of this vector are quadratic functions of m_1, m_2, m_3. We determine the mean values S_1, S_2, S_3 of these functions and obtain a vector

$$S = (S_1, S_2, S_3) \tag{48.61}$$

which we interpret as the vector of average flux of the unpolarized light wave. We find, for example, the expression for the component S_1

$$S_1 = \frac{c}{48\pi i k} \begin{bmatrix} \overline{F}_x F_{xx} + \overline{F}_y F_{xy} + \overline{F}_z F_{xz} - n^2 k^2 \overline{F}_x F \\ - F_x \overline{F}_{xx} - F_y \overline{F}_{xy} - F_z \overline{F}_{xz} + n^2 k^2 F_x \overline{F} \end{bmatrix} . \tag{48.62}$$

In order to simplify this result for the case of small wave lengths let us determine the flux through a finite section Γ of a plane, which is normal to the unit vector (a_1, a_2, a_3). The flux is given by

$$\iint_{\Gamma} (S_1 a_1 + S_2 a_2 + S_3 a_3) d\sigma = \iint_{\Gamma_1} S_1 \, dydz + \iint_{\Gamma_2} S_2 \, dzdx + \iint_{\Gamma_3} S_3 dxdy \tag{48.63}$$

where $\Gamma_1, \Gamma_2, \Gamma_3$ are the projections of Γ onto the three coordinate planes. With the aid of the relations

$$\iint_{\Gamma_1} (\overline{F}_y F_{xy} + \overline{F}_z F_{xz}) dydz = - \iint_{\Gamma_1} F_x (\overline{F}_{yy} + \overline{F}_{zz}) dydz + \int_{\nu_1} F_x \frac{\partial \overline{F}}{\partial \nu} ds$$

$$\iint_{\Gamma_1} (F_y \overline{F}_{xy} + F_z \overline{F}_{xz}) dydz = - \iint_{\Gamma_1} \overline{F}_x (F_{yy} + F_{zz}) dydz + \int_{\nu_1} \overline{F}_x \frac{\partial F}{\partial \nu} ds \tag{48.64}$$

we find

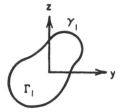

$$\iint_{\Gamma_1} S_1 \, dy \, dz = \frac{n^2 k^2 c}{24\pi i k} \iint_{\Gamma_1} (F_x \overline{F} - \overline{F}_x F) \, dy \, dz$$

$$+ \frac{c}{48\pi i k} \int_{\nu_1} \left(F_x \frac{\partial \overline{F}}{\partial \nu} - \overline{F}_x \frac{\partial F}{\partial \nu} \right) ds \ . \tag{48.65}$$

Figure 191   As before we conclude that the predominant contribution in (48.65) is given by the first integral on the right side. This allows us to consider the expression

$$S_1 = \frac{n^2 k^2 c}{24\pi i k} (F_x \overline{F} - \overline{F}_x F) = \frac{n^2 k^2 c}{24\pi i k} |F|^2 \frac{\partial}{\partial x} \left(\log \frac{F}{\overline{F}} \right) \tag{48.66}$$

to be a sufficiently good approximation to the component S_1. By assuming F in the form

$$F = |F| e^{ikX} \tag{48.67}$$

so that kX (x,y,z) is the phase angle of the complex function F (x,y,z) we can write (48.66) as follows

$$S_1 = \frac{n^2 k^2 c}{12\pi} |F|^2 \frac{\partial X}{\partial x} = \frac{c \, \mathcal{W}}{n^2} \frac{\partial X}{\partial x} \ . \tag{48.68}$$

Similar results are found for the components S_2 and S_3 so that the average flux vector S is given, with sufficient accuracy, by the relation

$$S = \frac{c \, \mathcal{W}}{n^2} \operatorname{grad} X \ . \tag{48.69}$$

Obviously this last relation is a direct generalization of the corresponding formula $S = \frac{c \, \mathcal{W}}{n^2} \operatorname{grad} \psi$ in geometrical optics. The vector grad X in (48.69) however does <u>not</u> in general coincide with the direction of the light ray. This illustrates that by diffraction light may penetrate into the regions of geometrical shadow.

48.7 The above results allow us to investigate the diffraction phenomena of unpolarized light waves with the aid of the scalar function F defined by the integral (48.33). In order to get the observable characteristics of such a wave it is not necessary to determine the individual polarized elements of the wave, i.e., the vectors u and v as functions of m_1, m_2, m_3. These wave characteristics are completely determined by the scalar function $F(x,y,z)$. The distribution of the light energy is given by $|F|^2$ and the energy flux by the phase angle X of this function

$$F = |F| e^{ikX}. \tag{48.71}$$

The following applications of the theory are carried out on the basis of this result.

§49. DIFFRACTION PATTERNS FOR DIFFERENT TYPES OF ABERRATIONS.

49.1 <u>Spherical aberration</u>. First we consider the case of a point source on the axis of a system of rotational symmetry. The mixed characteristic W is a function of

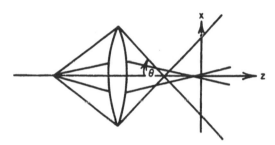

Figure 192

$$\rho = \sqrt{p^2 + q^2}.$$

We can assume that the function $\phi(p,q)$ in (48.33) also depends on this combination alone. We introduce polar coordinates

$$p = \rho \cos \varphi$$
$$q = \rho \sin \varphi \tag{49.11}$$

and obtain the integral

$$F(r) = \int_0^{\rho_0} \phi(\rho) J_0(k\rho r) e^{ik\left(W(\rho) + z\sqrt{n^2 - \rho^2}\right)} \rho d\rho \tag{49.12}$$

where $r = \sqrt{x^2 + y^2}$ and J_0 is Bessel's function of order zero. The limit of integration, ρ_0, is determined by the aperture of the objective. We have

$$\rho_0 = n \sin \theta \tag{49.13}$$

where θ is the maximum angle which is subtended by the refracted rays with the axis.

The function $W(\rho)$ can be obtained from the lateral spherical aberration $\ell(\rho)$ of the bundle with the aid of the formula (refer to §31.1)

$$W(\rho) = W(0) + \int_0^{\rho} \ell(\rho) d\rho. \tag{49.14}$$

If the Gaussian image plane is chosen as the plane $z = 0$ we have the development

$$W(\rho) \doteq W_0 + \alpha\rho^4 + \beta\rho^6 + \ldots . \tag{49.15}$$

The coefficients α, β, ... can be found by determining several points of the curve $\ell(\rho)$ by tracing rays.

In practice one generally assumes $\phi(\rho) = 1$ and considers the integral

$$F(r) = \int_0^{\rho_0} J_0 \, (k\rho r) e^{ik\left[W(\rho) + z \sqrt{n^2 - \rho^2}\right]} \rho \, d\rho \tag{49.16}$$

sufficient for calculating the light distribution of the refracted axial bundle.

The diffraction pattern in the Gaussian image plane is given by

$$F(r) = \int_0^{\rho_0} J_0 \, (k\rho r) e^{ikW} \rho \, d\rho . \tag{49.17}$$

For a bundle which is free from spherical aberration, so that $W = W_0$, we obtain

$$F(r) = e^{ikW_0} \int_0^{\rho_0} J_0 \, (k\rho r) \rho \, d\rho$$

or

$$F(r) = \frac{\rho_0}{kr} J_1 (k\rho_0 r) e^{ikW_0} \tag{49.18}$$

where J_1 is Bessel's function of order one. The light energy $\mathcal{W}$ of the diffraction pattern in this ideal case is given by

$$\mathcal{W}(r) = \frac{n^4 \pi}{12\lambda^2} \rho_0^4 \left(\frac{2J_1\left(\frac{2\pi r}{\lambda}\rho_0\right)}{\frac{2\pi r}{\lambda}\rho_0} \right)^2 \tag{49.19}$$

which follows from letting $k = \frac{2\pi}{\lambda}$ in (48.55). At the center $r = 0$ we have

$$\mathcal{W}(0) = \frac{n^4 \pi}{12} \frac{\rho_0^4}{\lambda^2} \tag{49.191}$$

which shows that the energy increases proportionally to the inverse square of λ and proportionally to the fourth power of the aperture ρ_0. The function

$$D(r) = \left[\frac{2J_1\left(\frac{2\pi}{\lambda}\rho_0 r\right)}{\frac{2\pi}{\lambda}\rho_0 r} \right]^2 \tag{49.192}$$

measures the relative energy distribution $D(r) = \dfrac{\mathcal{W}(r)}{\mathcal{W}(0)}$. It attains the value zero at

$$\frac{2\pi}{\lambda}\rho_0 r = (0.61)\,2\pi$$

i.e., at

$$r = 0.61\,\frac{\lambda}{n\,\sin\theta}. \tag{49.193}$$

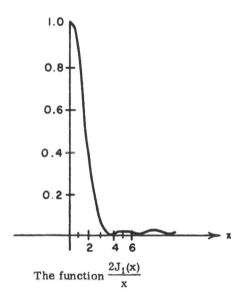

The function $\dfrac{2J_1(x)}{x}$

Figure 193

This value of r is often considered as the limit of resolution for the reason that two point sources are difficult to separate if the diffraction patterns of their images are located so that the central maximum of the one lies over the first minimum of the other.

The concentration of the energy of the diffraction pattern deteriorates rapidly in the presence of spherical aberration. This is caused by the fact that the function e^{ikW} oscillates several times in the interval of integration if $W(\rho) - W(0)$ reaches values greater than λ. The effect on $F(r)$ is to decrease its values in the neighborhood of $r = 0$. This effect however is not noticeable in practical observation if $W(\rho)$ does not exceed the value $1/4\lambda$. This gives us the so called Rayleigh Limit: An objective can be considered as sufficiently corrected for spherical aberration if, in the interval $0 \leq \rho \leq \rho_0$, the function $W(\rho)$ satisfies the inequality

$$|W(\rho) - W(0)| < \frac{1}{4}\,\lambda\,.$$

If this condition is not satisfied it is still possible to obtain satisfactory definition by changing the location of the image plane to a position $z \neq 0$. This leads to a generalization of (49.194): <u>If there exists a value z such that</u>

$$\left| W(\rho) - W(0) + z \sqrt{n^2 - \rho^2} \right| < \frac{1}{4} \lambda \qquad (49.195)$$

<u>then we can consider the correction of the spherical aberration satisfactory</u>.

We mention that in photographic objectives one can allow a considerably greater tolerance, namely 4 or 5 wave lengths, without impairing the suitability of the objective for its purpose.

49.2 <u>Coma</u>: We consider next a light wave which originates at a point source at the point $(x_0, 0)$ of the object plane. For small values of x_0 we have, according to §31.34, the expression

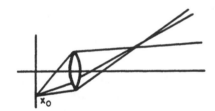

Figure 194

$$W = W_0(\rho) - x_0 p\, M(\rho) \qquad (49.21)$$

where $W_0(\rho) = \int_0^\rho \ell(\rho)\, d\rho$

and where $M(\rho)$ is the zonal magnification. By introducing (49.21) in (48.33) we find

$$F = \int_0^{\rho_0} J_0 \left(k\rho \ \sqrt{(x - x_0 M(\rho))^2 + y^2} \right) e^{ik\left[W_0(\rho) + z\sqrt{n^2 - \rho^2} \right]} \rho\, d\rho$$

$$(49.22)$$

assuming $\phi(p,q) = 1$. If the axial bundle is free of spherical aberration we obtain in the plane $z = 0$ the <u>diffraction pattern of pure coma</u>:

$$F = e^{ikW_0} \int_0^{\rho_0} J_0 \left(k\rho \ \sqrt{(x - x_0 M(\rho))^2 + y^2} \right) \rho\, d\rho \ . \qquad (49.23)$$

The function $M(\rho)$ has the development

$$M(\rho) = M_0 + \alpha\rho^2 + \beta\rho^4 + \ldots \qquad (49.24)$$

and can be determined numerically by tracing axial rays.

49.3 <u>Astigmatism</u>. We finally apply the formula (48.33) to an anastig-
matic bundle of rays. Let us assume that a narrow manifold of rays about a

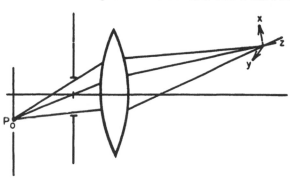

principal ray is selected
by a small diaphragm.
The principal ray of the
bundle shall be an oblique
meridional ray in a
system of revolution. We
consider the wave which
originates at a point
source located at a point
P_n on the principal ray.
The mixed characteristic
W in this case can be
developed in the following
form (refer to §36.1)

Figure 195

$$W = W_0 + \frac{1}{2}(C_1 p^2 + C_2 q^2) + \dots \tag{49.31}$$

where the dots indicate polynomials of p and q of an order higher than 2. With
a suitable choice of the origin of the coordinate system on the refracted
principal ray we can assume that $C_1 = -C_2$ and thus we have without loss of
generality

$$W = W_0 + \frac{1}{2}C(p^2 - q^2) . \tag{49.32}$$

We introduce this expression in (48.33) and, letting $\phi(p,q) = 1$, we find the
integral

$$F = \frac{1}{2\pi} e^{ikW_0} \iint e^{ik\left[\frac{1}{2}C(p^2-q^2)+xp+yq+zr\right]} dp\, dq . \tag{49.33}$$

We assume n = 1 and in the chosen approximation we may replace r by
$1 - \frac{1}{2}(p^2 + q^2)$. It follows that

$$F = \frac{1}{2\pi} e^{ik(W_0+z)} \iint e^{ik\left[-\frac{1}{2}(z-C)p^2-\frac{1}{2}(z+C)q^2+xp+yq\right]} dp\, dq . \tag{49.34}$$

This integral can be reduced to a well known type if the diaphragm is rectan-
gular. In this case we find that the light distribution is proportional to the
product of two functions:

$$|F|^2 = |A(x)|^2\, |B(y)|^2 \tag{49.35}$$

where

$$A(x) = \frac{1}{\sqrt{2\pi}} \int_{-p_0}^{p_0} e^{ik\left[xp-\frac{1}{2}(z-C)p^2\right]} dp$$

$$B(y) = \frac{1}{\sqrt{2\pi}} \int_{-q_0}^{q_0} e^{ik\left[yq-\frac{1}{2}(z+C)q^2\right]} dq \ .$$

(49.36)

In the plane z = C, i.e., at the position of the primary focal line we have

$$A(x) = \sqrt{\frac{2}{\pi}} \frac{\sin kp_0 x}{kx}$$

$$B(y) = \frac{1}{\sqrt{2\pi}} \int_{-q_0}^{q_0} e^{ik\left[yq-Cq^2\right]} dq \ .$$

(49.37)

Similarly at the position z = - C of the secondary focal line:

$$A(x) = \frac{1}{\sqrt{2\pi}} \int_{-p_0}^{p_0} e^{ik\left[xp + Cp^2\right]} dp$$

$$B(y) = \sqrt{\frac{2}{\pi}} \frac{\sin kq_0 y}{ky} \ .$$

(49.38)

The remaining integrals in (49.37) and (49.38) and the integrals (49.36) in case z ≠ ± C can be easily evaluated with the aid of the function

$$X(\sigma) + iY(\sigma) = Z(\sigma) = \int_0^\sigma e^{is^2} ds \ .$$

(49.39)

The curve $x = x(\sigma)$; $y = y(\sigma)$ which is determined by (49.39) is known as Cornu's Spiral.

§50. RESOLUTION OF TWO LUMINOUS POINTS OF EQUAL INTENSITY.

50.1   The diffraction pattern of a wave which corresponds to a given system of wave fronts depends on the function $\phi(p,q)$ in (48.33), i.e., on the intensity and phase of the wave in the direction p,q. In this section we shall study the influence of the function $\phi(p,q)$ on the form of the diffraction pattern. Many ways exist in practice to change $\phi(p,q)$ without influencing the light rays or the wave fronts essentially, for example, by evaporating thin films of metallic or dielectric substances onto the lenses.

Let us therefore assume that $W(p,q)$ is a given function. We choose a plane z = const. as the plane of observation and write our integral (48.33) in the form

$$F(\xi,\eta) = \frac{1}{2\pi} \iint_D A(p,q) \, e^{i(p\xi + q\eta)} \, dp \, dq \tag{50.11}$$

where

$$A(p,q) = \phi(p,q)e^{ik \left(W+z \sqrt{n^2-p^2-q^2} \right)} \tag{50.12}$$

and where ξ and η are defined by

$$\xi = kx = 2\pi \frac{x}{\lambda}$$
$$\eta = ky = 2\pi \frac{y}{\lambda} \ . \tag{50.13}$$

Since $\phi(p,q)$ can be assigned quite arbitrarily in practice we can consider $A(p,q)$ itself as a function which we are free to choose in the domain D of integration. This domain D is determined by the aperture of the optical instrument; we know that it must lie inside the circle $p^2 + q^2 \leq n^2$.

We may define $A(p,q)$ as a function in the whole p,q plane which is zero outside D. The function $F(x,y)$ then is obtained by a Fourier transformation of $A(p,q)$. The manifold of all possible diffraction patterns of an optical instrument is thus given by the Fourier adjoints of all functions $A(p,q)$ which are zero outside D.

From a general theorem for Fourier integrals we conclude that

$$\iint_{-\infty}^{\infty}|F|^2 \, d\xi \, d\eta = \iint_D |A|^2 \, dp \, dq \tag{50.14}$$

which we can interpret as follows: <u>The total light energy of the diffraction pattern is equal to the energy radiated through the aperture of the instrument.</u>

50.2 In the following we shall consider only the case of rotational symmetry, i.e., the diffraction patterns in the immediate neighborhood of the axis of an optical instrument. The domain D becomes a circle $p^2 + q^2 \leq \rho_0^2$ and $A(p,q)$ is a function $A(\rho)$. In this case we obtain F as a function of $r = \sqrt{\xi^2 + \eta^2}$ namely

$$F(r) = \int_0^{\rho_0} \rho A(\rho)J_0(\rho r)d\rho \ . \tag{50.21}$$

The energy relation (50.14) is replaced by

$$\int_0^\infty r|F|^2 dr = \int_0^{\rho_0} \rho\,|A|^2 d\rho \ . \tag{50.22}$$

The light distribution in the diffraction pattern is proportional to $D = |F|^2$; we define the relative light distribution by the ratio

$$R(r) = \frac{D(r)}{D(0)} \ . \tag{50.23}$$

In the special case $A(\rho) = 1$ the function $R(r)$ is given by the expression

$$R(r) = \left(\frac{2J_1(\rho_0 r)}{\rho_0 r}\right)^2 . \tag{50.24}$$

50.3 Let us now consider two independent, i.e., incoherent point sources which are located symmetrically to the axis at two points of the object plane. Let us assume that both point sources are of equal intensity so that the energy distributions $D\left(\sqrt{(\xi-\delta)^2+\eta^2}\right)$ and $D\left(\sqrt{(\xi+\delta)^2+\eta^2}\right)$ are determined by the same function $D(r)$. The assumption that the point sources are incoherent has the consequence that the combined light distribution of both point sources is given by the sum of the energies:

$$D\left(\sqrt{(\xi-\delta)^2 + \eta^2}\right) + D\left(\sqrt{(\xi+\delta)^2 + \eta^2}\right) \tag{50.31}$$

i.e., the sum of the squares of the functions $|F|$.

In the case of two coherent point sources, which may be obtained by illuminating two pin holes by a small source one can show that the combined light distribution has to be found from the expression

$$\left|F\left(\sqrt{(\xi-\delta)^2 + \eta^2}\right) + F\left(\sqrt{(\xi+\delta)^2 + \eta^2}\right)\right|^2 \tag{50.32}$$

i.e., by the square of the sum of the functions F.

On the line $\eta = 0$ which connects the two central maxima of the two individual diffraction patterns we have the light distributions

$$D(\xi-\delta) + D(\xi+\delta)$$

and   (50.33)

$$|F(\xi-\delta) + F(\xi+\delta)|^2$$

respectively.

In both cases these curves have the form illustrated in Figure 196, namely a curve with two maxima if δ is large enough and with only one maximum at $\xi = 0$ if δ is smaller than a certain value δ_0. We can consider $2\delta_0$ as the limit of resolution. Indeed if $\delta > \delta_0$ then the two bright maxima of the distributions (50.31) and (50.32) are separated by a strip of lower intensity. If $\delta < \delta_0$ however, this strip is no longer present as is shown in Figure 196. The intensity curves (50.33) at $\xi = 0$ are concave in case $\delta > \delta_0$ and convex if $\delta < \delta_0$. It follows that the second derivative at $\xi = 0$ must be zero for the curve which belongs to $\delta = \delta_0$. This leads to an equation for δ_0. For

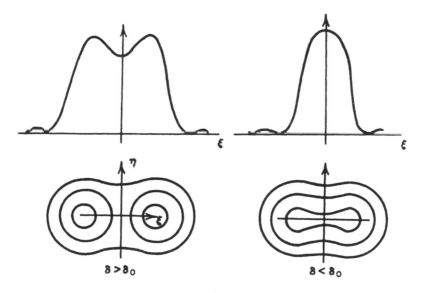

Figure 196

incoherent sources we find

$$D''(\delta_0) + D''(-\delta_0) = 0$$

or since $D(\xi)$ and $D''(\xi)$ are even functions we obtain the equation

$$D''(\delta_0) = 0. \tag{50.34}$$

In other words: the limit of resolution, $2\delta_0$, is given by the separation of the two inflection points of the intensity curve $D = D(r)$ of a single point source. (Fig. 197)

In the case of coherent light sources we find by a similar consideration the condition

$$F''(\delta_0)\overline{F}(\delta_0) + F(\delta_0)\overline{F}''(\delta_0) = 0 . \tag{50.35}$$

If $F(r)$ is a real function then (50.35) reduces to $F''(\delta_0) = 0$, i.e., $2\delta_0$ is the

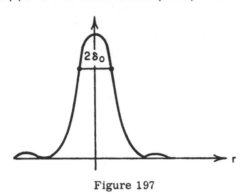

distance of the two inflection points on the curve $F = F(r)$. Since $D = F^2$ it is quite clear that the distance of the inflection points for $D = D(r)$ is smaller than for $F = F(r)$. This demonstrates that it will be more difficult to resolve coherent light sources than incoherent sources, i.e., self luminous points. We shall verify this result in the next section by another approach.

Figure 197

If, for example, $F(r)$ is given by the normal function

$$F(r) = \frac{2J_1(\rho_0 r)}{\rho_0 r} \tag{50.36}$$

we obtain

a) $2\delta_0 = 0.8\dfrac{2\pi}{\rho_0}$ in the case of non–self–luminous points

b) $2\delta_0 = 0.52\,\dfrac{2\pi}{\rho_0}$ in the case of self–luminous points.

This leads to the following formulae for the limits of resolution:

a) $r_0 = 0.8\,\dfrac{\lambda}{n\sin\theta}$

b) $r_0 = 0.52\,\dfrac{\lambda}{n\sin\theta}$ (50.37)

where $r_0 = \sqrt{x^2 + y^2} = \dfrac{1}{k}\sqrt{\xi^2 + \eta^2}$. The last formula shows that the present limit of resolution is smaller than the one given in (49.193) which was based upon the distance of the first minimum of $D(r)$ from the center.

50.4 We continue the investigation of the integral (50.21). Let us assume that the functions $A(\rho)$ satisfy the condition

$$\int_0^{\rho_0} \rho\,|A(\rho)|^2 d\rho = 1 \tag{50.41}$$

so that the total energy in our diffraction patterns has a given fixed value. By Schwartz's inequality it follows that

$$|F(r)|^2 \leq \left(\int_0^{\rho_0} \rho J_0^2(\rho r)\, d\rho \right) \left(\int_0^{\rho_0} \rho |A|^2\, d\rho \right)$$

or on account of (50.41):

$$|F(r)|^2 \leq \int_0^{\rho_0} \rho J_0^2(\rho r)\, d\rho \ . \tag{50.42}$$

This yields with the aid of the identity

$$\int_0^1 t J_0^2(xt)\, dt = \frac{1}{2}\left(J_0^2(x) + J_1^2(x) \right) \tag{50.43}$$

the inequality

$$0 \leq |F(r)|^2 \leq \frac{1}{2}\rho_0^2 \left(J_0^2(\rho_0 r) + J_1^2(\rho_0 r) \right) \tag{50.44}$$

i.e., all curves $D = |F|^2$ must lie in the domain of the D,r plane which is determined by the r-axis and the curve $\frac{1}{2}\rho_0^2 \left(J_0^2(\rho_0 r) + J_1^2(\rho_0 r) \right)$. For $r = 0$ we have

$$|F(0)|^2 \leq \frac{1}{2}\rho_0^2 \ . \tag{50.45}$$

An equality can only be obtained if A = a constant. This means: <u>Among all diffraction patterns (50.21) of equal total energy the highest central maximum is obtained by the normal pattern (50.24) i.e. for $A(\rho)$ = a constant.</u>

50.5 The normal curve (50.24) reaches its first zero at the point $r_0 = 0.61 \frac{2\pi}{\rho_0}$. Let us attempt to increase the resolving power of our instrument by applying a "coat" $A(\rho)$ to the aperture so that the corresponding function $|F|^2$ reaches its first zero at a point $r = \alpha < r_0$. From the above result it follows that such a function $D = |F|^2$ must have a lower central maximum than the normal curve. We are of course interested in keeping the central maximum as high as possible. Thus we are led to the following problem of variation:

<u>To find a function $A(\rho)$ which satisfies the conditions</u>

$$\int_0^{\rho_0} \rho |A(\rho)|^2\, d\rho = 1 \qquad \int_0^{\rho_0} \rho A(\rho) J_0(\rho\alpha)\, d\rho = 0 \tag{50.51}$$

such that the integral

$$|F(0)|^2 = \left| \int_0^{\rho_0} \rho A(\rho) \, d\rho \right|^2 \tag{50.52}$$

assumes a maximal value.

One can see easily that $A(\rho)$ can be assumed to be real. All other solutions are then given in the form $cA(\rho)$ where c is a constant of absolute value $|c| = 1$.

We introduce two Lagrangian multipliers λ and μ and consider the problem of variation

$$V = \left(\int_0^{\rho_0} \rho A(\rho) \, d\rho \right)^2 + \lambda \int_0^{\rho_0} \rho A(\rho) J_0 (\rho\alpha) d\rho + \mu \int_0^{\rho_0} \rho A^2 d\rho = \text{Max.} \tag{50.53}$$

Let us assume that $A(\rho)$ is the solution of this problem. We choose an arbitrary function $\zeta(\rho)$ and introduce in (50.33) the variation function $A(\rho) + \epsilon\zeta(\rho)$. The integral V becomes a function $V = V(\epsilon)$ which has a maximum for $\epsilon = 0$. The necessary condition $V'(0)$ leads to the following equation.

$$\int_0^{\rho_0} \rho \zeta(\rho) \left\{ 2 \int_0^{\rho_0} \rho A \, d\rho + \lambda J_0 (\rho\alpha) + 2\mu A(\rho) \right\} d\rho = 0 \tag{50.54}$$

which must be satisfied for every function $\zeta(\rho)$. It follows that the bracket in (50.54) is zero. This means that $A(\rho)$ has the form

$$A(\rho) = \sigma - \tau J_0 (\alpha\rho) \tag{50.55}$$

where σ and τ are constants, which can be found by introducing (50.55) in the equations (50.51). The second equation (50.51) yields

$$\frac{\sigma}{\tau} = \frac{\displaystyle\int_0^{\rho_0} \rho J_0^2 (\rho\alpha) d\rho}{\displaystyle\int_0^{\rho_0} \rho J_0 (\rho\alpha) d\rho} = \frac{1}{2}\rho_0 \, \alpha \, \frac{J_0^2 (\rho_0\alpha) + J_1^2 (\rho_0\alpha)}{J_1 (\rho_0\alpha)} \, . \tag{50.56}$$

Hence letting

$$\sigma_0 = J_0^2 (\rho_0\alpha) + J_1^2 (\rho_0\alpha)$$

$$\tau_0 = \frac{2J_1 (\rho_0\alpha)}{\rho_0\alpha} \tag{50.57}$$

we obtain for $A(\rho)$ the expression

$$A(\rho) = \Gamma \left(\sigma_0 - \tau_0 J_0 (\alpha \rho) \right) \tag{50.58}$$

where the constant Γ has to be found by the first condition (50.51). The diffraction pattern $F(r)$ is determined by the integral

$$F(r) = \Gamma \int_0^{\rho_0} \rho \left(\sigma_0 - \tau_0 J_0 (\rho \alpha) \right) J_0 (\rho r) d\rho . \tag{50.59}$$

By a suitable choice of α it is indeed possible to decrease the limit of resolution $2\delta_0$ of our instrument. A numerical investigation has shown however that for $\alpha < 0.31 \dfrac{2\pi}{\rho_0}$ the functions $|F|^2$ have too low a central maximum.

50.6 We can apply a more direct method of decreasing the limit of resolution $2\delta_0$. Again let us consider only real functions $A(\rho)$. We have seen that δ_0 is a solution of the equation

$$D''(\delta_0) = F(\delta_0)F''(\delta_0) + \left(F'(\delta_0) \right)^2 = 0 . \tag{50.61}$$

By introducing the integral expressions for F we obtain the equation

$$D''(\delta_0) = \left(\int_0^{\rho_0} \rho A(\rho) J_0 (\rho \delta_0) d\rho \right) \left(\int_0^{\rho_0} \rho^3 A(\rho) J_0 ''(\rho \delta_0) d\rho \right) + \left(\int_0^{\rho_0} \rho^2 A(\rho) J_0' (\rho \delta_0) d\rho \right)^2 = 0 . \tag{50.62}$$

This is a condition for $A(\rho)$ if δ_0 is pre-assigned. Again we have to expect a decrease of the central maximum of $|F(r)|^2$ if δ_0 is chosen smaller than the corresponding value in the case of the normal curve (50.35). Therefore we try to find the function $A(\rho)$ which satisfies (50.62) and which makes $|F(0)|^2$ as great as possible. This leads us to the following problem of variation:

To find a function $A(\rho)$ which satisfies the equation (50.62) and the condition

$$\int_0^{\rho_0} \rho |A|^2 d\rho = 1 \tag{50.63}$$

such that the integral

$$|F(0)|^2 = \left| \int_0^{\rho_0} \rho A(\rho) d\rho \right|^2 \tag{50.64}$$

assumes a maximal value.

We introduce two Lagrangian multipliers λ and μ and consider the problem

$$V = \left(\int_0^{\rho_0} \rho A(\rho) d\rho \right)^2 + \lambda D''(\delta_0) + \mu \int_0^{\rho_0} \rho A^2 d\rho = \text{Extr.} \qquad (50.65)$$

where $D''(\delta_0)$ is the quadratic form (50.62).

We assume that $A(\rho)$ is the solution and introduce the variation function $A(\rho) + \epsilon \zeta(\rho)$. The condition $V'(0) = 0$ leads to the equation

$$\int_0^{\rho_0} \rho \zeta(\rho) \left\{ 2 \int_0^{\rho_0} \rho A \, d\rho + \lambda \left(J_0(\rho\delta_0) \int_0^{\rho_0} \rho^3 A J_0''{}' d\rho + \rho^2 J_0' \int_0^{\rho_0} \rho A J_0 \, d\rho \right. \right.$$

$$\left. \left. + 2\rho J_0' \int_0^{\rho_0} \rho^2 A J_0' d\rho \right) + 2\mu A \right\} d\rho = 0 \qquad (50.66)$$

which must hold for every function $\zeta(\rho)$. It follows that the $\{\}$ bracket is identically zero thus yielding the following expression for $A(\rho)$:

$$A(\rho) = \sigma_0 + \sigma_1 J_0(\rho\delta_0) + \sigma_2 \rho J_0'(\rho\delta_0) + \sigma_3 \rho^2 J_0''(\rho\delta_0) \qquad (50.67)$$

with constant values of σ_1.

We conclude from (50.66) that $\sigma_1, \sigma_2, \sigma_3$ satisfy the equations

$$\frac{\sigma_1}{\sigma_3} = \frac{\displaystyle\int_0^{\rho_0} \rho^3 A J_0''(\rho\delta_0) d\rho}{\displaystyle\int_0^{\rho_0} \rho A J_0(\rho\delta_0) d\rho}$$

$$\frac{\sigma_2}{\sigma_3} = \frac{\displaystyle 2\int_0^{\rho_0} \rho^2 A J_0'(\rho\delta_0) d\rho}{\displaystyle\int_0^{\rho_0} \rho A J_0(\rho\delta_0) d\rho} \qquad . \qquad (50.68)$$

By introducing the expression (50.67) in these equations and in (50.62) and (50.63) we obtain four quadratic equations which allow us to calculate the constants $\sigma_0, \sigma_1, \sigma_2, \sigma_3$ for any given δ_0 and ρ_0.

50.7 Finally we mention a third method of finding functions $A(\rho)$ which give an increased resolution. We consider a circle $r < \delta$ in the ξ, η plane.

The total energy contents of this circle is proportional to the integral

$$\mathcal{W} = \int_0^\delta |F|^2 \, r \, dr \; .$$

(50.71)

We introduce the integral expression (50.21) for F and obtain the quadratic form

$$\mathcal{W} = \iint_0^{\rho_0} K(\rho,\rho')A(\rho)\overline{A}(\rho') \, d\rho \, d\rho'$$

(50.72)

where the Kernel $K(\rho,\rho')$ is the symmetrical function

$$K(\rho,\rho') = \rho\rho' \int_0^\delta rJ_0(\rho r)J_0(\rho' r) \, dr \; .$$

(50.73)

Instead of requiring that the value $|F^2(0)|$ of the central maximum is as high as possible we can ask that the energy contents of a given circle $r < \delta$ shall be as great as possible. We thus obtain the following maximum problem: To find among the functions $A(\rho)$ which satisfy the condition

$$\int_0^{\rho_0} \rho|A|^2 \, d\rho = 1$$

(50.74)

a function for which the quadratic form (50.72) assumes its maximum value.

We introduce the Lagrangian multiplier λ and consider the problem

$$V = \iint_0^{\rho_0} K(\rho,\rho') A(\rho) \, \overline{A}(\rho') \, d\rho d\rho' - \lambda \int_0^{\rho_0} \rho |A|^2 \, d\rho = \text{Extr.}$$

(50.75)

By applying our method of variation we obtain $A(\rho)$ as the solution of the homogeneous integral equation

$$\lambda\rho A(\rho) = \int_0^{\rho_0} K(\rho,\rho') A(\rho') \, d\rho'$$

(50.76)

where λ is the greatest characteristic value of the Kernel $K(\rho,\rho')$ and $A(\rho)$ the corresponding characteristic solution.

§51. RESOLUTION OF OBJECTS OF PERIODIC STRUCTURE.

51.1 We have found in the preceding section that the limit of resolution can be decreased by "coating" the aperture with a function $A(\rho)$. This result was obtained for an object which consists of two separated luminous points. With the aid of the "improved" function $D(r) = |F|^2$ we can now calculate the light distribution which belongs to a larger number of object points, lined up, for example, on the ξ - axis. If the points are equidistant we have the light distribution on the ξ - axis

$$\sum_{\nu} D(\xi - \nu\delta) . \tag{51.11}$$

Let us assume that the distance δ of the image points is considerably smaller than the limit of resolution obtained from the normal function (50.35). Let $D(r)$, however, be a function which allows us to resolve two points of distance δ if no other points are present. By calculating (51.11) we would find that the improvement of the resolution becomes practically unnoticeable if a greater number of points are considered. The strips of low intensity which separate the bright maxima disappear gradually when more points are added.

51.2 Therefore in this section we shall consider the problem of resolution from a more general point of view. We assume a distribution of self luminous objects in the object plane which is characterized by a function $U_0(x_0,y_0)$. In the image plane we obtain a perfect reproduction of the object if the light distribution in this plane is proportional to the function $U(x,y) = U_0\left(\dfrac{x}{M} , \dfrac{y}{M}\right)$ where M is the magnification.

Let us assume that the image of a self luminous point (x_0,y_0) is given by the light distribution

$$D(x - Mx_0, y - My_0) = |F(x - Mx_0, y - My_0)|^2 \tag{51.21}$$

where $F(x,y)$ is the integral

$$F(x,y) = \frac{1}{2\pi} \iint A(p,q)e^{ik(xp+yq)}dpdq . \tag{51.22}$$

It follows that the image $V(x,y)$ of the object $U_0(x_0,y_0)$ is represented by the integral

$$V(x,y) = \iint U(\xi,\eta) D (x - \xi, y - \eta)d\xi d\eta . \tag{51.23}$$

51.3 If the object $U_0(x_0,y_0)$ is not self luminous but illuminated by a wave from a point source, then we do not obtain the image $V(x,y)$ by superposition of the energy functions $D = |F|^2$ from the different points but by superposition of the corresponding functions F. The light from the different object points is coherent in this case and interference of the wavelets F is to be expected. We obtain

$$V(x,y) = \left| \iint U(\xi,\eta)\, F\,(x - \xi, y - \eta) d\xi\, d\eta \right|^2 \tag{51.31}$$

i.e., the "square of the sum" of the wavelets F instead of the "sum of the squares" as in (51.23).

51.4 In many cases we have to deal with objects which consist of similar elements distributed regularly or at random. A network of cells for example or a group of stars. We may consider the function $U(x,y)$ in these cases as periodic or at least as almost periodic. In the following let us therefore study the images $V(x,y)$ which belong to objects of periodic structure. We develop $U(x,y)$ in a Fourier series

$$U(x,y) = \sum_{\nu,\mu} U_{\nu\mu} e^{i\frac{2\pi}{\ell}(\nu x + \mu y)} \tag{51.41}$$

where ℓ is the period of the structure. First we treat the case of self luminous objects. We introduce (51.41) in (51.23) and obtain

$$V(x,y) = \sum_{\nu,\mu} U_{\nu\mu} D_{\nu\mu} e^{i\frac{2\pi}{\ell}(\nu x + \mu y)} \;. \tag{51.42}$$

The quantities $D_{\nu\mu}$ are given by the integrals

$$D_{\nu\mu} = \iint D(\xi,\eta) e^{-i\frac{2\pi}{\ell}(\nu\xi + \mu\eta)} d\xi\, d\eta \;. \tag{51.43}$$

The equation (51.42) demonstrates that a complete reproduction of the object $U(x,y)$ can only be obtained if all the quantities $D_{\nu\mu}$ are equal to unity. This however is never the case. In fact we will find that only a finite number of the $D_{\nu\mu}$'s are different from zero so that (51.42) represents a trigonometric polynomial of finitely many members.

By applying Fourier's inversion theorem to the expression (51.22) we obtain

$$A(p,q) = \frac{k^2}{2\pi} \iint F(x,y) e^{ik(xp + yq)}\, dx\, dy \;. \tag{51.44}$$

In order to determine the quantities $D_{\nu\mu}$ we make use of the following relation which is a direct consequence of (51.44):

$$\iint A(p + s, q + t)\overline{A}(p,q)dpdq = \frac{k^2}{2\pi} \iint |F|^2 e^{-ik(xs+yt)} dxdy \; . \qquad (51.45)$$

Since $D(x,y) = |F(x,y)|^2$ we obtain

$$D_{\nu\mu} = \frac{2\pi}{k^2} \iint A\left(p + \frac{2\pi}{k\ell}\nu, \; q + \frac{2\pi}{k\ell}\mu\right) \overline{A}(p,q)dpdq$$

or letting $k = \frac{2\pi}{\lambda}$ we obtain the equation

$$D_{\nu\mu} = \frac{\lambda^2}{2\pi} \iint A\left(p + \frac{\lambda}{\ell}\nu, \; q + \frac{\lambda}{\ell}\mu\right) \overline{A}(p,q)dpdq \; . \qquad (51.46)$$

The function $A(p,q)$ is zero outside a certain domain Γ of the (p,q) plane which is given by the aperture of the optical instrument. Therefore we can interpret the integration (51.46) as follows. We shift the domain without rotation to a new position, $\Gamma_{\nu\mu}$, in the p,q plane such that the point $p = q = 0$ goes over into the point $p = \frac{\lambda}{\ell}\nu, q = \frac{\lambda}{\ell}\mu$. We define the function $A_{\nu\mu}(p,q)$ by

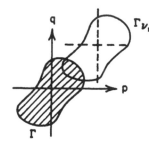

$$\Gamma_{\nu,\mu} \quad A_{\nu\mu}(p,q) = A\left(p + \frac{\lambda}{\ell}\nu, \; y + \frac{\lambda}{\ell}\mu\right) \qquad (51.47)$$

so that $A_{\nu\mu}$ in $\Gamma_{\nu\mu}$ has the same values as A at the corresponding points in Γ. The integral (51.46) becomes

$$D_{\nu\mu} = \frac{\lambda^2}{2\pi} \iint A_{\nu\mu}(p,q)\overline{A}(p,q)dpdq \qquad (51.48)$$

Figure 198

and is different from zero only if the domains Γ and $\Gamma_{\nu\mu}$ overlap. This is possible, however, only for a finite number of integers ν and μ.

51.5 The smaller ℓ is the fewer of the domains $\Gamma_{\nu\mu}$ overlap the domain Γ, i.e., the more coefficients are zero. This means that the image $V(x,y)$ given by (51.42) shows less and less detail of the structure of the object. If finally ℓ is so small that no domain $\Gamma_{\nu\mu}$ overlaps Γ then we have $V = U_{00}D_{00}$, i.e., a constant light distribution. The optical instrument cannot resolve any detail in the object. The limiting value ℓ_0 determines the limit of resolution. It depends only on the aperture Γ and cannot be decreased by changing the function $A(p,q)$ <u>inside</u> Γ. This means that: <u>It is impossible to increase the</u>

resolving power with regard to periodic structures by coating the aperture of
an optical instrument.

Let us now assume that the domain Γ is a circle of radius $\rho_0 = n \sin \theta$.

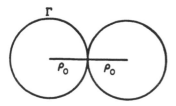

A circle $\Gamma_{\nu\mu}$ can overlap the circle Γ only if

$$\frac{\lambda}{\ell} \sqrt{\nu^2 + \mu^2} < 2\rho_0 . \qquad (51.51)$$

It follows that those $D_{\nu\mu}$ are zero for which we have

$$\sqrt{\nu^2 + \mu^2} > \frac{2\rho_0 \ell}{\lambda} . \qquad (51.52)$$

Figure 199

No detail will be observed if this last inequality holds for $\nu = 1; \mu = 0$ and
$\nu = 0; \mu = 1$, i.e., if

$$\ell < \frac{1}{2} \frac{\lambda}{\rho_0} = \frac{1}{2} \frac{\lambda}{n \sin \theta} . \qquad (51.53)$$

The limit of resolution, ℓ_0, of objects of periodic structure is thus given by
the formula

$$\ell_0 = 0.5 \frac{\lambda}{n \sin \theta} . \qquad (51.54)$$

Let us consider ℓ as given but $\rho_0 = n \sin \theta$ as variable, i.e., let us observe a given object with an optical instrument of variable aperture (diaphragm).

For $\rho_0 < \frac{1}{2} \frac{\lambda}{\ell}$ we obtain $V = U_{00}D_{00}$, i.e., uniformity without any detail.

For $\frac{1}{2} \frac{\lambda}{\ell} < \rho_0 < \frac{1}{2} \sqrt{2} \frac{\ell}{\lambda}$ we have

$$V = U_{00}D_{00} + U_{10}D_{10}e^{\frac{2\pi i}{\ell}x} + U_{01}D_{01}e^{\frac{2\pi i}{\ell}y} . \qquad (51.55)$$

This is a sinusoidal light distribution which indicates the periodicity of the
object.

By increasing ρ_0 we observe gradually more and more detail since more
terms appear in the trigonometric polynomial. However, since $\rho_0 \leq n_1$ we can
never obtain infinitely many terms which demonstrates that it is impossible

by optical means to observe details of the structure which are smaller than a certain constant.

By choosing different "coating functions" A(p,q) we can vary those coefficients $D_{\nu\mu}$ which are different from zero. The resolving power of the instrument is not affected by this, the <u>contrast of the image however can be varied considerably</u>.

51.6 <u>Non self luminous objects</u>. We treat the case of objects which are not self luminous in a similar manner. The image distribution is given by $V = |G|^2$ where G is the integral

$$G = \iint U(\xi,\eta) F(x-\xi, y-\eta) d\xi\, d\eta .$$
(51.61)

We introduce the Fourier series (51.41) and obtain

$$G = \sum U_{\nu\mu} F_{\nu\mu}\, e^{i\frac{2\pi}{\ell}(\nu x + \mu y)}$$
(51.62)

where

$$F_{\nu\mu} = \iint F(\xi,\eta) e^{-i\frac{2\pi}{\ell}(\nu\xi + \mu\eta)}\, d\xi\, d\eta .$$
(51.63)

On account of (51.44) this yields

$$F_{\nu\mu} = \frac{\lambda^2}{2\pi} A\left(\frac{\lambda}{\ell}\nu, \frac{\lambda}{\ell}\mu\right) .$$
(51.64)

Only finitely many quantities $F_{\nu\mu}$ are different from zero, namely those for

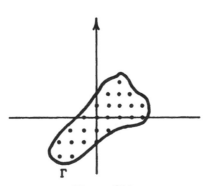

Figure 200

which the lattice points $\frac{\lambda}{\ell}\nu, \frac{\lambda}{\ell}\mu$ lie inside the domain Γ. For a circular domain Γ of radius ρ_0 we have $F_{\nu\mu} = 0$ if

$$\frac{\lambda}{\ell}\sqrt{\nu^2 + \mu^2} > \rho_0 .$$
(51.65)

The function G becomes a constant if

$$\ell < \frac{\lambda}{\rho_0} = \frac{\lambda}{n \sin\theta}$$
(51.66)

i.e., <u>the limit of resolution of</u>

non-self-luminous objects is given by the formula

$$\ell_0 = \frac{\lambda}{n \sin \theta} \quad .$$

(51.67)

It is twice as great as the limit of resolution for self-luminous objects.

This limit cannot be decreased by a different choice of A, i.e., by coating the aperture. The only effect of changing A inside the domain is to produce images of different contrast.

APPENDIX I

VECTOR ANALYSIS: DEFINITIONS AND THEOREMS

I.1 Let A be a vector in the three dimensional space. We write

$$A = (A_1, A_2, A_3) \tag{I.11}$$

where A_1, A_2, A_3 are the components of the vector A in a given Cartesian coordinate system.

The scalar product of two vectors $A = (A_1, A_2, A_3)$ and $B = (B_1, B_2, B_3)$ is defined by

$$A \cdot B = AB = A_1 B_1 + A_2 B_2 + A_3 B_3 . \tag{I.12}$$

It follows $A \cdot B = B \cdot A$. The length or absolute value of a vector A is given by the expression

$$|A| = \sqrt{A \cdot A} = \sqrt{A^2} = \sqrt{A_1^2 + A_2^2 + A_3^2} . \tag{I.13}$$

The angle between two vectors A and B follows from the equation

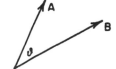

$$A \cdot B = |A| \, |B| \cos \vartheta ; \; 0 \leqq \vartheta \leqq \pi . \tag{I.14}$$

Two vectors A and B are orthogonal if

$$A \cdot B = 0 \tag{I.15}$$

The vector product A x B of two vectors A and B is a vector whose components are equal to the sub-determinants of the matrix of the components:

$$\begin{pmatrix} A_1 & A_2 & A_3 \\ B_1 & B_2 & B_3 \end{pmatrix}$$

i.e.

$$A \times B = (A_2 B_3 - A_3 B_2, \; A_3 B_1 - A_1 B_3, \; A_1 B_2 - A_2 B_1) . \tag{I.16}$$

This vector A x B is orthogonal to both A and B:

$$A \cdot (A \times B) = B \cdot (A \times B) = 0 . \qquad (I.161)$$

Its length is given by

$$|A \times B| = |A| \, |B| \sin \vartheta . \qquad (I.162)$$

It follows from (I.16):

$$A \times B = - B \times A . \qquad (I.163)$$

Three vectors (i,j,k) of unit length, which point in the direction of the three axes of a Cartesian coordinate system, form an <u>orthogonal system of unit vectors</u>. They satisfy the relations:

$$i^2 = j^2 = k^2 = 1$$

$$i \cdot j = j \cdot k = k \cdot i = 0$$

$$i \times j = k$$
$$j \times k = i \qquad (I.17)$$
$$k \times i = j .$$

Every vector A can be represented by a linear combination of these three unit vectors:

$$A = A_1 i + A_2 j + A_3 k . \qquad (I.18)$$

I.2 <u>Vector identities</u>. The scalar product of the two vectors A x B and C is equal to the determinant of the components of the three vectors A, B, and C:

$$(A \times B) \cdot C = \begin{vmatrix} A_1 & A_2 & A_3 \\ B_1 & B_2 & B_3 \\ C_1 & C_2 & C_3 \end{vmatrix} \qquad (I.21)$$

Hence:

$$(A \times B) \cdot C = (C \times A) \cdot B = (B \times C) \cdot A . \qquad (I.22)$$

The vector product of the two vectors A x B and C is a linear combination of the vectors A and B:

$$(A \times B) \times C = (AC)B - (BC)A . \qquad (I.23)$$

The vector $(A \times B) \times C$ is normal to the vector $A \times B$. Thus it must lie in the plane determined by A and B; i.e., it must be a linear combination of A and B. We write

$$(A \times B) \times C = \alpha A + \beta B . \tag{I.24}$$

Scalar multiplication of C with equation (I.24) gives

$$\alpha (A \cdot C) + \beta (B \cdot C) = 0$$

and hence

$$\alpha = - \lambda(B \cdot C)$$

$$\beta = + \lambda(A \cdot C)$$

where λ is a certain factor. It follows

$$(A \times B) \times C \equiv \lambda \left[(A \cdot C)B - (B \cdot C)A \right] . \tag{I.25}$$

The left side and the bracket on the right side represent three homogeneous polynomials of the components A_y, B_y, C_y. The identity (I.25) thus is possible only if λ is a constant, independent of A_y, B_y, C_y. We can find its value by introducing special vectors A, B, C, for example, $A = i$, $B = j$, $C = i$. It follows $\lambda = 1$.

The square of the vector product $A \times B$ can be written in the form

$$(A \times B)^2 = A^2 B^2 - (A \cdot B)^2 . \tag{I.26}$$

Indeed, on account of (I.22):

$$(A \times B) \cdot (A \times B) = \left((A \times B) \times A \right) \cdot B .$$

Hence, by applying (I.25):

$$(A \times B) \cdot (A \times B) = \left(A^2 B - (A \cdot B)A \right) \cdot B = A^2 B^2 - (A \cdot B)^2 .$$

From (I.26) follows Schwartz's inequality,

$$(A \cdot B)^2 \leqq A^2 B^2 . \tag{I.27}$$

I.3 Vector fields. A manifold of vectors A whose components are functions of x,y,z is called a vector field:

$$A(x,y,z) = \left(A_1(x,y,z), \ A_2(x,y,z), \ A_3(x,y,z) \right) . \tag{I.31}$$

The <u>divergence of a vector</u> $A(x,y,z)$ is the scalar function,

$$\text{div } A = i \cdot \frac{\partial A}{\partial x} + j \cdot \frac{\partial A}{\partial y} + k \cdot \frac{\partial A}{\partial z}$$

$$= \frac{\partial A_1}{\partial x} + \frac{\partial A_2}{\partial y} + \frac{\partial A_3}{\partial z} \ . \tag{I.32}$$

The <u>curl of a vector</u> $A(x,y,z)$ is a vector defined by the equations,

$$\text{curl } A = i \times \frac{\partial A}{\partial x} + j \times \frac{\partial A}{\partial y} + k \times \frac{\partial A}{\partial z}$$

$$= \left(\frac{\partial A_3}{\partial y} - \frac{\partial A_2}{\partial z}, \frac{\partial A_1}{\partial z} - \frac{\partial A_3}{\partial x}, \frac{\partial A_2}{\partial x} - \frac{\partial A_1}{\partial y} \right) . \tag{I.33}$$

A vector field $A(x,y,z)$ is called

$$\begin{aligned} &\text{solenoidal, if div } A = 0 \ ; \\ &\text{lamellar, if curl } A = 0 \ . \end{aligned} \tag{I.34}$$

A solenoidal field A is obtained if $A(x,y,z)$ is the curl of a vector W:

$$A = \text{curl } W$$

where $W(x,y,z)$ is an arbitrary vector field with continuous second derivatives. One proves easily

$$\text{div curl } W = 0 \ . \tag{I.35}$$

Let $\phi(x,y,z)$ be a scalar function with continuous derivatives. The <u>gradient</u> of this function at a point x,y,z is the vector

$$\text{grad } \phi = (\phi_x, \phi_y, \phi_z) = i \frac{\partial \phi}{\partial x} + j \frac{\partial \phi}{\partial y} + k \frac{\partial \phi}{\partial z} \ . \tag{I.36}$$

This vector grad ϕ is a function of x,y,z and thus determines a vector field; this vector field is <u>lamellar</u>. From a simple calculation it follows:

$$\text{curl grad } \phi = 0 \tag{I.37}$$

provided that ϕ has continuous second derivatives.

The divergence of a gradient field is given by the <u>Laplacian operator</u>,

$$\Delta \phi = \phi_{xx} + \phi_{yy} + \phi_{zz} = \text{div grad } \phi \ . \tag{I.38}$$

Hence: A gradient field $A = \operatorname{grad} \phi$ is both solenoidal and lamellar if ϕ is a solution of Laplace's differential equation $\Delta \phi = 0$.

I.4 Vector identities. Let $f(x,y,z)$ be a scalar function and $A(x,y,z)$ a vector field with continuous first derivatives. Then we have the following identities:

$$\operatorname{div} fA = f \operatorname{div} A + (\operatorname{grad} f) \cdot A \tag{I.411}$$

$$\operatorname{curl} fA = f \operatorname{curl} A + (\operatorname{grad} f) \times A \ . \tag{I.412}$$

Indeed, from the definition (I.32) it follows:

$$\operatorname{div} f A = f \left(\frac{\partial A_1}{\partial x} + \frac{\partial A_2}{\partial y} + \frac{\partial A_3}{\partial z} \right) + A_1 \frac{\partial f}{\partial x} + A_2 \frac{\partial f}{\partial y} + A_3 \frac{\partial f}{\partial z}$$

$$= f \operatorname{div} A + (\operatorname{grad} f) \cdot A \ .$$

From (I.33):

$$\operatorname{curl} fA = i \times (fA)_x + j \times (fA)_y + k \times (fA)_z$$

$$= f \left[i \times A_x + j \times A_y + k \times A_z \right] + (if_x + jf_y + kf_z) \times A$$

$$= f \operatorname{curl} A + (\operatorname{grad} f) \times A \ .$$

Let $f(x,y,z)$ be a scalar function with continuous first derivatives and $A(x,y,z)$ a vector field with continuous second derivatives. Then

$$\operatorname{curl} (f \operatorname{curl} A) = -f \Delta A + f \operatorname{grad} \operatorname{div} A + (\operatorname{grad} f) \times \operatorname{curl} A \ . \tag{I.42}$$

In order to prove this identity, we first apply (I.412):

$$\operatorname{curl} (f \operatorname{curl} A) = f \operatorname{curl} \operatorname{curl} A + (\operatorname{grad} f) \times \operatorname{curl} A \ .$$

This leaves us to show that

$$\operatorname{curl} \operatorname{curl} A = -\Delta A + \operatorname{grad} \operatorname{div} A \ . \tag{I.43}$$

On account of the definition (I.33):

$$\operatorname{curl} \operatorname{curl} A = i \times (i \times A_{xx} + j \times A_{yx} + k \times A_{zx})$$

$$+ j \times (i \times A_{yx} + j \times A_{yy} + k \times A_{zy})$$

$$+ k \times (i \times A_{xz} + j \times A_{yz} + k \times A_{zz}) \ .$$

Finally with the aid of the identity (I.23):

$$\text{curl curl A} = - A_{xx} + i\frac{\partial^2 A_1}{\partial x^2} + j\frac{\partial^2 A_1}{\partial x\,\partial y} + k\frac{\partial^2 A_1}{\partial x\,\partial z}$$

$$- A_{yy} + i\frac{\partial^2 A_2}{\partial x\,\partial y} + j\frac{\partial^2 A_2}{\partial y^2} + k\frac{\partial^2 A_2}{\partial y\,\partial z}$$

$$- A_{zz} + i\frac{\partial^2 A_3}{\partial x\,\partial z} + j\frac{\partial^2 A_3}{\partial y\,\partial z} + k\frac{\partial^2 A_3}{\partial z^2} \ .$$

This however can be written as follows:

$$\text{curl curl A} = - \Delta A + i\frac{\partial}{\partial x}(\text{div A}) + j\frac{\partial}{\partial y}(\text{div A}) + k\frac{\partial}{\partial z}(\text{div A})$$

$$= - \Delta A + \text{grad div A} \ .$$

If $A(x,y,z)$ and $B(x,y,z)$ are two vector fields with continuous first derivatives, then the following three identities are satisfied:

$$\text{div } (A \times B) = B \cdot \text{curl A} - A \cdot \text{curl B} \tag{I.44}$$

$$\text{curl } (A \times B) = \frac{\partial A}{\partial \beta} - \frac{\partial B}{\partial \alpha} + A \text{ div B} - B \text{ div A} \tag{I.45}$$

$$A \times \text{curl B} + B \times \text{curl A} = - \left(\frac{\partial B}{\partial \alpha} + \frac{\partial A}{\partial \beta}\right) + \text{grad } (A \cdot B) \ . \tag{I.46}$$

In these formulae the differential operators $\frac{\partial}{\partial \alpha}$ and $\frac{\partial}{\partial \beta}$ are defined by

$$\frac{\partial}{\partial \alpha} = A_1 \frac{\partial}{\partial x} + A_2 \frac{\partial}{\partial y} + A_3 \frac{\partial}{\partial z} \ ,$$

$$\frac{\partial}{\partial \beta} = B_1 \frac{\partial}{\partial x} + B_2 \frac{\partial}{\partial y} + B_3 \frac{\partial}{\partial z} \ . \tag{I.47}$$

Hence $\frac{\partial}{\partial \alpha}$ denotes differentiation in the direction of the vector A and $\frac{\partial}{\partial \beta}$ differentiation in the direction of B.

We prove first the identity (I.44). We have

$$B \cdot \text{curl A} = (i \times A_x) \cdot B + (j \times A_y) \cdot B + (k \times A_z) \cdot B \ .$$

Hence by applying (I.22):

$$B \cdot \text{curl } A = (A_x \times B) \cdot i + (A_y \times B) \cdot j + (A_z \times B) \cdot k \ .$$

We interchange A and B:

$$A \cdot \text{curl } B = (B_x \times A) \cdot i + (B_y \times A) \cdot j + (B_z \times A) \cdot k \ .$$

Hence

$$B \cdot \text{curl } A - A \cdot \text{curl } B = (A \times B)_x \cdot i + (A \times B)_y \cdot j + (A \times B)_z \cdot k$$

$$= \text{div } (A \times B) \ .$$

The second identity (I.45) follows from the definition (I.33):

$$\text{curl } (A \times B) = i \times (A_x \times B) + i \times (A \times B_x)$$

$$+ j \times (A_y \times B) + j \times (A \times B_y)$$

$$+ k \times (A_z \times B) + k \times (A \times B_z)$$

with the aid of the relation (I.23). We find

$$\text{curl } (A \times B) = B_1 A_x + B_2 A_y + B_3 A_z + (\text{div } B)A$$

$$- A_1 B_x - A_2 B_y - A_3 B_z - (\text{div } A)B$$

which yields (I.45) if the notation (I.47) is introduced.

The last identity (I.46) can be found in a similar way. We have

$$A \times \text{curl } B = A \times (i \times B_x) + A \times (j \times B_y) + A \times (k \times B_z)$$

or by applying (I.23):

$$A \times \text{curl } B = (A \cdot B_x)i + (A \cdot B_y)j + (A \cdot B_z)k$$

$$- A_1 B_x - A_2 B_y - A_3 B_z \ .$$

We interchange A and B and add both equations. It follows:

$$A \times \text{curl } B + B \times \text{curl } A = -\left(\frac{\partial A}{\partial \beta} + \frac{\partial B}{\partial \alpha} \right) + (A \cdot B)_x i + (A \cdot B)_y j + (A \cdot B)_z k$$

which is the identity (I.46).

For the purpose of a later application, by adding (I.45) and (I.46), we obtain the following identity:

curl (A x B) + A x curl B + B x curl A

$$= -2\frac{\partial B}{\partial \alpha} - B \ div \ A + \left(A \ div \ B + grad \ (A \cdot B)\right) \ . \qquad (I.48)$$

APPENDIX II

TRACING OF LIGHT RAYS IN A SYSTEM
OF PLANE REFLECTING OR REFRACTING SURFACES

II.1 The vector form of the laws of reflection and refraction which we have derived in §14 can be used with advantage in practical problems of ray tracing. Both laws have the same mathematical form

$$S' = S + \Gamma M . \qquad (\text{II}.11)$$

The vectors S and S' are defined by $S = nT$ and $S' = n'T'$ where n and n' are the indices of refraction and T and T' are unit vectors in the direction of the rays. M is a unit vector normal to the surface and Γ is a scalar function of the angle of incidence, ϑ. By introducing the variable

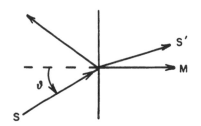

$$\rho = n \cos \vartheta = (S \cdot M) \qquad (\text{II}.12)$$

instead of ϑ we find from (14.191) and (14.28) that

$$\Gamma(\rho) = -2\rho \qquad , \quad \text{in case of reflection}$$

$$\Gamma(\rho) = \sqrt{n'^2 - n^2 + \rho^2} - \rho , \quad \text{in case of refraction} . \qquad (\text{II}.13)$$

II.2 Let us now consider the problem of tracing a ray through a number, k, of plane surfaces $L_0, L_1, L_2, \ldots, L_{k-1}$. These surfaces may be either reflecting or refracting surfaces. Any prism system is an example of this case. Let us assume that the normals, M_i, of these surfaces are known. On L_i we denote the incident vector by S_i and the reflected or refracted vector by S_{i+1}. The index of refraction of the medium with which the vector S_i is associated is called n_i. If L_i is a reflecting surface we have, of course, $n_i = n_{i+1}$.

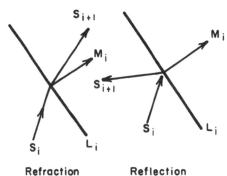

Refraction   Reflection

368

The vectors S_i are then related by the system of recursion formulae

$$S_{i+1} = S_{i-1} + \Gamma_i(\rho_i)M_i \qquad (\text{II}.21)$$

in which $\rho_i = S_i \cdot M_i$ and

$$\Gamma_i(\rho_i) = -2\rho_i ,$$

if L_i is a reflecting surface,

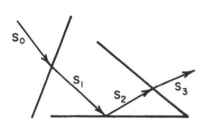

$$\Gamma_i(\rho_i) = \sqrt{n_{i+1}^2 - n_i^2 + \rho_i^2} - \rho_i , \qquad (\text{II}.22)$$

if L_i is a refracting surface.

The equations II.21 allow us to find the vectors S_i successively if the initial vector S_0 is known.

II.3 Of special interest is, of course, the last vector S_k which determines the direction of the ray in the final medium. From (II.21) it follows

$$S_k = S_0 + \sum_{i=0}^{k=1} \Gamma_i(\rho_i)M_i \qquad (\text{II}.31)$$

which demonstrates that the vector S_k is completely known if we have found the scalar quantities $\rho_i = S_i M_i$.

Our aim is to show that the quantities ρ_i can be found by a system of scalar recursion formulae. We form the scalar product of a vector M_ν with the equation (II.21). We find

$$S_{i+1} \cdot M_\nu - S_i \cdot M_\nu = \Gamma_i(\rho_i)M_i \cdot M_\nu \qquad (\text{II}.32)$$

and hence by summation from $i = 0$ to $i = \nu - 1$ we obtain

$$S_\nu \cdot M_\nu = S_0 \cdot M_\nu + \sum_{i=0}^{\nu-1} \Gamma_i(\rho_i)(M_i \cdot M_\nu) \qquad (\text{II}.33)$$

or

$$\rho_\nu = (S_0 \cdot M_\nu) + \sum_{i=0}^{\nu-1} \Gamma_i(\rho_i)(M_i \cdot M_\nu) . \qquad (\text{II}.34)$$

This is the desired recursion formula; it determines the value ρ_ν if the preceding quantities $\rho_1, \rho_2, \ldots, \rho_{\nu-1}$ are known.

II.4 The equations (II.21) and (II.34) are especially simple if all the surfaces L_i are reflecting surfaces. Since $\Gamma_i(\rho_i) = -2\rho_i$ we have

$$\rho_\nu = (S_0 \cdot M_\nu) - 2 \sum_{i=0}^{\nu-1} (M_i \cdot M_\nu)\rho_i \qquad \text{(II.41)}$$

i.e., a system of <u>linear recursion formulae</u> for the quantities ρ_ν. The final vector S_k is then given by

$$S_k = S_0 - 2 \sum_{i=0}^{k-1} \rho_i M_i . \qquad \text{(II.42)}$$

II.5 As an example we consider a set of three mirrors at right angles to each other. We choose a cartesian coordinate system so that the normals M_i of the mirrors have the direction of the unit vectors $\vec{i}, \vec{j}, \vec{k}$, i.e.,

$$M_0 = \vec{i}, \qquad M_1 = \vec{j}, \qquad M_2 = \vec{k} . \qquad \text{(II.51)}$$

Since $M_i \cdot M_\nu = 0$ it follows that

$$\rho_\nu = (S_0 \cdot M_\nu) \qquad \text{(II.52)}$$

and hence

$$S_3 = S_0 - 2\left[(S_0 \cdot \vec{i})\vec{i} + (S_0 \cdot \vec{j})\vec{j} + (S_0 \cdot \vec{k})\vec{k}\right]$$

or

$$S_3 = S_0 - 2S_0 = -S_0 . \qquad \text{(II.53)}$$

Thus any ray which enters this sytem of three mirrors comes back on itself, except for a lateral shift.

II.6 We consider next a 90° roof prism of the type illustrated in the following figure. Let us assume that a parallel bundle of rays enters the glass body normal to the first surface and therefore is unchanged in its direction. The bundle then is reflected on the surfaces L_0 and L_1 of the roof. Let M_0

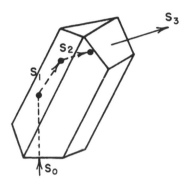

and M_1 be the normals of these surfaces. One part of the bundle is first reflected on L_0 and then on L_1, another part first on L_1 and then on L_0. Let us consider the rays of the first part. We have

$$\rho_0 = S_0 \cdot M_0$$

$$\rho_1 = S_0 \cdot M_1 - 2\rho_0(M_0 \cdot M_1)$$ (II.61)

and

$$S_2 = S_0 - 2(\rho_0 M_0 + \rho_1 M_1)$$

or

$$S_2 = S_0 - 2(S_0 \cdot M_0)M_0 - 2(S_0 \cdot M_1)M_1 + 4(S_0 \cdot M_0)(M_0 \cdot M_1)M_1 \ . \quad \text{(II.62)}$$

The vector S_2 determines the direction of the rays in the glass after the two reflections.

The corresponding vector S_2' of the other part of the bundle, which is reflected first on L_1 and then on L_0 is obtained by interchanging M_0 and M_1 in (II.62). The difference $S_2' - S_2$ therefore is given by the expression

$$S_2' - S_2 = 4(M_0 \cdot M_1)\left[(S_0 \cdot M_1)M_0 - (S_0 \cdot M_0)M_1\right] \ . \quad \text{(II.63)}$$

It follows that the two reflected bundles are parallel to each other only if

$$M_0 \cdot M_1 = 0 \ . \quad \text{(II.64)}$$

i.e., if the two roof surfaces are at right angles. In any other case these bundles are not parallel; if a distant object is viewed through such a prism it will appear as double.

II.7 For practical purposes it is important to know the departure from parallelism of the two bundles if the roof angle is not exactly 90°. Let us therefore assume that

$$M_0 \cdot M_1 = \cos(90 + \epsilon) = -\sin \epsilon = -\epsilon \quad \text{(II.71)}$$

where ϵ is a small quantity. From (II.63) it follows that

$$S_2' - S_2 = -4\epsilon\left[(S_0 \cdot M_1)M_0 - (S_0 \cdot M_0)M_1\right] \ . \quad \text{(II.72)}$$

In the bracket we may introduce the data M_0 and M_1 of the exact roof prism. By orienting the roof prism as illustrated in the following figure we have

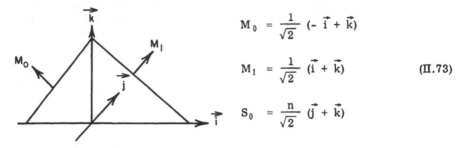

$$M_0 = \frac{1}{\sqrt{2}} (- \vec{i} + \vec{k})$$

$$M_1 = \frac{1}{\sqrt{2}} (\vec{i} + \vec{k}) \qquad\qquad (\text{II}.73)$$

$$S_0 = \frac{n}{\sqrt{2}} (\vec{j} + \vec{k})$$

and hence by (II.72):

$$S_2' - S_2 = \frac{4n}{\sqrt{2}} \, \epsilon \, \vec{i} \, . \qquad\qquad (\text{II}.74)$$

The angular deviation $\left| S_2' - S_2 \right|$ is thus given by

$$\left| S_2' - S_2 \right| = \sqrt{2} \, n \, \epsilon \quad . \qquad\qquad (\text{II}.75)$$

Assuming $n = 1.5$ we obtain

$$\left| S_2' - S_2 \right| = 4.3 \, \epsilon \qquad\qquad (\text{II}.76)$$

which demonstrates the accuracy to which the roof angle has to be held if noticeable doubling is to be avoided. The formula (II.76) refers to the rays inside the prism. However, one proves readily that no additional deviation is introduced by refraction on the last prism surface. This follows from the fact that in the exact roof prism the rays pass at right angles through this surface. For small values of ϵ this surface is passed <u>almost</u> at right angles. The refracted rays S_3' and S_3 differ then from S_2' and S_2 only by terms of order ϵ^2. Thus for the final bundle we have the same formula:

$$\left| S_3' - S_3 \right| = 4.3 \, \epsilon \quad . \qquad\qquad (\text{II}.77)$$

SUPPLEMENTARY NOTE. NO. I

ELECTRON OPTICS

INTRODUCTION

In electron optics it will be found that much the same problems occur as in ordinary optics and similar methods may be used to treat them. It is the purpose of this appendix to develop the theory of electron optics through the close parallelism of the two fields.

The subject deals with the behavior of electrons in an electromagnetic field. To discuss the theory rigorously, it would be necessary to consider Schroedinger's wave equation, or for fast electrons, Dirac's. However, as in geometrical optics, certain simplifying approximations can be made. The electron is considered to be a particle with an associated charge and mass and the "ray" will be its path in a given electromagnetic field.

It is possible to obtain the approximation of Geometrical Electron Optics from the wave equations of Quantum Mechanics by letting Planck's Constant $h \rightarrow 0$.[†] We may then treat Dirac's wave equations in precisely the same way as Maxwell's Equations in §16. Furthermore, it is then possible to construct a diffraction theory analagous to that of Chapter VI for electron optical instruments. Since the wave lengths of electrons are much smaller than those of visible light, the limits of resolution obtained will be much smaller. It is for this reason that electron optics takes on such importance in microscopy. The diffraction theory of such instruments will not be developed here, but we shall consider only the geometrical theory of electron optics.

The discussion will begin with a derivation of a variation principle analogous to Fermat's problem in optics:

$$V = \int n \, ds = \text{Extremum.}$$

However, because of the presence of a magnetic field, the electron optical "index of refraction" will be found to depend on the direction as well as the position. This corresponds to the non-isotropic, non-homogeneous case in ordinary optics.

[†] See W. Pauli, Die Allgemeinen Principien der Wellenmechanik, p. 240 ff. Handbuch der Physik, Bd 24.1, 1933

§1. THE EQUATIONS OF MOTION

1.1 We consider an electron with a mass, m, and a charge, e, moving in an electromagnetic field. The field is defined by its potentials, the electric potential $\phi(x,y,z)$, a scalar, and the magnetic potential $\vec{A}(x,y,z)$, a vector. The field vectors are given by the equations

$$\vec{E} = (E_1, E_2, E_3) = - \text{grad } \phi$$

$$\vec{H} = (H_1, H_2, H_3) = \text{curl } \vec{A} \ .$$

(1.11)

From Maxwell's equations it is found that

$$\Delta\phi = - 4\pi\rho \qquad \text{(Poisson's Equation)}$$

$$\text{and } \Delta\vec{A} = \frac{4\pi}{\tau} \vec{i} \qquad \text{div } \vec{A} = 0$$

(1.12)

where $\rho = \rho(x,y,z)$ is the electric charge density and $\vec{i}$ is the vector of current density and direction. We are mostly concerned, however, with parts of the electromagnetic field in which there is no charge and no current.† The equations (1.12) then become

$$\Delta\phi = 0$$

$$\Delta\vec{A} = 0 , \qquad \text{div } \vec{A} = 0 \ .$$

(1.13)

The force, $F = (F_1, F_2, F_3)$, acting upon the electron is the sum of the electric force $e\vec{E}$, and the Lorentz force of the magnetic field, $\frac{e}{c}(\dot{\vec{X}} \times H)$.

$$F = e\left[\vec{E} + \frac{1}{c}(\dot{\vec{X}} \times \vec{H})\right]$$

(1.14)

where $\dot{\vec{X}} = (\dot{x},\dot{y},\dot{z})$ is the velocity vector of the electron and the dot denotes differentiation with respect to the time parameter, t.

The kinetic energy of the electron may be written in the form $T(x,y,z; \dot{x},\dot{y},\dot{z}) = T(x_i,\dot{x}_i)$. (i = 1, ... ,3) where x_1,x_2,x_3 are written in place of x, y, z. For this problem T is a function of $\dot{x}, \dot{y}, \dot{z}$ only. For slow electrons we have the classical expression

$$T = \frac{m}{2} \Sigma \dot{x}_i^2$$

(1.15)

---

† Since the charge density is not zero in the neighborhood of the cathode of an instrument, this assumption is not always valid. ϕ must then be determined as a solution of Poisson's Equation.

whereas for high velocities we must apply the relativistic mechanics of a particle[†] in which

$$T = mc^2 \left(1 - \sqrt{1 - \beta^2}\right)$$

$$\beta^2 = \frac{1}{c^2} \Sigma \dot{x}_i^2 .$$

(1.16)

For small values of β this reduces to (1.15).

1.2 Knowing the kinetic energy we may write Lagrange's equations of motion

$$\frac{d}{dt} \frac{\partial T}{\partial \dot{x}_i} - \frac{\partial T}{\partial x_i} = F_i$$

(1.21)

which are valid for any field of force. The kinetic potential of the system

$$L(x_i, \dot{x}_i) = T - U$$

(1.22)

is determined if the system has a generalized potential U such that

$$F_i = -\frac{\partial U}{\partial x_i} + \frac{d}{dt} \frac{\partial U}{\partial \dot{x}_i} .$$

(1.23)

The scalar expression $e\phi - \frac{e}{c}(\vec{A} \cdot \dot{\vec{X}})$

(1.24)

is a generalized potential, U, since it does give the vector F as can be verified by direct computation. From (1.21) and (1.23) we obtain Lagrange's equations in the form

$$\frac{d}{dt} \frac{\partial L}{\partial \dot{x}_i} - \frac{\partial L}{\partial x_i} = 0 .$$

(1.25)

Hence, the equations of motion of an electron in an electromagnetic field may be given in terms of the Lagrangian function or kinetic potential, $L(x_i, \dot{x}_i)$, where

$$L = mc^2 \left(1 - \sqrt{1 - \beta^2}\right) - e\phi + \frac{e}{c}(\vec{A} \cdot \dot{\vec{X}}) .$$

(1.26)

[†] See O. Halpern, Relativitätsmechanik, Handbuch der Physik, Bd. 5, 1927

1.3 Now the equation (1.25) is precisely Euler's equation for the problem of variation

$$W = \int_{t_0}^{t_1} L(x_i, \dot{x}_i)dt = \text{Extremum} . \tag{1.31}$$

Hence the existence of L means that we may interpret the functions $x_i(t)$ as extremals of the problem where only those $x_i(t)$ which pass through two given points, $P_0 = (\alpha_1, \alpha_2, \alpha_3)$ and $P_1 = (\beta_1, \beta_2, \beta_3)$ are admitted. In other words we admit only those curves satisfying the boundary conditions

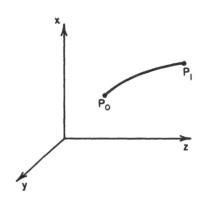

$$x_i(t_0) = \alpha_1$$
$$\tag{1.32}$$
$$x_i(t_1) = \beta_1 .$$

The extremals of this problem give not only the path of the electron but also its motion as a function of t.

1.4 The function L in (1.26) does not depend explicitly upon the time but only on x_i and $\dot{x}_i$. From this it is shown that the expression

$$\sum \dot{x}_i \frac{\partial L}{\partial \dot{x}_i} - L \tag{1.41}$$

is a constant along the path of a moving electron, for we have

$$\frac{d}{dt}\left[\sum \dot{x}_i \frac{\partial L}{\partial \dot{x}_i} - L\right] = \sum \left(\ddot{x}_i \frac{\partial L}{\partial \dot{x}_i} + \dot{x}_i \frac{d}{dt}\frac{\partial L}{\partial \dot{x}_i}\right) - \sum \dot{x}_i \frac{\partial L}{\partial x_i} - \sum \ddot{x}_i \frac{\partial L}{\partial \dot{x}_i}$$

$$= \dot{x}_i \left[\frac{d}{dt}\frac{\partial L}{\partial \dot{x}_i} - \frac{\partial L}{\partial x_i}\right]$$

$$= 0 \text{ by } (1.25) .$$

Hence, $$\sum \dot{x}_i \frac{\partial L}{\partial \dot{x}_i} - L = C . \tag{1.42}$$

This is the general principle of <u>conservation of energy</u>. The special case $L = \frac{m}{2}\sum \dot{x}_i^2 - U(x_i)$ yields the familiar equation, $\frac{m}{2}\sum \dot{x}_i^2 + U = C$, namely the sum of the kinetic and potential energies is constant.

From this general result we will derive a problem of Fermat's type in the next section.

§2. THE ASSOCIATED FERMAT PROBLEM.

2.1 Consider a mechanical problem which has a Lagrangian function $L(x_i, \dot{x}_i)$ that does not depend explicitly on t. The expression (1.41) is a constant along any actual path. We shall consider only those paths for which

$$\Sigma \dot{x}_i \frac{\partial L}{\partial x_i} - L = C \qquad (2.11)$$

where C is a given constant. We want to pick out the path from among these which passes through two given points P_0 and P_1. In other words a solution of Lagrange's equations (1.25) is sought which satisfies the condition (2.11) and passes through the points P_0 and P_1. This problem differs from that of (1.31) in that it is not required that the points P_0 and P_1 be passed at the particular times t_0 and t_1, but that the particle has the given energy C.

This new problem can be characterized as a variation problem as we now show. Let $x_i = x_i(s)$ be a curve passing through the points P_0 and P_1. The parameter, s, is chosen so that $x_i(0) = \alpha_i$ and $x_i(1) = \beta_i$ where α_i and β_i are the respective coordinates of P_0 and P_1. The time when the electron reaches a point $x_i(s)$ on the curve is expressed as $t = t(s)$. The Lagrangian function $L(x_i, \dot{x}_i)$ of the electron becomes a function of s

$$L = L\left(x_i(s), \frac{x_i'(s)}{t'(s)} \right)$$

where the prime denotes differentiation with respect to s. The solution of this problem is the solution of problem of variation: To find functions $x_i(s)$ satisfying the boundary conditions

$$x_i(0) = \alpha_i ; \quad x_i(1) = \beta_i \qquad (2.12)$$

for which the integral

$$V = \int_0^1 \left[L\left(x_i, \frac{x_i'}{t'} \right) + C \right] t' ds \qquad (2.13)$$

is an extremum. Note, in particular, that there are no boundary conditions upon t(s).

Variation with respect to x_i yields

$$\frac{d}{ds} L_i \left(x_i, \frac{x_i'}{t'} \right) - t' \frac{\partial L}{\partial x_i} = 0 \qquad (2.14)$$

where $L_1(x_1, \dot{x}_1) = \frac{\partial}{\partial \dot{x}_1} L_1(x_1, \dot{x}_1)$. If we divide by t' we recognize the equations (2.11).

Let us take the variation function of t, $t + \epsilon \zeta$. This gives

$$\int \zeta' \frac{\partial}{\partial t'} t' \left[L\left(x_1, \frac{x_1'}{t'}\right) + C \right] ds = 0 \tag{2.15}$$

for any function $\zeta(s)$. Integration by parts yields the equation

$$\left[\zeta \left\{ \frac{\partial}{\partial t'} (t'L) + C \right\} \right]_0^1 - \int \zeta \frac{d}{ds} \left[\frac{\partial}{\partial t'} (t'L) + C \right] ds = 0 . \tag{2.16}$$

But ζ is arbitrary, hence

$$\frac{d}{ds} \left[\frac{\partial}{\partial t'} (t'L) \right] = 0 \tag{2.17}$$

and $\frac{\partial}{\partial t'} (t'L) + C = 0$ for $s = 0$ and $s = 1$, therefore

$$\frac{\partial}{\partial t'} (t'L) + C = L - \Sigma \frac{x_1'}{t'} L_1 + C = 0 \text{ for all values of } s. \tag{2.18}$$

However, $\frac{x_1'}{t'} = \dot{x}_1$ which gives the energy condition

$$\Sigma \dot{x}_1 \frac{\partial L}{\partial \dot{x}_1} - L = C .$$

The special condition (2.11) is therefore obtainable from (2.13) by variation with respect to the time $t(s)$.

2.2 It will be shown now that the variation problem (2.13) can be transformed into an integral of Fermat's type

$$V = \int_0^1 F(x_1, x_1') ds \tag{2.21}$$

where $F(x_1, x_1')$ is homogeneous of the first order in x_1'. In order to do this the function $t(s)$ must be eliminated in (2.13).

Consider the equivalent problem of variation

$$V = \int_0^1 \left[L\left(x_1, \frac{x_1'}{\rho}\right) + p(s) (\rho - t') + Ct' \right] ds = \text{Extremum} \tag{2.22}$$

where p(s) is a Lagrangian multiplier.[†] Variation with respect to ρ gives

$$L\left(x_1, \frac{x_1'}{\rho}\right) - \Sigma \frac{x'}{\rho} L_1\left(x_1, \frac{x_1'}{\rho}\right) + p(s) = 0 . \tag{2.23}$$

Variation with respect to t gives

$$p'(s) = 0 \quad\text{and}\quad p(1) = p(0) = C . \tag{2.24}$$

Variation with respect to p(s) gives

$$\rho - t' = 0 . \tag{2.25}$$

The conditions (2.23),...,(2.25) may be placed upon (2.22) without affecting the solution of the problem. The condition (2.25) gives the original problem (2.13), however we obtain a new problem by imposing the other conditions. From (2.24)

$$p(s) = C \tag{2.26}$$

and hence by (2.23)

$$L\left(x_1, \frac{x_1'}{\rho}\right) + C - \Sigma \frac{x_1'}{\rho} L_1\left(x_1, \frac{x_1'}{\rho}\right) = \frac{\partial}{\partial\rho} \rho \left[C + L_1\left(x_1, \frac{x_1'}{\rho}\right)\right] = 0 . \tag{2.27}$$

This gives

$$V = \int_0^1 \rho \left[L\left(x_1, \frac{x_1'}{\rho}\right) + C\right] ds \tag{2.28}$$

where ρ may be eliminated by means of the condition (2.27).

2.3 The problem of variation has been expressed in the equivalent form

$$V = \int_0^1 F(x_1, x_1')ds = \text{Extremum} \tag{2.31}$$

where $F(x_1, x_1') = G\left(x_1, x_1', \rho(x_1, x_1')\right) = \rho\left[C + L\left(x_1, \frac{x_1'}{\rho}\right)\right] \tag{2.32}$

and where $\rho = \rho(x_1, x_1')$ is determined by the equation

---

[†] Compare the similar procedure for the elimination of $\theta(z)$ in §30.2.

$$\frac{\partial}{\partial \rho} G(x_i, x_i', \rho) = 0 .$$ (2.33)

It will now be shown that $F(x_i, x_i')$ is homogeneous in x_i' to the first order.

$G(x_i, x_i', \rho)$ satisfies the relation

$$\Sigma x_i' G_{x_i'} = G - \rho G_\rho$$ (2.34)

as we may easily verify. Now from (2.32) and (2.33) it follows that

$$F x_i' = G_{x_i'} + G_\rho \frac{\partial \rho}{\partial x_i'} = G_{x_i'} .$$

Hence $\Sigma x_i' F_{x_i'} = \Sigma x_i' G_{x_i'} = G - \rho G_\rho$. Since $G_\rho = 0$ and $G = F$, we obtain

$$\Sigma x_i' F_{x_i'} = F$$ (2.35)

which is Euler's condition for homogeneity of the first order in x_i'.

In summary: The paths of particles of a given energy C in a mechanical system which possesses a Lagrangian function of the form $L(x_i, \dot{x}_i)$ are extremals of the Fermat problem

$$V = \int F(x_i, \dot{x}_i') ds$$ (2.36)

where the function $F(x_i, x_i')$ is homogeneous of the first order in x_i' and can be obtained by eliminating ρ from the equations

$$F = \rho \left[C + L(x_i, x_i') \right] ; \quad \frac{\partial}{\partial \rho} \rho \left[C + \left(L \ x_i \ \frac{x_i'}{\rho} \right) \right] = 0 .$$ (2.37)

We note that this definition of F is equivalent to the definition by the Legendre transformation $\frac{\partial}{\partial \rho} \left[\rho L \left(x_i, \frac{x_i}{\rho} \right) \right] = -C; \quad \rho L + F = -C\rho$.

2.4 Consider the example of the case where the kinetic energy is a quadratic form in the velocities $T = \Sigma g_{ik} \dot{x}_i \dot{x}_k$, and the potential is a function only of the position, $U(x_i)$. Then $L = T - U$ or

$$L = \Sigma g_{ik} \dot{x}_i \dot{x}_k - U(x_i)$$ (2.41)

which gives the function F through the equations (2.37)

$$F = \frac{1}{\rho} \Sigma g_{ik} x_i' x_k' + \rho(C - U)$$

$$(2.42)$$

$$- \frac{1}{\rho^2} \Sigma g_{ik} x_i' x_k' + (C - U) = 0 .$$

We obtain the Jacobi principle of mechanics

$$F = 2\sqrt{C - U} \sqrt{\Sigma g_{ik} x_i' x_k'} \qquad (2.43)$$

2.5 The case of electron optics. The general result immediately yields the electron optical Fermat problem. From (1.26) we have the Lagrangian function

$$L = mc^2 \left(1 - \sqrt{1 - \beta^2}\right) - e\phi + \frac{e}{c} (\vec{A} \cdot \dot{\vec{X}})$$

where $\beta^2 = \frac{1}{c^2} \Sigma \dot{x}_i^2$. The substitution of $\frac{x_i'}{\rho}$ for $\dot{x}_i$ gives

$$G(x_i, x_i', \rho) = mc^2 \rho \left(1 - \sqrt{1 - \frac{1}{\rho^2 c^2} \Sigma x_i^2}\right) + \frac{e}{c} (\vec{A} \cdot \vec{X}') + \rho(C - e\phi) \qquad (2.52)$$

which we write in the form

$$\frac{G}{mc} = c\rho \left(1 - \sqrt{1 - \frac{1}{\rho^2 c^2} \Sigma x_i'^2}\right) + (\vec{a} \cdot \vec{X}') + \frac{1}{2} \rho c \varphi . \qquad (2.53)$$

The vector $\vec{a}$ and the scalar $\varphi(x,y,z)$ are defined by

$$\vec{a} = \frac{e}{mc^2} \vec{A} = (a_1, a_2, a_3) \qquad (2.54)$$

$$\varphi = \frac{2(C - e\phi)}{mc^2} .$$

In these expressions the right sides are dimensionless.

The extremals of the variation problem (2.26) are not changed by multiplying the function $F(x_i, x_i')$ by a constant. So that the index of refraction may be a dimensionless number, the function F is defined by the equations

$$F = \frac{1}{mc} G(x_i, x_i', \rho)$$

$$(2.55)$$

$$\frac{\partial}{\partial \rho} \frac{1}{mc} G(x_i, x_i', \rho) = 0 ;$$

through the elimination of ρ we obtain the equation

$$F(x_i, x_i') = \sqrt{\varphi + \frac{1}{4}\varphi^2} \sqrt{\Sigma x_i'^2 + (\vec{a} \cdot \vec{x'})} \, . \tag{2.56}$$

F is, in fact, homogeneous of first order in x_i'.

The paths of electrons of energy C are the extremals of the variation problem

$$V = \int F(x_i, x_i') ds = \int \left[\sqrt{\varphi + \frac{1}{4}\varphi^2} \sqrt{x'^2 + y'^2 + z'^2} + (a_1 x' + a_2 y' + a_3 z') \right] ds \tag{2.57}$$

in which φ and a_i are the dimensionless quantities of (2.54). If the parameter s is chosen as the length of the curves $\vec{X}(s)$ we can write

$$V = \int n \, ds = \text{Extremum} \tag{2.58}$$

where the "index of refraction", n, is given by the formula

$$n = \sqrt{\varphi + \frac{1}{4}\varphi^2} + (a_1 \xi + a_2 \eta + a_3 \zeta) \tag{2.581}$$

and (ξ, η, ζ) is a unit vector in the direction of the path.

For slow electrons we obtain a new formula by means of the classical expression for the kinetic energy $T = \frac{1}{2} m \Sigma \dot{x}_i^2$, namely

$$n = \sqrt{\varphi} + (a_1 \xi + a_2 \eta + a_3 \zeta) \tag{2.582}$$

which is obtained from (2.581) by dropping the $\frac{1}{4}\varphi^2$ from the radical.

2.6 If a vector of length $\frac{1}{n}$ is drawn in any direction (ξ, η, ζ) we obtain a surface closely related to Fresnel's ray surface in crystal optics. An electromagnetic field can thus be interpreted as an optical crystal whose properties vary from point to point. To each point there is then an associated ray surface which is given by

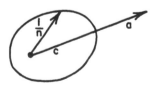

$$\sqrt{\varphi} \sqrt{x^2 + y^2 + z^2} + a_1 x + a_2 y + a_3 z = 1 \, . \tag{2.61}$$

This is the equation of an ellipsoid which is symmetric with respect to the vector $\vec{a}$. The center is offset in the same direction. In case $\vec{a} = 0$ the ellipsoid becomes a sphere.

§3. THE CANONICAL EQUATIONS OF ELECTRON OPTICS.

3.1 The parameter in the variation problem is now taken to be the coordinate z. The problem has become

$$V = \int_{z_0}^{z_1} F(x,y,z,\dot{x},\dot{y})dz = \int_{z_0}^{z_1} \left[\sqrt{\varphi + \frac{1}{4}\varphi^2} \sqrt{1 + \dot{x}^2 + \dot{y}^2} + (a_1\dot{x} + a_2\dot{y} + a_3) \right] dz$$

(3.11)

where $\dot{x} = \dfrac{dx}{dz}$ and $\dot{y} = \dfrac{dy}{dz}$.

Our aim is to transform this problem into canonical form. In §18.37 it has been shown in general that the canonical form is obtained by introducing the quantities x,y,p,q,H in (3.11). In place of x,y,x',y',F through the Legendre transformation,

$$p = \frac{\partial F}{\partial \dot{x}}, \qquad q = \frac{\partial F}{\partial \dot{y}}$$

(3.12)

$$F + H = \dot{x}p + \dot{y}q .$$

3.2 Let us adopt the notation

$$N(x,y,z) = \sqrt{\varphi + \frac{1}{4}\varphi^2}$$

(3.21)

giving $F = N\sqrt{1 + \dot{x}^2 + \dot{y}^2} + a_1\dot{x} + a_2\dot{y} + a_3$

whence

$$p = \frac{N\dot{x}}{\sqrt{1 + \dot{x}^2 + \dot{y}^2}} + a_1$$

(3.22)

$$q = \frac{N\dot{y}}{\sqrt{1 + \dot{x}^2 + \dot{y}^2}} + a_2 .$$

From this it follows that

$$\dot{x} = \frac{p - a_1}{\sqrt{N^2 - (p - a_1)^2 - (q - a_2)^2}} \qquad \dot{y} = \frac{q - a_2}{\sqrt{N^2 - (p - a_1)^2 - (q - a_2)^2}} .$$

(3.23)

If these expressions are introduced in $H = \dot{x}p + \dot{y}q - F$ they yield the Hamiltonian function

$$H = -\sqrt{N^2 - (p - a_1)^2 - (q - a_2)^2} - a_3 \tag{3.24}$$

or

$$H = -\sqrt{\varphi + \frac{1}{4}\varphi^2 - (p - a_1)^2 - (q - a_2)^2} - a_3 \quad . \tag{3.25}$$

For slow electrons this reduces to

$$H = -\sqrt{\varphi - (p - a_1)^2 - (q - a_2)^2} - a_3 \quad . \tag{3.26}$$

The quantities x, y, p, q satisfy the canonical equations

$$\dot{x} = H_p \qquad\qquad \dot{p} = -H_x$$
$$\dot{y} = H_q \qquad\qquad \dot{q} = -H_y \quad . \tag{3.27}$$

These equations will be used to derive the theory of first and third order electron optics by methods identical to those of Chapter IV and V to obtain similar results.

No use will be made of Hamilton's characteristic functions V, W, and T since the results of the following sections may be derived without them. The general theory of these functions developed in §§19-21 may be applied directly to the present problem since we based the theory upon canonical equations in general.

§4. ELECTROMAGNETIC FIELDS OF ROTATIONAL SYMMETRY.

4.1 Most electron optical instruments are, like ordinary optical instruments, symmetric about an axis of rotation. Let this axis be the z-axis of the coordinate system. The electric potential, $\phi(x,y,z)$, must then be a function of z and of $\rho = \sqrt{x^2 + y^2}$. Hence this is true for the function

$$\varphi(\rho,z) = \frac{2(C - e\phi)}{mc^2} \tag{4.11}$$

and for

$$N(\rho,z) = \sqrt{\varphi + \frac{1}{4}\varphi^2} \quad . \tag{4.12}$$

It is easy to verify that in a magnetic field of rotational symmetry the components (A_1, A_2, A_3) satisfy the conditions

$$A_1 x + A_2 y = 0; \quad A_3 = 0 \tag{4.13}$$

and that $A_1^2 + A_2^2$ is a function of ρ and z. In other words the vectors $\vec{A}$ are tangent to the circles $x^2 + y^2 = \rho^2$ of the planes $z = $ const. It follows that the same conditions are satisfied by the dimensionless vectors

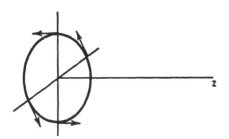

$a_1 = \dfrac{e}{mc^2} A_1$. If these conditions are satisfied, the vector $\vec{a}$ has the form

$$\vec{a} = (-ya, xa, 0) \tag{4.14}$$

where $a = a(\rho, z)$ is a scalar function.

4.2 The Hamiltonian (3.24) in systems of rotational symmetry is given by the expression

$$H = -\sqrt{N^2(\rho, z) - (p + ya)^2 - (q - xa)^2}$$

or   (4.21)

$$H = -\sqrt{N^2(\rho, z) - p^2 a^2 + 2a(xq - yp) - p^2 - q^2} \ .$$

But, since $N(\rho, z)$ and $a(\rho, z)$ are even functions of $\rho = \sqrt{x^2 + y^2}$, this means that H may be written as a function $H = H(z; u, v, w)$ of the three combinations

$$u = x^2 + y^2$$

$$v = p^2 + q^2 \tag{4.22}$$

$$w = 2(xq - yp) \ .$$

If no magnetic field is present H does not depend on w but only on u and v. This corresponds to the ordinary optical case. The canonical equations become

$$\dot{x} = 2(H_v p - H_w y) \ ; \qquad \dot{p} = -2(H_u x + H_w q)$$

$$\dot{y} = 2(H_v q + H_w x) \ ; \qquad \dot{q} = -2(H_u y - H_w p) \tag{4.23}$$

whence we obtain

$$\dot{x} q - \dot{y} p = -2H_w(xp + yq)$$

$$x\dot{q} - y\dot{p} = +2H_w(xp + yq) \tag{4.24}$$

and hence

$$\frac{d}{dz}(xq - yp) = 0 . \tag{4.25}$$

This yields the same result as in optics. The expression $xq - yp = \frac{1}{2}w$ is a constant along any given electron ray.

4.3 It is assumed in the following that the charge density $\rho = 0$. Thus ϕ and hence φ are solutions of Laplace's equation $\Delta\varphi = 0$. We have seen in §38.1 that the solution $\varphi = \varphi(\rho,z)$ which assumes the boundary values

$$\varphi(0,z) = f(z) \tag{4.31}$$

on the z-axis is expressed by the integral [†]

$$\varphi(\rho,z) = \frac{1}{2\pi}\int_0^{2\pi} f(z + i\rho\cos \varphi)d\varphi . \tag{4.32}$$

The function, f(z) may be expressed in terms of the potential on the axis, $\phi(0,z)$

$$f(z) = \frac{2}{mc^2}\left[C - e\phi(0,z)\right] \tag{4.33}$$

where the constant C is the kinetic energy of the electron $C = \frac{1}{2}mv_0^2$ in those parts of the field where the electric potential is zero. Hence from (4.33)

$$f(z) = \frac{v_0^2}{c^2} - \frac{2e}{mc^2}\phi(0,z) . \tag{4.34}$$

The formula (4.32) may be used to derive a power series for $\varphi(\rho,z)$. With the aid of the relations

$$\frac{1}{2\pi}\int_0^{2\pi} \cos^{2\nu}\varphi \, d\varphi = \frac{(2\nu)!}{2^{2\nu}(\nu!)^2}$$

$$\tag{4.35}$$

$$\frac{1}{2\pi}\int_0^{2\pi} \cos^{2\nu+1}\varphi \, d\varphi = 0$$

† Scherzer, Z. Physik 80, 193, 1933.

we obtain

$$\varphi(\rho,z) = \sum_{\nu=0}^{\infty} \frac{(-1)^{\nu}}{(\nu!)^2} \left(\frac{\rho}{2}\right)^{2\nu} f^{(2\nu)}(z) \qquad (4.36)$$

or

$$\varphi(\rho,z) = f(z) - \frac{1}{4}\rho^2 f''(z) + \frac{1}{64}\rho^4 f^{(4)}(z) + \ldots \quad . \qquad (4.37)$$

4.4 The vector $\vec{a} = \left(- ya(\rho,z), xa(\rho,z), 0\right)$ must satisfy the conditions

$$\Delta\vec{a} = 0, \qquad \text{div } \vec{a} = 0 \qquad (4.41)$$

by (1.13). By computation it is easy to see that div $\vec{a} = 0$ for any choice of the scalar function $a(\rho,z)$ and that $\Delta\vec{a} = 0$ if $a(\rho,z)$ satisfies the differential equation

$$a_{\rho\rho} + \frac{3}{\rho}a_{\rho} + a_{zz} = 0 \quad . \qquad (4.42)$$

In a way similar to that for $\varphi(\rho,z)$ (compare §38.1) it may be verified that the integral

$$a(\rho,z) = \frac{1}{2\pi} \int_0^{2\pi} g(z + i\rho \cos \varphi)\sin^2 \varphi \, d\varphi \qquad (4.43)$$

is the solution of (4.42) which satisfies the boundary condition

$$2a(0,z) = g(z) \qquad (4.44)$$

on the z-axis.

To interpret the function $g(z)$ physically let us derive the magnetic field of the system

$$\frac{e}{mc^2} \vec{H} = \text{curl } \vec{a} = (-xa_z, -ya_z, 2a + \rho a_{\rho}) \quad . \qquad (4.45)$$

It follows that on the z-axis

$$\frac{e}{mc^2} \vec{H} = \left(0, 0, 2a(0,z)\right) = \left(0, 0, g(z)\right) \quad . \qquad (4.46)$$

This means that $g(z)$ depends on the component H_3 of the magnetic field along the z-axis, namely

$$g(z) = \frac{e}{mc^2} H_3(0,z) .$$ (4.47)

Note that $g(z)$ is not dimensionless but has the dimension $1/cm$.

A power series for $a(\rho,z)$ may be derived from the integral formula

$$\frac{1}{2\pi} \int_0^{2\pi} \cos^{2\nu} \varphi \sin^2 \varphi \, d\varphi = \frac{(2\nu)!}{2^{2\nu+1} \nu!(\nu+1)!} .$$ (4.48)

This gives

$$a(\rho,z) = \frac{1}{2} \sum_0^\infty \frac{(-1)^\nu}{\nu!(\nu+1)!} \left(\frac{\rho}{2}\right)^{2\nu} g^{(2\nu)}(z)$$ (4.49)

or

$$a(\rho,z) = \frac{1}{2}g(z) - \frac{1}{16}\rho^2 g''(z) + \dots .$$ (4.491)

The principal result of these last two sections is that the electromagnetic field is completely determined by two analytic functions $f(z)$ and $g(z)$ which are given by the electric potential $\phi(0,z)$ and the magnetic vector component $H_3(0,z)$ along the axis of rotation.

§5. FIRST ORDER ELECTRON OPTICS IN SYSTEMS OF ROTATIONAL SYMMETRY.

5.1 As in optics the first order equations may be obtained by taking the first order development of the Hamiltonian

$$H(z;u,v,w) = -\sqrt{N^2 - a^2u + aw - v}$$ (5.11)

in powers of u, v, and w. N and a are functions of z and $u = x^2 + y^2$. Now, to the first order we have

$$H = H_0 + H_1u + H_2v + H_3w$$ (5.12)

where H_0, H_1, H_2, H_3 are certain functions of z. Consider the canonical equations for the Hamiltonian (5.12)

$$\dot{x} = 2(H_2p - H_3y) \qquad \dot{p} = -2(H_1x + H_3q)$$

$$\dot{y} = 2(H_2q + H_3x) \qquad \dot{q} = -2(H_1y - H_3p) .$$ (5.13)

The solutions $x(z)$, $y(z)$, $p(z)$, $q(z)$ of these equations are called the paraxial electron rays. Now for reasons which will become obvious later we introduce the notation

$$D(z) = 2H_1$$

$$\frac{1}{n(z)} = 2H_2 \qquad (5.14)$$

$$\frac{d\omega}{dz} = \dot{\omega}(z) = 2H_3 \ .$$

The Hamiltonian (5.12) assumes the form

$$H = H_0 + \frac{1}{2}\left(Du + \frac{v}{n} + \dot{\omega}w\right) \qquad (5.15)$$

and the canonical equations become

$$\dot{x} = \frac{1}{n}p - \dot{\omega}y \ , \qquad\qquad \dot{p} = -Dx - \dot{\omega}q$$

$$\qquad (5.16)$$

$$\dot{y} = \frac{1}{n}q + \dot{\omega}x \ , \qquad\qquad \dot{q} = -Dy + \dot{\omega}p \ .$$

5.2   The equations (5.16) may be written as follows

$$\frac{d}{dz}\left[(x + iy)e^{-i\omega(z)}\right] = \frac{1}{n}(p + iq)e^{-i\omega(z)}$$

$$\qquad (5.21)$$

$$\frac{d}{dz}\left[(p + iq)e^{-i\omega(z)}\right] = -D(x + iy)e^{-i\omega(z)} \ .$$

This result suggests the introduction of the complex notation

$$X(z) = (x + iy)e^{-i\omega(z)}$$

$$\qquad (5.22)$$

$$P(z) = (p + iq)e^{-i\omega(z)} \ .$$

Hence we may express the canonical equations (5.21) in the form

$$\dot{X} = \frac{1}{n}P$$

$$\qquad (5.23)$$

$$\dot{P} = -DX$$

which is a system of equations of the same type as the paraxial equations of geometrical optics (37.46). Therefore in the investigation of these equations the methods of §37 may be applied.

5.3 The coefficients $D(z)$, $n(z)$ and $\dot{\omega}(z)$ are determined by the values of the functions $f(z)$ and $g(z)$ defined in §4. From (5.11) we obtain

$$H_0 \;=\; -\,N(0,z) \;=\; -\sqrt{f(z) + \tfrac{1}{4}f^2(z)} \;. \tag{5.31}$$

Furthermore differentiation yields

$$H_0 D(z) \;=\; 2H_0 H_1 \;=\; \frac{\partial}{\partial u}\left[N^2(0,z)\right] \;-\; a^2(0,z)$$

$$H_0 \,\frac{1}{n} \;=\; 2H_0 H_2 \;=\; -\,1 \tag{5.32}$$

$$H_0 \dot{\omega} \;=\; 2H_0 H_3 \;=\; a(0,z) \;.$$

Now, since $N^2(u,z) = \varphi(u,z) + \tfrac{1}{4}\varphi^2(u,z)$ we obtain $\left(\dfrac{\partial N^2}{\partial u}\right)_{u=0} = \varphi_u(0,z) + \tfrac{1}{2}\varphi(0,z)\,\varphi_u(0,z)$ and hence by (4.37)

$$\left(\frac{\partial N^2}{\partial u}\right)_{u=0} \;=\; -\frac{1}{4}\left(1 + \tfrac{1}{2}f(z)\right)f''(z) \;. \tag{5.33}$$

The function $\tfrac{1}{2}g(z) = a(0,z)$ is introduced, thus giving the following result:

$$D(z) \;=\; \frac{1}{4}\,\frac{f'' + g^2 + \tfrac{1}{2}ff''}{\sqrt{f + \tfrac{1}{4}f^2}}$$

$$n(z) \;=\; \sqrt{f + \tfrac{1}{4}f^2} \tag{5.34}$$

$$\dot{\omega}(z) \;=\; -\frac{1}{2}\,\frac{g(z)}{\sqrt{f + \tfrac{1}{4}f^2}} \;.$$

For slow electrons (non-relativistic motion) the second order terms f^2 and ff'' may be neglected to give the simplified formulae:

$$D(z) = \frac{1}{4} \frac{f'' + g^2}{\sqrt{f}}$$

$$n(z) = \sqrt{f} \tag{5.35}$$

$$\dot{\omega}(z) = -\frac{1}{2} \frac{g(z)}{\sqrt{f}} \ .$$

The formulae for purely electrical fields (g = 0) have already been derived in §38.

5.4 The general solution of the canonical equations (5.23) can be written in the form of a linear combination of any two particular linearly independent solutions. As in §37.5 the particular solutions chosen are:

The axial ray h(z), ϑ (z) defined by the boundary conditions

$$h(z_0) = 0$$
$$\vartheta(z_0) = 1 \tag{5.41}$$

and the field ray H(z), θ(z) defined by the conditions

$$H(z_0) = 1$$
$$\theta(z_0) = 0 \ . \tag{5.42}$$

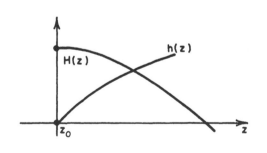

The general solution of (5.23) can be expressed in terms of these two rays, namely

$$X(z) = X_0 H(z) + P_0 h(z) \tag{5.43}$$

$$P(z) = X_0 \theta(z) + P_0 \vartheta(z) \ .$$

This ray satisfies the boundary conditions

$$X(z_0) = X_0$$
$$P(z_0) = P_0 \tag{5.44}$$

where X_0 and P_0 are arbitrary complex numbers.

To obtain the solution of the original paraxial equations (5.16) from (5.43), let us define $\omega(z)$ in accordance with (5.34) by the integral

$$\omega(z) = -\frac{1}{2} \int_{z_0}^{z} \frac{g(z)dz}{\sqrt{f + \frac{1}{4}f^2}} \tag{5.45}$$

so that $\omega(z_0) = 0$.

Now $\qquad X_0 = x_0 + iy_0, \qquad P_0 = p_0 + iq_0$

and $\qquad X = (x + iy)e^{-i\omega(z)}, \qquad P = (p + iq)e^{-i\omega(z)}$

and hence by (5.43)

$$x + iy = e^{i\omega}\left[(x_0 + iy_0)H(z) + (p_0 + iq_0)h(z)\right]$$
$$p + iq = e^{i\omega}\left[(x_0 + iy_0)\theta(z) + (p_0 + iq_0)\vartheta(z)\right]. \tag{5.46}$$

Thus these equations represent the ray $x(z)$, $y(z)$, $p(z)$, $q(z)$ which has the coordinates x_0, y_0, p_0, q_0 at $z = z_0$.

5.5 <u>Conjugate planes.</u> Assume that the axial ray $h(z)$ intersects the z-axis at the point $z = z_1$ so that $h(z_0) = 0$ and $h(z_1) = 0$. The coordinates x_1, y_1, p_1, q_1 of an arbitrary ray are then related to those at $z = z_0$ by the equations

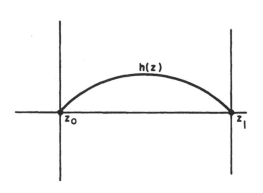

$$x_1 + iy_1$$
$$= e^{i\omega(z_1)}\left[(x_0 + iy_0)H(z_1)\right] \tag{5.51}$$

$$p_1 + iq_1$$
$$= e^{i\omega(z_1)}\left[(x_0 + iy_0)\theta(z_1)\right.$$
$$\left. + (p_0 + iq_0)\vartheta(z_1)\right].$$

As in §37.5 we introduce the quantities

$$H(z_1) = M, \qquad \text{(Magnification)}$$
$$-\theta(z_1) = \frac{1}{F}, \qquad \text{(Equivalent focal length)} \tag{5.52}$$
$$\vartheta(z_1) = 1/M.$$

This yields the relations

$$x_1 + iy_1 = Me^{i\omega(z_1)}(x_0 + iy_0)$$

$$p_1 + iq_1 = e^{i\omega(z_1)}\left[-\frac{1}{F}(x_0 + iy_0) + \frac{1}{M}(p_0 + iq_0)\right] .$$

(5.53)

The first equation shows that all the paraxial rays from a point (x_0, y_0) of the plane $z = z_0$ intersect in a point on the plane $z = z_1$, namely

$$x_1 = M(x_0 \cos \omega - y_0 \sin \omega)$$

$$y_1 = M(x_0 \sin \omega + y_0 \cos \omega) .$$

(5.54)

This means that, in the paraxial region, <u>the instrument produces an image of the plane $z = z_0$ on the plane $z = z_1$ which is enlarged M times and rotated through an angle $\omega(z_1)$.</u>

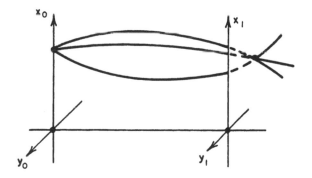

The planes z_0 and z_1 are <u>conjugate planes</u> of first order optics and we may make the statement of §37: <u>If a real solution h(z) of the equations</u> (5.23) <u>intersects the z-axis at two different points z_0 and z_1, then the planes $z = z_0$ and $z = z_1$ are conjugate planes.</u>

The existence of a solution h(z) of this type depends on the functions D(z) and n(z), that is, on the functions f(z) and g(z) which characterize the electromagnetic field. A proper choice of either f or g or of a combination of f and g can be made so that solutions h(z) of the desired type are always obtained. Thus the focusing action of fields of rotational symmetry may be demonstrated.

The magnetic field not only contributes to the focusing effect but also produces the rotation of the image. The angle of rotation is, in fact, given by the integral

$$\omega(z_1) = -\frac{1}{2}\int_{z_0}^{z_1}\frac{g(z)dz}{\sqrt{f + \frac{1}{4}f^2}}$$

(5.55)

which vanishes with the magnetic field.

§6. THE GAUSSIAN CONSTANTS OF AN ELECTRON OPTICAL INSTRUMENT.

6.1 In this section formulae will be derived which determine the first order imagery of an electron optical instrument. These formulae are perhaps of less interest mathematically than as a means of practical computation.

Consider an electromagnetic field which practically vanishes outside a given region $0 \leq z \leq \ell$. This means that $g(z) = 0$ and $f(z)$ is a constant in the regions $0 > z$ and $z > \ell$. Since the angular rotation of the image for all conjugate planes is given by the integral

$$\omega = -\frac{1}{2} \int_0^\ell \frac{g(z)}{\sqrt{f + \frac{1}{4}f^2}} \, dz \qquad (6.11)$$

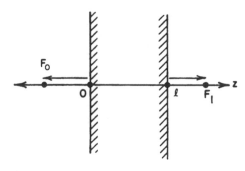

all other first order characteristics can then be derived with the aid of the real solutions of the differential equations

$$\dot{X} = \frac{1}{n} P$$

$$\dot{P} = - DX \, . \qquad (6.12)$$

The theory of these equations is the same as in ordinary Gaussian optics.

Denote the constant values of the function $n(z) = \sqrt{f + \frac{1}{4}f^2}$ in the regions $z < 0$ and $z > \ell$ by n_0 and n_1 respectively. From the general theory of Gaussian optics we know that the first order imagery can be characterized by three constants: the equivalent focal length F, and the positions F_0 and F_1 of the focal points of the instrument. The position of conjugate planes is given by Newton's lens equation

$$Z_0 Z_1 = - n_0 n_1 F^2 \qquad (6.13)$$

where Z_0 and Z_1 are the positions of the conjugate planes relative to the focal points. The associated magnification M of these planes is obtained from

$$M = \frac{n_0}{Z_0} F = - \frac{Z_1}{n_1} \frac{1}{F} \, . \qquad (6.14)$$

6.2 Denote the distance of the focal point in the object space from $z = 0$ by F_0, and let F_1 denote the distance of the focal point in the image space from $z = \ell$. In general, we have $F_0 < 0$ and $F_1 > 0$. Now consider a ray which has the coordinates X_0 and P_0 at $z = 0$. With the aid of the axial ray and the field ray this ray can be represented in the form

$$X(z) = X_0 H(z) + P_0 h(z)$$

$$P(z) = X_0 \theta(z) + P_0 \vartheta(z) \ . \tag{6.21}$$

The coordinates of this ray at $z = \ell$ are therefore given by the linear equations

$$X(\ell) = X_0 H(\ell) + P_0 h(\ell)$$

$$P(\ell) = X_0 \theta(\ell) + P_0 \vartheta(\ell) \ . \tag{6.22}$$

The determinant of these equations, $H(\ell)\vartheta(\ell) - h(\ell)\theta(\ell)$, is equal to one as was shown in §37.75. By means of this result the equations (6.22) may be written in the form

$$X(\ell) = \frac{F_1}{n_1 F} X_0 + \left(F + \frac{F_0 F_1}{n_0 n_1 F}\right) P_0$$

$$P(\ell) = -\frac{1}{F} X_0 - \frac{F_0}{n_0 F} P_0 \ . \tag{6.23}$$

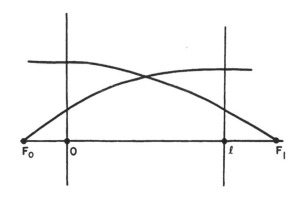

These equations may be verified by considering two special rays

a) $X_0 = 1$; $P_0 = 0$

b) $X(\ell) = 1$; $P(\ell) = 0$.

The intersections of these rays with the axis are the focal points F_0 and F_1. In addition we have

$P(\ell) = -\dfrac{1}{F}$ for the first

ray and $P_0 = \dfrac{1}{F}$ for the

second. These conditions give the coefficients of the above equations directly.

6.3 A similar representation of the ray coordinates $X(\ell)$ and $P(\ell)$ may be obtained from the canonical equations (6.12). By integration

$$X(z) - X_0 = \int_0^z \frac{1}{n(s)} P(s)ds \qquad P(z) - P_0 = \int_0^z \big(- D(s)\big) X(s)ds \tag{6.31}$$

which gives the following integral equation for $X(z)$ and $P(z)$:

$$X = X_0 + P_0 \int_0^z \frac{ds}{n} + \int_0^z \frac{ds_1}{n} \int_0^{s_1} (-D)X \, ds_2$$

$$P = P_0 + X_0 \int_0^z (-D)ds + \int_0^z (-D)ds_1 \int_0^{s_1} \frac{1}{n} P \, ds_2 \quad .$$

(6.32)

We will try to solve these equations by the method of iteration. Consider the sequence of functions $X_\nu(z)$, $P_\nu(z)$ defined by the recursion formulae

$$X_{\nu+1}(z) = X_0 + P_0 \int_0^z \frac{ds}{n} + \int_0^z \frac{ds_1}{n} \int_0^{s_1} (-D)X_\nu(s_2)ds_2$$

$$P_{\nu+1}(z) = P_0 + X_0 \int_0^z (-D)ds + \int_0^z (-D)ds_1 \int_0^{s_1} \frac{1}{n} P_\nu(s_2)ds_2$$

(6.33)

and the conditions $X_0(z) = X_0$, $P_0(z) = P_0$. The convergence of the sequence X_ν, P_ν may be demonstrated under quite general assumptions about $D(z)$ and $n(z)$.

The functions $X_\nu(z)$ and $P_\nu(z)$ may be seen to have the form

$$X_\nu(z) = X_0(\tau_0 + \tau_2 + \ldots + \tau_{2\nu}) + P_0(\sigma_1 + \sigma_3 + \ldots + \sigma_{2\nu-1})$$

$$P_\nu(z) = X_0(\tau_1 + \tau_3 + \ldots + \tau_{2\nu-1}) + P_0(\sigma_0 + \sigma_2 + \ldots + \sigma_{2\nu})$$

(6.34)

where the functions τ_i and σ_i are defined by the iterated integrals

$$\sigma_0 = 1 \qquad\qquad\qquad\qquad \tau_0 = 1$$

$$\sigma_1 = \int_0^z \frac{ds_1}{n} \qquad\qquad\qquad \tau_1 = \int_0^z (-D)ds_1$$

$$\sigma_2 = \int_0^z (-D)ds_2 \int_0^{s_2} \frac{ds_1}{n} \qquad \tau_2 = \int_0^z \frac{ds_2}{n} \int_0^{s_2} (-D)ds_1$$

$$\sigma_3 = \int_0^z \frac{ds_3}{n} \int_0^{s_3} (-D)ds_2 \int_0^{s_2} \frac{ds_1}{n} \qquad \tau_3 = \int_0^z (-D)ds_3 \int_0^{s_3} \frac{ds_2}{n} \int_0^{s_2} (-D)ds_1$$

. . . .

(6.35)

Assuming convergence, the coordinates $X(\ell)$ and $P(\ell)$ may be written in the form

$$X(\ell) = X_0(\tau_0 + \tau_2 + \ldots) + P_0(\sigma_1 + \sigma_3 + \ldots)$$

$$(6.36)$$

$$P(\ell) = X_0(\tau_1 + \tau_3 + \ldots) + P_0(\sigma_0 + \sigma_2 + \ldots) \; .$$

By comparison of these formulae with (6.23), the quantities F, F_0, F_1 are obtained from the infinite series

$$-\frac{1}{F} = \tau_1 + \tau_3 + \tau_5 + \ldots$$

$$\frac{F_1}{n_1 F} = \tau_0 + \tau_2 + \tau_4 + \ldots$$

$$(6.37)$$

$$-\frac{F_0}{n_0 F} = \sigma_0 + \sigma_2 + \sigma_4 + \ldots \; .$$

The quantities σ_i and τ_i are expressed in terms of the integrals (6.35) for $z = \ell$.

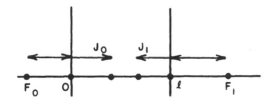

It is often preferable to use the <u>unit points</u> as reference points instead of the focal points. The position of the conjugate planes is then determined by the lens equation

$$\frac{n_1}{z_1} - \frac{n_0}{z_0} = \frac{1}{F} \; .$$

$$(6.38)$$

Let J_0 and J_1 be the distances of the unit points from $z = 0$ and $z = \ell$ respectively. This gives the relations

$$F_1 - J_1 = n_1 F$$

$$(6.381)$$

$$J_0 - F_0 = n_0 F$$

and hence from (6.37) we obtain the equations

$$\frac{J_1}{n_1 F} = \tau_2 + \tau_4 + \ldots$$

$$-\frac{J_0}{n_0 F} = \sigma_2 + \sigma_4 + \ldots$$

$$(6.39)$$

$$-\frac{1}{F} = \tau_1 + \tau_3 + \tau_5 + \ldots \; .$$

6.4 For a first approximation we take only the first member of these series. The smaller the electron optical lens, that is, the smaller l, then the better is the approximation. The first approximation gives the formulae

$$\frac{1}{F} = \int_0^l D(z)dz$$

$$\frac{J_1}{n_1 F} = -\int_0^l \frac{dz}{n} \int_0^z D(s)ds \qquad (6.41)$$

$$\frac{J_0}{n_0 F} = \int_0^l D(z)dz \int_0^z \frac{ds}{n(s)}$$

where the functions $n(z)$ and $D(z)$ are given in (5.34).

$$D(z) = \frac{1}{4} \frac{f'' + g^2 + \frac{1}{2}ff''}{\sqrt{f + \frac{1}{4}f^2}}$$

$$\qquad (6.42)$$

$$n(z) = \sqrt{f + \frac{1}{4}f^2} \quad .$$

§7. THIRD ORDER ELECTRON OPTICS IN FIELDS OF ROTATIONAL SYMMETRY.

7.1 The third order theory of electron optics will be developed here by considerations like those of §42.

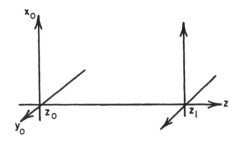

Let us assume that the two planes $z = z_0$ and $z = z_1$ are conjugate planes in the first order theory. Let x_0, y_0, p_0, q_0 be the coordinates of an electron ray at $z = z_0$ and $x(z), y(z), p(z), q(z)$ its coordinates at an arbitrary z. Consider a development of these functions into series of the form

$$x = x_1 + x_3 + x_5 + \ldots$$

$$y = y_1 + y_3 + y_5 + \ldots$$

$$p = p_1 + p_3 + p_5 + \ldots \qquad (7.11)$$

$$q = q_1 + q_3 + q_5 + \ldots$$

where the functions x_i, y_i, p_i, q_i are homogeneous polynomials of order i in the initial coordinates x_0, y_0, p_0, q_0. The coefficients of these polynomials are functions of z such that at $z = z_0$ the conditions

$$x_1(z_0) = x_0, \quad y_1(z_0) = y_0, \quad p_1(z_0) = p_0, \quad q_1(z_0) = q_0, \quad (i = 1)$$

$$(7.12)$$

and the conditions

$$x_i(z_0) = y_i(z_0) = p_i(z_0) = q_i(z_0) = 0 \quad (i \geq 3) \qquad (7.13)$$

are satisfied.

7.2 The canonical equations in the system under consideration have the form

$$\dot{x} = 2(H_v p - H_w y); \qquad \dot{p} = - 2(H_u x + H_w q)$$
$$\dot{y} = 2(H_v q + H_w x); \qquad \dot{q} = - 2(H_u y - H_w p) .$$

$$(7.21)$$

From the first order development of the Hamiltonian

$$H = H_0 + H_1 u + H_2 v + H_3 w + \ldots = H_0 + \frac{1}{2}\left(Du + \frac{1}{n}v + \dot{\omega}w\right) + \ldots .$$

These equations may be written in the form

$$\dot{x} + i\dot{y} - \dot{\omega}(x + iy) - \frac{1}{n}(p + iq) = 2(H_v - H_2)(p + iq) + 2i(H_w - H_3)(x + iy)$$

$$(7.22)$$

$$\dot{p} + i\dot{q} - \dot{\omega}(p + iq) + D(x + iy) = - 2(H_u - H_1)(x + iy) + 2i(H_w - H_3)(p + iq) .$$

By introducing the complex functions

$$X = (x + iy)e^{-i\,\omega(z)}$$

$$P = (p + iq)e^{-i\,\omega(z)} .$$

$$(7.23)$$

These yield the complex equations

$$\dot{X} - \frac{1}{n}P = 2(H_v - H_2)P + 2i(H_w - H_3)X$$

$$(7.24)$$

$$\dot{P} + DX = - 2(H_u - H_1)X + 2i(H_w - H_3)P .$$

The quantities u,v,w can be expressed directly in terms of the complex functions X and P, namely

$$u = x^2 + y^2 = |X|^2$$

$$v = p^2 + q^2 = |P|^2 \qquad (7.25)$$

$$w = 2(xq - yp) = i(X\overline{P} - \overline{X}P)$$

so that the equations (7.24) may be considered as a pair of differential equations for the complex functions $X(z)$ and $P(z)$.

Now, as we have shown in §4.2 the quantity w does not depend upon z. Hence

$$w = 2(x_0 q_0 - y_0 p_0) \ . \qquad (7.26)$$

Now the angle $\omega(z)$ is defined by the integral (5.45) so that $\omega(z_0) = 0$ and therefore

$$X_0 = x_0 + iy_0, \quad P_0 = p_0 + iq_0 \ .$$

The development in (7.11) of the functions x,y,p,q may be replaced by a corresponding development of the complex functions

$$X = X_1 + X_3 + X_5 + \ldots$$

$$(7.27)$$

$$P = P_1 + P_3 + P_5 + \ldots$$

where the X_i and P_i are homogeneous polynomials of the ith order in the quantities x_0, y_0, p_0, q_0 with coefficients which are complex functions of z. Further, we have

$$X_1(z_0) = X_0 \ , \qquad\qquad P_1(z_0) = P_0$$

and $\qquad\qquad\qquad\qquad\qquad\qquad\qquad\qquad\qquad (7.28)$

$$X_i(z_0) = 0 \ , \qquad\qquad P_i(z_0) = 0 \text{ for } i \geq 3 \ .$$

7.3 Let us introduce the development (7.27) into the equations (7.24) and equate the polynomials of the same order. This yields a set of equations for the polynomials X_i, P_i. For $i = 1$ we find

$$\dot{X}_1 - \frac{1}{n}P_1 = 0 \ ,$$

$$(7.31)$$

$$\dot{P}_1 + DX_1 = 0$$

and for $i = 3$

$$\dot{X}_3 - \frac{1}{n}P_3 = 2(H_{12}u_1 + H_{22}v_1 + H_{23}w_1)P_1 + 2i(H_{13}u_1 + H_{23}v_1 + H_{33}w_1)X_1$$
$$(7.32)$$

$$\dot{P}_3 + DX_3 = -2(H_{11}u_1 + H_{12}v_1 + H_{13}w_1)X_1 + 2i(H_{13}u_1 + H_{23}v_1 + H_{33}w_1)P_1$$

where the coefficients H_{ik} are functions of z which are defined by the development of the Hamiltonian to the second order in u,v,w:

$$H = H_0 + H_1u + H_2v + H_3w$$

$$+ \frac{1}{2}\left[H_{11}u^2 + H_{22}v^2 + H_{33}w^2 + 2H_{12}uv + 2H_{13}uw + 2H_{23}vw\right] . \qquad (7.33)$$

The coefficients will be determined later.

The right hand members of the equations (7.32) are known functions of z if we have the first order polynomials X_1 and P_1. In fact, we have

$$u_1 = |X_1|^2$$

$$v_1 = |P_1|^2 \qquad (7.34)$$

$$w_1 = i(X_1\overline{P}_1 - \overline{X}_1P_1) .$$

The solutions X_1 and P_1 may be expressed in terms of the two particular first order rays:

The axial ray: $h(z), \vartheta(z)$

and the field ray: $H(z), \theta(z)$

which are defined by the boundary conditions (5.42) and (5.43) and hence satisfy the relation

$$\begin{vmatrix} H(z) & h(z) \\ \theta(z) & \vartheta(z) \end{vmatrix} = 1 . \qquad (7.35)$$

The planes z_0 and z_1 are conjugate, hence

$$H(z_1) = M , \qquad\qquad h(z_1) = 0 ,$$
$$(7.36)$$
$$\theta(z_1) = -\frac{1}{F} , \qquad\qquad \vartheta(z_1) = \frac{1}{M} .$$

The polynomials X_1 and P_1 are then given by the functions

$$X_1 = X_0 H(z) + P_0 h(z) ,$$

$$P_1 = X_0 \theta(z) + P_0 \vartheta(z)$$

(7.37)

whence it follows that

$$u_1 = (x_0^2 + y_0^2)H^2 + (p_0^2 + q_0^2)h^2 + 2(x_0 p_0 + y_0 q_0)Hh ,$$

$$v_1 = (x_0^2 + y_0^2)\theta^2 + (p_0^2 + q_0^2)\vartheta^2 + 2(x_0 p_0 + y_0 q_0)\theta\vartheta ,$$

(7.38)

$$w_1 = 2(x_0 q_0 - y_0 p_0) .$$

7.4 We wish to determine the third order departure x_3, y_3 of the intersection of the ray with the image plane from the ideal first order intersection by means of the quantity

$$X_3(z_1) = \left(x_3(z_1) + iy_3(z_1) \right) e^{-i\omega(z_1)} .$$

(7.41)

It is even preferable to determine the departures $\Delta\xi$ and $\Delta\eta$ given by

$$X_3(z_1) = \Delta\xi + i\Delta\eta$$

(7.42)

where these are the departures measured in a coordinate system which is rotated through the same angle $\omega(z_1)$ as the first order image.

Now X_3 may be found by the method of §42.3. From the two equations

$$\dot{X}_3 - \frac{1}{n} P_3 = 2(H_{12}u_1 + H_{22}v_1 + H_{32}w_1)P_1 + 2i(H_{13}u_1 + H_{23}v_1 + H_{33}w_1)X_1 ,$$

(7.43)

$$\dot{h} - \frac{1}{n} \vartheta = 0$$

it follows that

$$\dot{X}_3 \vartheta - \dot{h}P_3 = 2(H_{12}u_1 + H_{22}v_1 + H_{23}w_1)P_1 \vartheta$$

$$+ 2i(H_{13}u_1 + H_{23}v_1 + H_{33}w_1)X_1 \vartheta .$$

(7.44)

Similarly, from the second equation (7.32) and the equation $\vartheta + Dh = 0$, it follows that

$$X_3 \dot{\vartheta} - \dot{h}P_3 = 2(H_{11}u_1 + H_{12}v_1 + H_{13}w_1)X_1 h - 2i(H_{13}u_1 + H_{23}v_1 + H_{33}w_1)P_1 h .$$

(7.45)

The sum of the left sides of (7.44) and (7.45) is $\frac{d}{dz}(X_3 \vartheta - hP_3)$. Hence using the relations $h(z_0) = h(z_1) = 0$, $X_3(z_0) = 0$, $\vartheta(z_1) = \frac{1}{M}$ it follows that

$$
\frac{X_3}{M} = 2 \int_{z_0}^{z_1} \left[(H_{12}u_1 + H_{22}v_1 + H_{23}w_1)P_1\vartheta \right.
$$

$$
+ (H_{11}u_1 + H_{12}v_1 + H_{13}w_1)X_1h \Big] dz
$$

$$
+ 2i \int_{z_0}^{z_1} (H_{13}u_1 + H_{23}v_1 + H_{33}w_1)(X_1\vartheta - P_1h)dz \quad . \quad (7.46)
$$

If the expressions (7.37) and (7.38) are introduced the right side becomes a cubic polynomial in x_0, y_0, p_0, q_0. Since w_1 and $X_1\vartheta - P_1h = X_0$ are independent of z the integral may be written in the form

$$
\frac{\Delta\xi + i\Delta\eta}{M} = 2 \int_{z_0}^{z_1} \left[(H_{12}u_1 + H_{22}v_1)P_1\vartheta + (H_{11}u_1 + H_{12}v_1)X_1h \right] dz
$$

$$
+ 2w_1 \int_{z_0}^{z_1} \left[H_{13}X_1h + H_{23}P_1\vartheta \right] dz + 2iX_0 \int_{z_0}^{z_1} (H_{13}u_1 + H_{23}v_1)dz
$$

$$
+ 2iX_0w_1 \int_{z_0}^{z_1} H_{33}dz \quad . \quad (7.47)
$$

7.5 The first of these integrals is the same as the integral (42.352) of the corresponding optical problem. Now, if no magnetic field is present all the coefficients H_{i3} vanish so that only the first integral remains. The cubic polynomial defined by this integral must therefore be of the same form as in geometrical optics. Without loss of generality it may be assumed that $y_0 = 0$ which gives this polynomial in the form

$$
\left[A(p_0^2 + q_0^2)p_0 + \frac{1}{3}Bx_0(3p_0^2 + q_0^2) + Cx_0^2p + Ex_0^2 \right]
$$

$$
+ i \left[A(p_0^2 + q_0^2) + \frac{1}{3}Bx_0 2p_0q_0 + Dx_0^2q_0 \right] \quad . \quad (7.51)
$$

The coefficients are given as the same integrals as (42.44) with $\Gamma = 1$, namely

$$A = 2 \int_{z_0}^{z_1} \left[H_{11}h^4 + 2H_{12}h^2\vartheta^2 + H_{22}\vartheta^4 \right] dz ,$$

$$B = 6 \int_{z_0}^{z_1} \left[H_{11}h^3H + H_{12}h\vartheta(H\vartheta + h\theta) + H_{22}\theta\vartheta^3 \right] dz ,$$

$$C = 6 \int_{z_0}^{z_1} \left[H_{11}h^2H^2 + 2H_{12}Hh\theta\vartheta + H_{22}\theta^2\vartheta^2 \right] dz + 2 \int_{z_0}^{z_1} H_{12}dz , \qquad (7.52)$$

$$E = 2 \int_{z_0}^{z_1} \left[H_{11}hH^3 + H_{12}H\theta(H\vartheta + h\theta) + H_{22}\theta^3\vartheta \right] dz ,$$

$$D = 2 \int_{z_0}^{z_1} \left[H_{11}H^2h^2 + H_{12}(H^2\vartheta^2 + h^2\theta^2) + H_{22}\theta^2\vartheta^2 \right] dz .$$

The other integrals in (7.47) represent aberrations which are introduced with the magnetic field. In the investigation of these integrals only results will be stated since the procedure requires only simple algebraic manipulation. Again let us take $y_0 = 0$. For the sum of these integrals we obtain the polynomial

$$\left[\frac{2}{3}\beta x_0 p_0 q_0 + \gamma x_0^2 q_0 \right] + i\left[\frac{\beta}{3}x_0(p_0^2 + 3q_0^2) + x_0^2(\gamma p_0 + \delta q_0) + \epsilon x_0^3 \right] \qquad (7.53)$$

where the coefficients β, γ, δ, ϵ are given by the integrals

$$\beta = 6 \int_{z_0}^{z_1} (H_{13}h^2 + H_{23}\vartheta^2)dz ,$$

$$\gamma = 4 \int_{z_0}^{z_1} (H_{13}Hh + H_{23}\theta\vartheta)dz ,$$

$$\qquad (7.54)$$

$$\epsilon = 2 \int_{z_0}^{z_1} (H_{13}H^2 + H_{23}\theta^2)dz ,$$

$$\delta = 4 \int_{z_0}^{z_1} H_{33} dz .$$

The complete third order polynomial $X_3 = \Delta\xi + i\Delta\eta$ is the sum of the polynomials (7.51) and (7.53). Hence we have the general expression for the third order aberration of an electron optical system with reference to an object point $(x, 0)$:

$$\frac{\Delta\xi}{M} = A(p_0^2 + q_0^2)p_0 + \frac{1}{3}x_0\left[B(3p_0^2 + q_0^2) + 2\beta p_0 q_0\right]$$

$$+ x_0^2\left[Cp_0 + \gamma q_0\right] + Ex_0^3 ,$$

$$\frac{\Delta\eta}{M} = A(p_0^2 + q_0^2)q_0 + \frac{1}{3}x_0\left[2Bp_0q_0 + \beta(p_0^2 + 3q_0^2)\right]$$

$$+ x_0^2\left[\gamma p_0 + D^* q_0\right] + \epsilon x_0^3 .$$

(7.55)

The quantity D^* is then defined as

$$D^* = D + \delta ,$$ (7.56)

that is

$$D^* = 2\int_{z_0}^{z_1}\left[H_{11}H^2h^2 + H_{12}(H^2\vartheta^2 + h^2\theta^2) + H_{22}\theta^2\vartheta^2\right]dz$$

$$+ 4\int_{z_0}^{z_1}H_{33}dz .$$ (7.57)

From the formula (7.55) we see that <u>an electron optical instrument has</u>, <u>in general, eight types of third order image aberrations given by the eight coefficients</u>

$$A, B, C, D^*, E; \beta, \gamma, \epsilon .$$ (7.58)

7.6 To complete the foregoing considerations we shall determine the explicit form of the functions $H_{1k}(z)$ in terms of the functions $f(z)$ and $g(z)$ which characterize the electromagnetic field. The functions H_{1k} are the second derivatives at the point $u = v = w = 0$

$$H_{uu}, H_{vv}, H_{ww}, H_{uv}, H_{uw}, H_{vw}$$ (7.61)

of the Hamiltonian (5.11). The function H is given by the expression

$$H = -\sqrt{\varphi + \frac{1}{4}\varphi^2} - ua^2 + aw - v$$ (7.62)

and the functions $a(u,z)$ and $\varphi(u,z)$ by the power series

$$a(u.z) = \frac{1}{2}g(z) - \frac{u}{16}g''(z) + \dots ,$$

$$\varphi(u,z) = f(z) - \frac{1}{4}uf''(z) + \frac{1}{64}u^2 f^{(4)}(z) \dots .$$

Elementary calculation will yield the result:

$$2H_{11} = \frac{1}{n}\left[\frac{1}{2}D^2 - \frac{1}{8}gg'' - \frac{1}{32}(f'''' + \frac{1}{2}ff'''' + f''^2)\right] ,$$

$$2H_{12} = \frac{1}{2n^2}D ,$$

$$2H_{22} = \frac{1}{2n^3} ,$$

$$2H_{13} = -\frac{1}{4n}\left[\frac{1}{n}Dg + \frac{1}{4}g''\right] ,$$

$$2H_{23} = -\frac{1}{4n^3}g ,$$

$$2H_{33} = \frac{1}{8n^3}g^2 .$$

(7.64)

The expressions for $n(z)$ and $D(z)$

$$n(z) = \sqrt{f + \frac{1}{4}f^2}, \quad D(z) = \frac{1}{4n}(f'' + g^2 + \frac{1}{2}ff'')$$

(7.65)

should be introduced in the formulae.

For slow electrons (non-relativistic motion) the terms $\frac{1}{2}ff''''$ and f''^2 in the formula for H_{11} and the terms $\frac{1}{4}f^2$ and $\frac{1}{2}ff''$ in the expressions for $D(z)$ and $n(z)$ may be dropped, that is, all quadratic combinations involving the function f vanish. The formulae obtained in this case for $g = 0$ are the same as in (42.75) from Chapter V.

7.7 <u>Petzval's Equation</u>: A generalization of Petzval's equation in geometrical optics (see §42.6) may be obtained from the foregoing results. From (7.52) and (7.57) it follows that

$$3D* - C = 4 \int_{z_0}^{z_1} H_{12} \, dz + 12 \int_{z_0}^{z_1} H_{33} dz \; . \tag{7.71}$$

Introducing the expressions (7.64) we find

$$3D* - C = \frac{1}{4} \int_{z_0}^{z_1} \frac{1}{n^3} \left(f'' + 4g^2 + \frac{1}{2} ff'' \right) dz \tag{7.72}$$

or explicitly

$$3D* - C = \frac{1}{4} \int_{z_0}^{z_1} \frac{f'' + 4g^2 + \frac{1}{2} ff''}{\left(f + \frac{1}{4} f^2 \right)^{3/2}} \, dz \; . \tag{7.73}$$

Hence in an electron optical system in which $D* = C = 0$ this integral must be zero. Note that the presence of a magnetic field makes a positive contribution to the integral and thereby decreases the chance of satisfying Petzval's condition.

§8. PHYSICAL DISCUSSION OF THE THIRD ORDER ABERRATIONS OF AN ELECTRON OPTICAL INSTRUMENT.

In conclusion we shall discuss the different types of aberrations and their physical appearance. The several aberrations will be considered in groups according to the exponent of the variable x_0 in the different terms of (7.55).

8.1 Spherical Aberration. The terms in (7.55) which do not depend on x_0 are

$$\frac{1}{M} \Delta \xi = A(p_0^2 + q_0^2) p_0 \; ,$$

$$\frac{1}{M} \Delta \eta = A(p_0^2 + q_0^2) q_0 \; . \tag{8.11}$$

By introducing

$$p_0 = \rho \cos \psi \; ,$$

$$q_0 = \rho \sin \psi \; . \tag{8.12}$$

the image of a zonal bundle of rays of aperture ρ is obtained as a circle

$$\Delta \xi = AM \rho^3 \cos \psi \; , \qquad \Delta \eta = AM \rho^3 \sin \psi \tag{8.13}$$

whose radius is proportional to the cube of the aperture ρ. Hence the entire
bundle of rays from an axial object point intersects the image plane in a
circular spot. This aberration is called spherical aberration as in optics.

8.2 <u>Coma</u>. The aberration which is proportional to x_0 is given by the
expressions

$$\Delta \xi = \frac{1}{3} Mx_0 \left[B(3p_0^2 + q_0^2) + 2\beta p_0 q_0 \right] ,$$

$$\Delta \eta = \frac{1}{3} Mx_0 \left[2Bp_0 q_0 + \beta(p_0^2 + 3q_0^2) \right] .$$

(8.21)

Using ρ and ψ we obtain

$$\Delta \xi = \frac{1}{3} Mx_0 \rho^2 \left[2B + B \cos 2\psi + \beta \sin 2\psi \right]$$

$$\Delta \eta = \frac{1}{3} Mx_0 \rho^2 \left[2\beta + B \sin 2\psi - \beta \cos 2\psi \right] .$$

(8.22)

By introducing

$$BMx_0 \rho^2 = R \cos 2\alpha ,$$

$$\beta Mx_0 \rho^2 = R \sin 2\alpha$$

(8.23)

the equations (8.22) are obtained in the form

$$\Delta \xi = 2R \cos 2\alpha + R \cos 2(\psi - \alpha) ,$$

$$\Delta \eta = 2R \sin 2\alpha + R \sin 2(\psi - \alpha) .$$

(8.24)

This is the equation of a circle of radius R with the point $2R \cos 2\alpha$,
$2R \sin 2\alpha$ as its center. The quantity

$$R = \sqrt{B^2 + \beta^2} \ Mx_0 \rho^2 \qquad (8.25)$$

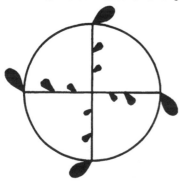

increases with the square of the aperture
ρ. The circles which belong to different
apertures ρ, have their centers on the
same straight line which includes an angle
α with the $\Delta \xi$-axis. The superposition of
all these circles produces a typical third
order coma flare of $\pm 30°$ angular opening.
However, if $\beta \neq 0$, that is, in the presence
of a magnetic field, the flare no longer
points to the center of the field. In

Electromagnetic Coma

general it includes an angle with the radius towards the center as is shown in the above figure. The magnetic field affects the coma flares by deviating them from a central orientation.

8.3 <u>Astigmatism</u>. Next we consider the terms

$$
\begin{aligned}
\Delta \xi &= Mx_0^2 \left[(Cp_0 + \gamma q_0) \right] , \\
\Delta \eta &= MX_0^2 \left[(\gamma p_0 + D^* q_0) \right]
\end{aligned}
\tag{8.31}
$$

which are proportional to x_0^2. In polar coordinates we have

$$
\begin{aligned}
\Delta \xi &= Mx_0^2 \rho \, (C \cos \psi + \gamma \sin \psi) \\
\Delta \eta &= Mx_0^2 \rho \, (\gamma \cos \psi + D^* \sin \psi) .
\end{aligned}
\tag{8.32}
$$

For a given value of ρ these equations represent an ellipse. This aberration is caused by the astigmatic character of the refracted bundle of rays. If

Electromagnetic Astigmatism

$\gamma = 0$, that is, in the absence of a magnetic field, these ellipses are symmetric to the radial lines of the field. The presence of a magnetic field, however, may cause the axes of the ellipse to take on any angle with the radius vector towards the center of the field. The ellipse (8.32) degenerates into a straight line if the determinant of the coefficients is zero, that is, when

$$
D^* C = \gamma^2 .
\tag{8.33}
$$

The line may include any angle with the radius toward the center. In optics and in electrostatic fields it must be either normal or parallel to the radius vector.

8.4 <u>Distortion</u>. The final terms in (7.55) give the aberration

$$
\begin{aligned}
\Delta \xi &= MEx_0^3 \\
\Delta \eta &= M\epsilon x_0^3 .
\end{aligned}
\tag{8.41}
$$

This aberration is obtained for bundles of extremely small aperture ρ. Therefore the definition is sharp but the resulting image is distorted. For

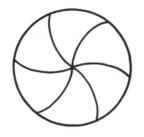

Electromagnetic Distortion

optical and electrostatic instruments we have $\epsilon = 0$ which leaves an object consisting of radial lines undistorted. This is no longer the case with magnetic fields as appears from (8.41). The appearance of the image is shown in the figure. This type of distortion is sometimes referred to as Spider distortion.

REFERENCES:

E. Brueche, O. Scherzer, Geom. Elektr. optik, Berlin 1934.

E. Brueche, W. Henneberg, Geom. Elektr. optik. Erg. d. exakt. Naturwiss. Bd 15.

J. Picht, Einf. in die Theorie der Elektr. optik, J. W. Edwards, 1944.

W. Glaser, Z.f. Physik 97, 177, 1935.

SUPPLEMENTARY NOTE NO. II

OPTICAL QUALITIES OF GLASS

M. Herzberger, Kodak Research Laboratories

Lecture given at Brown University, August 21, 1944

SUMMARY

Two graphs, one plotting reciprocal dispersion against ν-value and the other plotting red and violet partials against ν-value give all the data on glass useful to the optical designer. A short history of the development of optical glass is given and a dispersion formula developed which is satisfactory for all optical materials from $\lambda = 0.365\mu$ to $\lambda = 1.0\mu$.

Let us consider a single thin lens with a front curvature ρ_1 and a back curvature ρ_2. The power of the lens for the wave length D = 589.3 is

$$\varphi_D = (n_D - 1)K \tag{1}$$

where $K = \rho_1 - \rho_2$, and n_D is the index of the glass for the wave length D.

If we have two or more thin lenses, the power of the combination is the sum of the powers of the simple lenses. Consider especially two lenses and assume them to be corrected for two colors, for instance for the lines C and F of the solar spectrum, corresponding to wave lengths of 656.3 and 486.1 mμ. We then have

$$(n_{1D} - 1)K_1 + (n_{2C} - 1)K_2 = \varphi_D ,$$

$$(n_{1C} - n_{1F})K_1 + (n_{2C} - n_{2F})K_2 = 0 . \tag{2}$$

The solution is

$$K_1 = \frac{\dfrac{1}{n_{1C} - n_{1F}}}{\dfrac{n_{1D} - 1}{n_{1C} - n_{1F}} - \dfrac{n_{2D} - 1}{n_{2C} - n_{2F}}} \varphi_D \tag{3}$$

411

$$K_2 = \frac{\dfrac{1}{n_{2C} - n_{2F}}}{\dfrac{n_{2D} - 1}{n_{2C} - n_{2F}} - \dfrac{n_{1D} - 1}{n_{2C} - n_{2F}}} \varphi_D .$$

We introduce as abbreviations

$$N = \frac{1}{n_C - n_F} , \qquad \nu = \frac{n_D - 1}{n_C - n_F} , \tag{4}$$

and find

$$K_1 = \frac{N_1}{\nu_2 - \nu_1} \varphi_D , \qquad K_2 = \frac{N_2}{\nu_1 - \nu_2} \varphi_D . \tag{5}$$

For a third wave length λ we find the deviation

$$\varphi_\lambda - \varphi_F = (n_{1\lambda} - n_{1F})K_1 + (n_{2\lambda} - n_{2F})K_2 . \tag{6}$$

Abbreviating

$$P_\lambda = \frac{n_\lambda - n_F}{n_C - n_F} \tag{7}$$

we find finally

$$\varphi_\lambda - \varphi_F = \frac{P_{1\lambda} - P_{2\lambda}}{\nu_1 - \nu_2} \varphi_D . \tag{8}$$

Formulas (5) and (8) suggest a practical way to plot the characteristics of glasses so as to show immediately the facts which an optical designer would like to know.

If we plot N (the reciprocal mean dispersion) against the ν-value (introduced by E. Abbe), we can obtain immediately the values of K_1 and K_2 by an elementary construction (Fig. 1).[†] We find

$$K_1 = \varphi_D \, \text{tg} \, \alpha_1 ,$$
$$\tag{9}$$
$$K_2 = \varphi_D \, \text{tg} \, \alpha_2 .$$

[†]The figures are not consecutively numbered. The missing numbers are for projection slides which could not be reproduced.

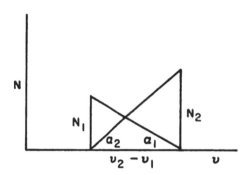

Figure 1

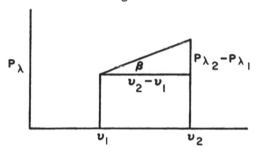

Figure 2

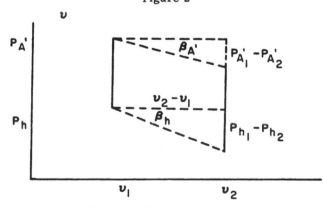

Figure 3

FUNDAMENTAL GLASS PLOT

In the same way, if we plot P_λ against ν, we find the deviation of the power for the wave length D again by a simple construction, (Fig. 2).

$$\varphi_\lambda - \varphi_F = \varphi_D \operatorname{tg} \beta \ . \tag{10}$$

It will be sufficient, as we shall see later, to make such a plot for two wave lengths, the line A' (λ = 768.2) at the red end of the spectrum, and the line h (λ = 404.7) at the violet end of the spectrum. Both plots can easily be made on the same sheet if we use a different scale for $P_{A'}$ and P_h. Under these circumstances a glass is given by a vertical line (Fig. 3) and the tg $\beta_{A'}$ characterizes the deviation at the red end, whereas tg β_h gives the deviation towards the violet end.

Let us now sketch briefly the development of our knowledge of optical glass, insofar as this knowledge is of special interest to the optical designer. The knowledge of glass as a substance transparent to light of all colors is frequently ascribed to the Phoenicians. It was applied in the late middle ages to the manufacture of spectacles. The sixteenth century saw the invention of different types of optical instruments. The glass used was, in the main, what we today call crown glass, a mixture of sand, lime, and soda, properly heated and cooled, with a refractive index of about 1.5 and a reciprocal dispersion of about ν = 55. In about 1666 Sir Isaac Newton found that glass had a different refractive index for light of different colors. His famous experiments in which a prism separates white light into light of the "seven colors of the spectrum" are well known to every scientist. Newton concluded from this that a simple lens must have chromatic aberrations.

Measuring the dispersion of water, (comp. Fig. 4) which has a ν-value of 55.6, and comparing it with the available crown glass, the equality of the ν-value for two such different materials made him surmise that all materials have the same relative dispersion, and he concluded from equation 2 that lenses could not be achromatized. For this reason he recommended and designed reflecting telescopes for astronomical purposes. However, in his lifetime prisms were made of glass of much higher dispersive power—the so-called flint glasses made by introducing lead into the usual mixture.

An English Justice of the Peace, Chester Moore Hall, in 1733 was the first to design an achromatic objective, i.e., a lens system corrected for two colors. The systematic manufacture of such lenses was undertaken by J. Dollond who acquired the first patent in 1758. Variations in the proportion of lead made it possible to change the refractive index and the dispersion at the same time, but not independently, so that up to about 1830 a one-dimensional manifold of glasses was known, the crown-flint series, such that to each refractive index belonged one, and only one, value of ν.

Joseph v. Fraunhofer (1787-1826) discovered the fixed dark lines in the solar spectrum†, and bright lines in the spectra of other light sources, which

---

† See Table 1.

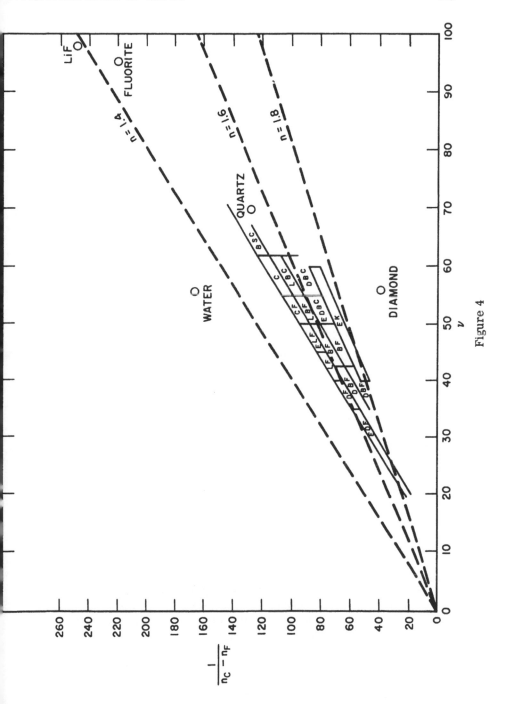

Figure 4

TABLE I

Spectrum Lines for Refractometry

Name	Element	Wave length (Å)	
A'	K	7699.0 7664.9	7681.9
b	He	7065.2	
C	H	6562.8	
D	Na	5895.9 5890.0	5892.9
d	He	5875.6	
e	Hg	5460.7	
F	H	4861.3	
g†	Hg	4358.4	
G'	H	4340.5	
h†	Hg	4046.6	
(uv)	Hg	3650.1	

†Unfortunately, g and h are each used for two different wave lengths. They
were originally applied to two absorption lines in the solar spectrum, g being
a calcium line at 4226.7 Å, and h a hydrogen line at 4101.7.

Kirchhoff and Bunsen soon tied up with the chemical constitution of the light
sources. This made possible exact measurements of the refractive indices,
and thus formed the basis for scientific measurement and production of glass.
Table I gives a list of the most important emission lines now used in the
optical industry, wave length, color, and the element producing it.

Fraunhofer, Utzschneider, and Reichenbach built the first optical factory
in Benedictbueren near Munich. One of their collaborators, the Swiss,
P. Guinand, whose sons later founded the glass industry in France, perfected
the art of glass melting to such a degree that disks for big telescope objectives
could be made with the necessary optical quality.

In the first half of the nineteenth century optical glass factories were
built in France, Switzerland, and England. However, besides Fraunhofer's
work, which we shall mention again, no significant process was made until
1879 when E. Abbe, the founder of the Zeiss works in Jena, hired the chemist

Otto Schott, with the intention of investigating systematically the influence of different chemical elements on the optical and physical characteristics of glass. This marks the beginning of the study of the nature of glass, a study in which significant steps have been made in recent years. (I refer to the papers of Berger, the books of Morey, and the recent work of Huggins and Sun.) From an optical standpoint the most significant discovery was that for a given index widely different dispersions can be obtained.

The use of boric acid and barium oxide proved especially advantageous. It lead to the borosilicate crowns, the barium crowns, and the barium flint types, and later on to the very important dense barium crown glasses (SK and SSK in the Schott catalogue). This development was crowned in recent years by the invention of the Eastman Kodak glasses which contain no silicon but certain rare-earth oxides.

The construction shown in Fig. 1 indicates the important effect of an increase in refractive index for constant ν-value. It means that all the curvatures of the lens system become smaller, i.e., we can use larger radii in the lenses. To every optical designer it is immediately evident that this means smaller zonal errors, or the possibility of increasing either field or aperture, or the quality of a lens system.

These fundamental types of glasses are, with the exception of the Kodak glasses, now manufactured in all the large countries and, because of their importance for military instruments, their manufacture is frequently subsidized by the government. In this country, Bausch and Lomb produce these types in sufficient varieties for all optical purposes, in England, G. B. Chance, in France, Parra-Mantois, in Germany, Schott, to name only the most important manufacturers. Each manufacturer issues a catalogue which includes a chart in which the position of a glass is given by its refractive index and its ν-value, a practice introduced by Abbe.

However, if we investigated the partial dispersions of most of these glasses, we should find our hope of correcting an optical system for more than two colors is difficult to fulfill. (See Fig. 5.)

For nearly all of the glasses, we find that (Fig. 14)

$$P_\lambda = A_\lambda \nu + B_\lambda . \tag{11}$$

Glasses which fulfill this relation may be called ordinary glasses. Equation (10) teaches us that for them

$$\varphi_\lambda - \varphi_F = \varphi_D A_\lambda \tag{12}$$

which means that if we achromatize a lens for C and F, the aberration for any other color is determined. This aberration is called the secondary spectrum.

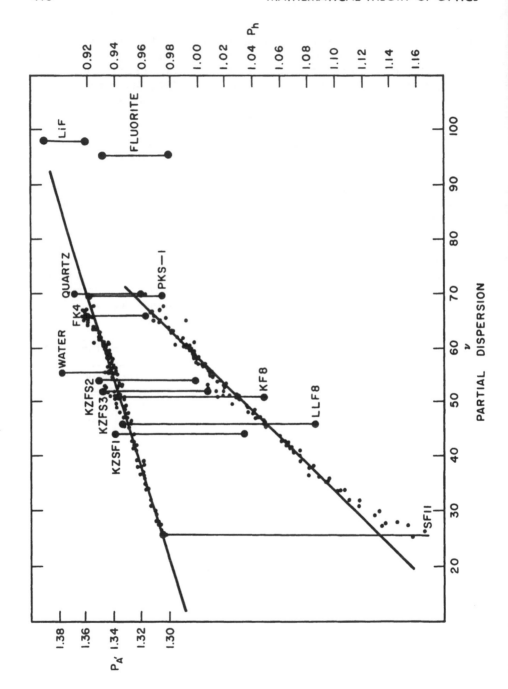

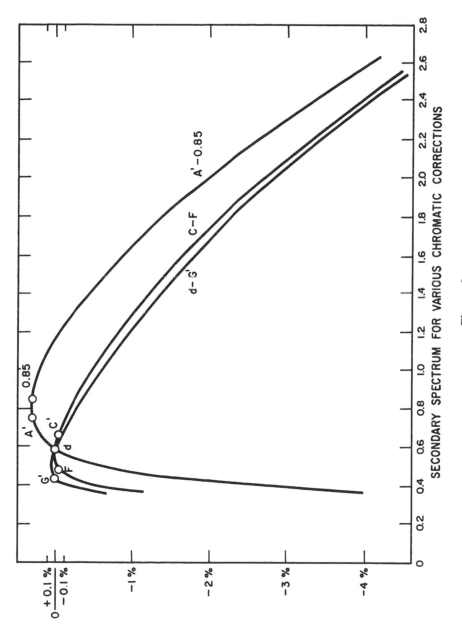

SECONDARY SPECTRUM FOR VARIOUS CHROMATIC CORRECTIONS

Figure 9

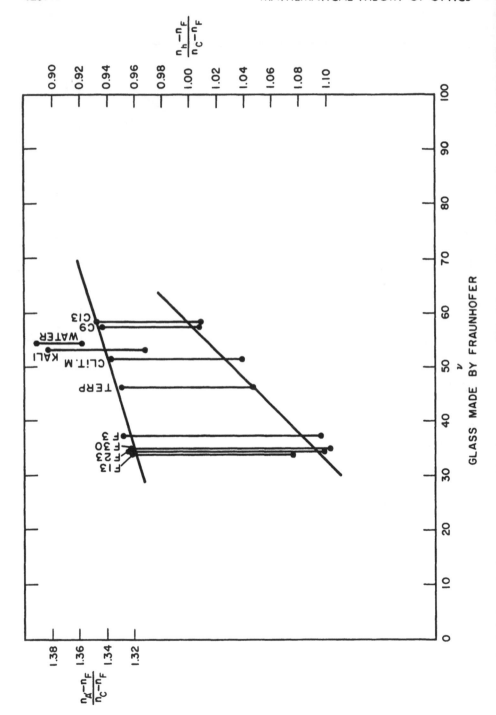

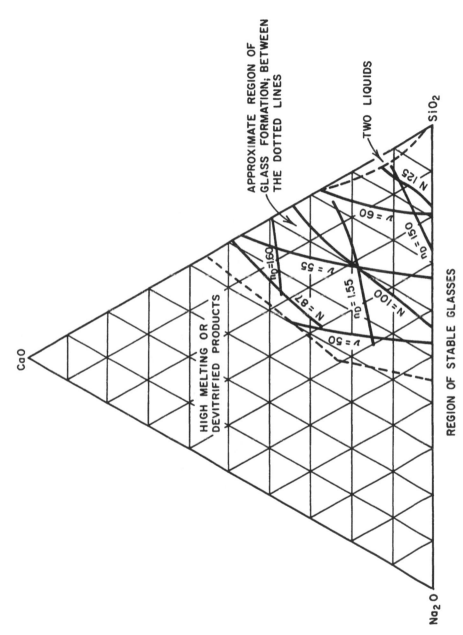

Figure 11

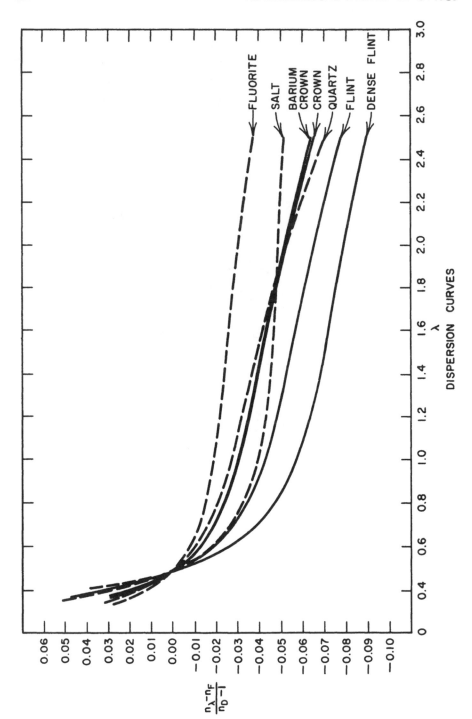

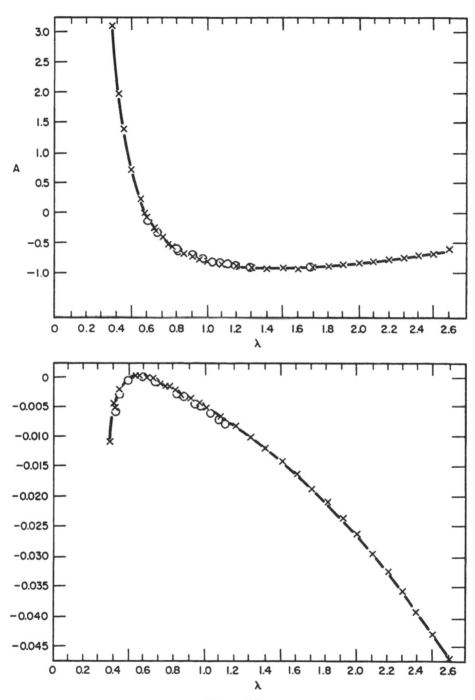

Figure 14

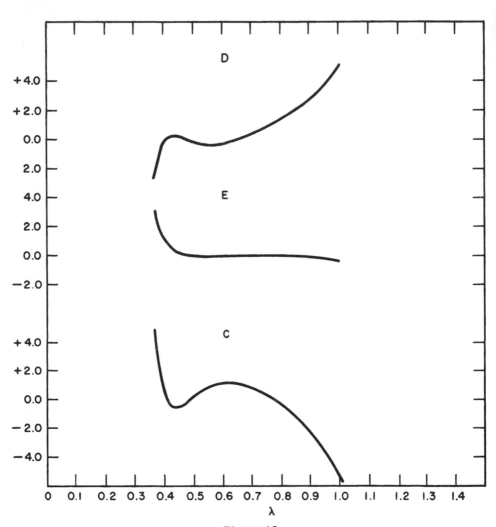

Figure 15

If we achromatize for two other colors, for instance λ_1 and λ_2, we find

$$\varphi_\lambda - \varphi_F = \frac{\begin{vmatrix} A_\lambda & A_1 \\ B_\lambda & B_1 \end{vmatrix}}{\begin{vmatrix} A_1 & B_1 \\ A_2 & B_2 \end{vmatrix}} \tag{13}$$

where A_1, A_2, B_1, B_2 designate the values of A and B for the two achromatized wave lengths.

Figure 9 shows the secondary spectrum for different chromatic corrections.

Frequently attempts have been made to find substances that permit a better chromatic correction of lenses, i.e., glasses which have different partial dispersion ratios. The first systematic experiments were made by Fraunhofer, and Fig. 10 shows the list of glasses which he investigated with their deviations from the normal. It is not known, however, whether these glasses were stable or not.

The first catalogues of Schott show a number of fluoride phosphate glasses with unusual dispersion, but most of them were dropped in later catalogues because they proved to be unstable; they were either water-soluble or changed into the crystalline state, or discolored. Careful investigations in the last thirty years have proved that, for any combination of three or more substances, only a certain percentage of the parts leads to stable glasses (Fig. 11) (Berger and Morey). In any case all these glasses show only small deviations from the normal partial dispersions. Thus a combination with a normal glass leads to small ν-differences, and therefore very short radii of curvature.

In 1883 Abbe introduced microscopic objectives containing fluorite. One glance at the partial dispersion plot shows immediately that here a significant step forward was made. A large difference in ν-value separates fluorite from comparable glasses, the only drawback being the large value N which demands relatively strong curvature for one of the elements.

Attempts to find other materials with extraordinary secondary spectra have not been successful until now. Lithium fluoride is out of the ordinary, but we know that only water matches it; quartz, fused or crystalline, follows closely the dispersion of glasses.

We saw in equation (11) that for ordinary glasses the ν-value is sufficient to give all partial dispersions. Investigation of the dispersion of a number of extraordinary glasses, fluorite, quartz, sylvin, water, etc., show that the knowledge of two partial dispersions, the red partial P_A, and the violet partial

P_h , are certainly sufficient to give the remaining partials with all necessary accuracy. This means the existence of three universal functions C_λ, D_λ, E_λ (Fig. 15) such that

$$P_\lambda = C_\lambda + D_\lambda P_{A'} + E_\lambda P_h \ . \tag{14}$$

Tables II and III give this function from $\lambda = 365 \ m\mu$ to $\lambda = 1.\mu$, and Tables IV and V show for a number of extraordinary glasses and minerals the deviation from formula (11) and formula (14). The latter are well within the necessary accuracy for optical calculations.

The question of a dispersion formula has been discussed considerably in the literature. We have plotted, for a number of typical glasses as well as for water and fluorite and rock salt, $\dfrac{n_\lambda - n_F}{n_D - 1}$ against wave length (Fig. 13). These curves of course are of the same type as n_λ against the wave length except that they go through the same point and are easier to compare.

It is obvious that the dispersion curves of all glasses belong to the same family. It is obvious, too, that the other materials are different.

It seems immediately apparent also that the famous Hartmann formula

$$n_\lambda = \frac{A}{\lambda - \lambda_0} \tag{15}$$

or its generalization

$$n_\lambda = \frac{A}{(\lambda - \lambda_0)^{1.2}} \tag{16}$$

can only be sufficient for a very small range of wave lengths. The dispersion curve is certainly not a hyperbola.

From a physical and a mathematical standpoint, we can say that the Helmholtz-Ketteler formula, or a modification of it, proves best fitted. We assume glasses to have at least one absorption band in the near ultra-violet and one in the far infra-red. That leads to

$$n = n_0 + \frac{A}{\lambda^2 - \lambda_0^2} + \frac{B}{\lambda^2 - \lambda_1^2} \ . \tag{17}$$

The red absorption band is, for the purposes of the part of the spectrum in which we are interested, far enough away to be replaced by the linear member of its development:

$$n = \frac{A}{\lambda^2 - \lambda_0^2} + C_0 + C_1\lambda^2 \ . \tag{18}$$

TABLE II

	A	B
.3650	+ 2.396	- .01090
.4047 h	+ 1.246	- .00427
.4340 G'	+ .679	- .00189
4358 g	+ .650	- .00179
.4861 F	+ .000	.00000
.5461 e	- .490	+ .00054
.5876 d	- .724	+ .00048
.5893 D	- .732	+ .00047
.6563 C	- 1.000	.00000
.7682 A'	- 1.275	- .00122
.8	- 1.329	- .00162
.85	- 1.399	- .00228
.9	- 1.455	- .00298
.95	- 1.500	- .00371
1.0	- 1.537	- .00447

$$P_\lambda = A_\lambda + B_\lambda \nu.$$

TABLE III

	D	E	C
.365015	− 2.5101	+ 3.3885	+ 5.0023
.404656 h	.0	+ 1.0	.0
.434047 G'	+ .2833	+ .3510	− .6002
.435834 g	+ .2828	− .3269	− .6008
.436103	.0	.0	.0
.546073	− .3019	− .0348	+ .8304
.587562 d	− .3067	− .0912	+ 1.0908
.589295 D	− .3033	− .0185	+ 1.0947
.656279 C	.0	.0	+ 1.0000
.706519 b	+ .4226	+ .0050	+ .6240
.768194 A'	+ 1.0	.0	.0
.8	+ 1.3664	− .0100	− .4257
.85	+ 2.0340	− .0457	− 1.2507
.9	+ 2.8501	− .1187	− 2.3253
.95	+ 3.8547	− .2407	− 3.7132
1.00	+ 5.0557	− .4135	− 5.4245

$$P_\lambda = DP_{A'} + EP_h + C.$$

TABLE IV

Extraordinary Glass

Residuals from $P_\lambda = A_\lambda + B_\lambda \nu$

		A'	C	D	e	G'	L
FK	4	.000	.000	-.001	.000	+.004	+.017
SF	1	+.002	.000	.000	.000	+.001	+.007
	3	+.001	-.001	.000	-.001	+.003	+.010
	4	+.002	.000	.000	.000	+.004	+.012
	6	+.003	.000	-.001	-.001	+.007	+.019
	10	.000	.000	.000	.000	+.007	+.020
	11	+.003	+.002	.000	.000	+.015	+.037
	13	+.001	+.001	.000	.000	+.010	+.027
	14	+.003	+.002	.000	.000	+.014	+.033
$K_z F$	1	-.004	-.003	-.002	-.001	-.008	-.012
	2	-.010	-.003	.000	.000	-.008	-.014
	3	-.010	-.003	-.001	.000	-.010	-.013
	5	-.010	-.003	-.001	.000	-.010	-.017
	6	-.009	-.003	-.001	+.001	-.009	-.018
PKS	1	+.003	+.001	.000	+.001	+.011	+.028
PSKS	1	+.002	+.001	.000	+.001	+.006	+.018
$K_z FS$	1	-.015	-.004	.000	.000	-.016	-.029
	2	-.014	-.003	.000	+.001	+.013	+.023
	3	-.014	-.004	.000	.000	-.012	-.020
SFS	1	+.008	+.003	.000	-.001	+.022	+.054
Quartz		-.008	-.005	.000	+.001	-.009	-.009
LiF		.000	+.005	.000	+.008	+.038	+.095
Fluorite		+.059	+.018	+.004	-.002	+.073	+.149
Sylvine		+.021	+.004	.000	+.002	-.002	+.149

TABLE V

Extraordinary Glass

Residuals from $P_\lambda = DP_A + EP_h + C$

		G'	g	e	d	D
FK	4	.000	+.001	.000	-.001	-.001
SF	1	+.001	-.001	.000	.000	-.001
	3	.000	.000	+.001	+.002	.000
	4	+.001	.000	.000	-.001	-.001
	6	+.001	.000	.000	-.001	-.001
	10	+.001	.000	-.001	.000	-.001
	11	+.001	.000	.000	.000	-.001
	13	+.002	+.001	-.001	.000	-.001
	14	+.002	.000	.000	-.001	+.002
	15	+.001	.000	.000	.000	-.001
K_z F	1	-.001	-.002	+.001	+.002	-.002
	2	+.001	.000	+.001	+.001	.000
	3	-.002	-.002	.000	+.001	-.001
	5	.000	-.001	.000	+.001	-.001
	6	+.001	.000	-.002	+.001	-.001
PKS	1	.000	-.001	+.002	+.001	+.001
PSKS	1	-.001	-.002	+.001	.000	.000
K_z FS	1	.000	+.001	.000	.000	.000
	2	.000	+.001	.000	.000	.000
	3	.000	.000	.000	.000	.000
SFS	1	.000	+.001	.000	.000	-.001
Amorphous SiO_2		+.002	-.001	-.004	-.001	-.003
LiF		.000	-.002	.000	-.001	.000
Fluorite		-.004	-.001	.000	+.001	.000

The absorption band λ_0 varies from glass to glass, but it is sufficiently near to $\lambda^2 = 0.035$ to be replaced by

$$n = \frac{A_1}{\lambda^2 - .035} + \frac{B_1}{(\lambda^2 - .035)^2} + C_0 + C_1\lambda^2 . \tag{19}$$

Formula (19) leads to a four-constant formula, which is in agreement with our theory. Four data—refractive index, ν-value, and the partials for red and violet—are sufficient to describe a glass. The formula has been tested with many materials and has proved to give sufficient information from 365 mμ up to 1 μ for all the substances used in designing optical systems.

Let us recapitulate. The lens designer has at his disposal a great variety of optical glasses of high quality. The aim of the glass manufacturer must still be to give him crowns with higher refractive index in order to reduce the curvatures of the lenses used.

There is a great need to reduce the secondary spectrum in optical systems. The discovery of the qualities of fluorite was a big step in the right direction. The attempts to make glass with special partial dispersions as yet have not been too successful. Here is an important field in which the technologist, chemist, physicist, and mathematician should work together. Success will mean a great step forward in the development of optical instruments.

SUPPLEMENTARY NOTE NO. III

MATHEMATICS AND GEOMETRICAL OPTICS

M. Herzberger, Kodak Research Laboratories

Lecture given at Brown University, August 23, 1944

Mathematical disciplines are models for certain relationships. They are not necessarily independent of each other, even though their development has been independent. The discovery that two or more of them say the same thing in different languages, and the ability to translate one into the other makes the enormous wealth of experience gained in one field available to the other. A significant example is the transfer of Hamilton's ideas in the calculus of variations to the theory of partial differential equations. The field of classical and modern theoretical physics is to the mathematician another example of such a series of models stating the same mathematical fact in different language. It is my feeling that science can be built up into a series of disciplines like a terrace where each level has a broader foundation than the preceding one and becomes identical with it, if we neglect certain quantities. Let us choose as an example the field of optics. The uppermost level is reserved for geometrical optics. Here we have light rays and waves along these light rays, and we investigate what happens to them in transversing all kinds of media. The light rays and waves form the fundamental skeleton of all optical problems. The next level could be assigned to Fresnel and Fraunhofer diffraction phenomena. We assume some impulses, periodic in time and space, moving along the light rays or waves. The investigation of what happens at a great distance from the source of disturbance is open to a mathematical analysis which is not too difficult. If the wave length λ were zero, then the laws of geometrical optics would be accurate. Assigning this case to a lower level signifies that laws governing the upper one are applicable wherever the wave length can be neglected.

The next lower level would be formed by the electromagnetic theory of light, which is able to explain the diffraction phenomena near the disturbing object; but this is not the bottom level. The problem of emission, the quantum theory of light, and psychophysics form even more basic levels, each larger then the former, none identical with the absolute truth.

I believe that a similar structure exists in all the physical sciences, and that mathematics, if it is really to help the physicist in his effort to understand nature, would do well to build the simplest and sharpest tools to deal with the problems in the different levels.

I hope this paper will be regarded as a first attempt to fashion such a tool for the highest and easiest of the fields to be conquered.

The history of geometrical optics up to the nineteenth century is the history of mathematics: Euclid, Hero of Alexandria, Ptolemy, the Arabs, Kepler, Galileo, Descartes, Fermat, Newton, Huyghens, the Bernouillis, D'Alembert, Clairaut, Euler, Lagrange, Gauss. This period ends with W. R. Hamilton, who has travelled far in the direction of our aim, much farther than is yet known to most comtemporary mathematicians.

In the nineteenth century, which future historians might call the century of specialization, mathematics was divided into branches with mathematical specialists. A few such specialists in the field of geometrical optics may be mentioned: G. B. Airy, L. Seidel, L. Schleiermacher, E. Abbe, T. Smith, A. Gullstrand, M. Boegehold. Occasionally some of the great mathematicians used geometrical optics as a field of application for their special field of interest.

Malus, Gergonne, Dupin, and Charles Sturm applied the differential geometry of Monge to the problems of optics. Mobius took the image of rays near the axis as a good model for collinear transformation and applied the theory of continued fractions to get the fundamental data of an optical system. W. R. Hamilton developed the geometrical calculus of variations using optics as a model. E. Abbe, R. Straubel, and especially A. Gullstrand investigated the invariants of optical imagery. Lie and W. Bruns thought geometrical optics a good application for contact and canonical transformations. H. Poincaré and G. Prange applied the theory of integral invariants. Caratheodory wrote a booklet on geometrical optics developing it as a parallel to his theory of partial differential equations, presenting both as applications of Huygens' principle of superposition of waves. A recent paper by Korringa applied the modern concept of groupoids to solving a special problem of optics. In my own papers I have made use of such diverse fields of mathematics as matrix algebra, vector algebra, Gaussian brackets, etc..

Geometrical optics thus owes a great debt of gratitude to mathematics. The correct understanding of the work of W. R. Hamilton and its extension here is a first attempt to try to repay this debt.

The fundamental problem of geometrical optics is to trace a manifold of rays through a surface dividing two media of different refracting indices. Let $\vec{a}$ be the vector from an arbitrary origin to a point on one of these rays. Let $\vec{n}$ be the vector along the ray, of length μ, μ being the refractive index of the first medium. Let $\vec{b}$ be the vector to the intersection point with the surface and $\vec{a}'$ the vector to an arbitrary point on the image ray. Let $\vec{0}$ be the unit vector in the direction of the surface normal, λ, λ' the length from the starting surface, to the initial and to the final points, respectively. We then have

$$\vec{a} = \vec{a} + \lambda \vec{n} + \lambda' \vec{n}'$$

$$(\vec{n}' - \vec{n}) \times \vec{0} = 0 \tag{1}$$

Let us now assume we have a two-dimensional manifold of rays, i.e., $\vec{a}$, $\vec{n}$, $\vec{0}$ and therefore $\vec{a}'$ $\vec{n}'$ are continuous functions of two variables u and v.

An easy calculation leads us now to the fundamental differential invariant

$$\vec{n}_u' \, \vec{a}_v' - \vec{n}_v' \, \vec{a}_u' = \vec{n}_u \vec{a}_v - \vec{n}_u \vec{a}_v \tag{2}$$

which, because of the generality of our derivation, is not only valid for one but any number of refractions. Formulas (1) and (2) contain formally also the laws of refraction in crystals, with the only difference that in crystal optics the vector $\vec{n}$, while well defined, no longer has the direction of the ray itself. If we consider media with continuously varying refractive index, the path of the light rays is no longer straight in any medium, but curved. If we choose a curve parameter t on each curve, we can easily see that equation (2) goes over into

$$\frac{d}{dt} \left(\vec{n}_u \vec{a}_v - \vec{n}_v \vec{a}_u \right) = 0 \ . \tag{3}$$

A vector field in n-dimensional space coordinates to each point (vector $\vec{a}$) a vector $\vec{n}$. Such vector fields have been studied to a considerable extent. The geometrical configuration which forms the basis of our problem is somewhat more complicated. We want to assume that at every point there belongs a vector to every direction through that point, i.e., we have through each point light rays in every direction, or in the language of mathematics, that to every fixed line element belongs one and only one vector $\vec{n}$. Such a vector manifold may be called a vector flux until a better name is found.

A system of curves such that for any two-dimensional manifold (3) is fulfilled along these curves, I have called a system of transversal curves.

That the condition (3) has a simple geometrical significance is obvious as soon as we restrict consideration to a vector field, i.e., we consider an n-dimensional submanifold of transversal curves so that to each point of a segment of space goes one and only one curve. Then (3) is equivalent to

$$\frac{d}{dt} (\text{curl } \vec{n}) = 0 \ . \tag{4}$$

For the benefit of those not quite familiar with vector analysis in n dimensions, it might be remembered that the curl in n-dimensional space is a $\binom{n}{2}$ vector characteristic for a two-dimensional area.

Let us now assume a manifold of transversal curves. The maximum number of parameters is $\binom{n}{2}$, therefore equation (2) can stand for $6 \times \binom{n+1}{4}$ equations. We call the equation (3) the Lagrange bracket of our problem or the Lagrange invariant, Lagrange having been the first to make use of its remarkable qualities. We carry out the differentiation and obtain, if we designate differentiation according to the curve parameter t by a point

$$\dot{n}_u \vec{a}_v + \vec{n}_u \dot{a}_v - \dot{n}_v \vec{a}_u - \vec{n}_v \dot{a}_u = 0 . \tag{5}$$

Equation (5) can be considered the integrability condition of four functions, of which, however, one could be identical to zero. We then write down in order

$$dL = \dot{n}\, d\vec{a} + \vec{n}\, d\dot{a} ,$$

$$dH = \dot{n}\, d\vec{a} - \dot{a}\, d\vec{n} ,$$

$$dK = \vec{a}\, d\dot{n} + \dot{a}\, d\vec{n} , \tag{6}$$

$$dJ = \dot{a}\, d\vec{n} - \vec{n}\, d\dot{a} .$$

Equation (6) shows that there exists a function (L) of object point $(\vec{a})$, and direction $\dot{\vec{a}}$, such that

$$\frac{\partial L}{\partial x_i} = n^i , \qquad \frac{\partial L}{\partial x_i} = n^i . \tag{7}$$

This function L is the refractive index μ in geometrical optics. If we know L, we find the components of the vector $\vec{n}$ and therefore the vector $\vec{n}$ itself, simply by the process of differentiation. $\vec{n}$ can be eliminated from (7) and we obtain Euler's differential equation characterizing a variation problem with L as the function under the integral, and with the transversal curves as extremals.

From (6) we conclude immediately

$$dL = \frac{d}{dt}\, (\vec{n}\, d\vec{a}) . \tag{8}$$

Integration of (8) along a series of extremals between two curves $\vec{a}(u)$ and $\vec{a}'(u)$ leads to

$$E = \int_{\vec{a}(u)}^{\vec{a}'(u)} L\, dt \tag{9}$$

with

$$dE = \vec{n}'\, d\vec{a}' - \vec{n}\, d\vec{a} . \tag{10}$$

Equation (10), Hamilton's fundamental equation, is valid if $\vec{a}$, $\vec{a}'$ and therefore E depend on any number of parameters.

E in formula (10) is the characteristic function of Hamilton. Restricting $\vec{a}$ and $\vec{a}'$ to be two (n-1)-dimensional surfaces so that one and only one ray goes through a given pair of points $(\vec{a}, \vec{a}')$ leads to the Eiconal of Bruns. In this case the two-point function is a function of $2(n-1)$ variables and completely characterizes our problem. We then obtain the vectors $\vec{n}, \vec{n}'$ simply by differentiation.

A manifold for which $(\vec{n}\,\vec{da}) = d\Phi$ is a total differential, is called a normal system. The reason for it is that making in (1) $E \equiv \Phi$ + const. leads to

$$\vec{n}'\,\vec{da}' = 0 \; , \tag{11}$$

i.e., we have found a system of surfaces $\vec{a}'$ which are perpendicular to the vector $\vec{n}$ of the transversal curve at the corresponding point. These surfaces are called the wave surfaces of the system of rays. This justifies the name normal system. We see immediately the rays coming from a point $(\vec{da} = 0)$ always form a normal system. Moreover E determines a certain kind of metric in our normal system. We can say that in the sense of this metric two wave surfaces are equidistant on all rays $(E_2 - E_1) = $ const..

Equation (10) is the starting point of Lie's theory of contact transformation.

If

then
$$\vec{n}\,\vec{da} = d\Phi$$
$$\vec{n}'\,\vec{da}' = d(E + \Phi) \; . \tag{12}$$

Let us now consider the second equation in (6). We have a function H of two vectors $\vec{a}$ and $\vec{n}$, from which we can derive directly the change of $\vec{a}$ and $\vec{n}$. We find

$$\frac{\partial H}{\partial x_i} = \frac{d}{dt}(n^i) \; ,$$

$$\frac{\partial H}{\partial n_i} = \frac{d}{dt}(x_i) \; . \tag{13}$$

Equations (13) are probably known to all present as the canonical differential equations, connected with the names Lagrange or Charpit. They show that

$$\frac{dH}{dt} = \frac{\partial H}{\partial x_i} \dot{x}_i + \frac{\partial H}{\partial n^i} \dot{n}^i = \dot{n}^i \dot{x}_i - \dot{x}_i \dot{n}^i \equiv 0 . \tag{14}$$

H is thus seen to remain constant along the transversal curves. It is what in physics usually is called the energy integral. Equations (13) are a system of (2n) ordinary differential equations. The solutions of these differential equations are the characteristics of a series of partial differential equations.

This can be shown in the following way. Consider a normal system. For such a normal system, we have a function Φ such that

$$\vec{n} \, d\vec{a} = d\Phi \tag{15}$$

or

$$n^i = \frac{\partial \Phi}{\partial x_i} . \tag{16}$$

H is then a function of x_i and $n^i = \frac{\partial \Phi}{x_i}$, which stays constant along the extremals. If a manifold had been chosen with less than n parameters, for which H was constant, then H would remain constant along each extremal starting from such a manifold. This construction corresponds to Cauchy's construction of the solution of a partial differential equation for given initial conditions.

Fundamental equation (3) constitutes a differential invariant. A corresponding integral invariant results by integrating over u and v

$$\frac{d}{dt} \int \int (\vec{n}_u \vec{a}_v - \vec{n}_v \vec{a}_u) du \, dv = 0 . \tag{17}$$

The integral in (17) does not depend on the surface which intersects the manifold, but the boundary of the second surface must be chosen such that it contains the same rays that go through the first. If (17) is taken over a boundary, it can be transformed into a line integral under the same precautionary measures as are usually connected with Green's theorem. We find

$$\int \int_S (\vec{n}_u \vec{a}_v - \vec{n}_v \vec{a}_u) du \, dv = \oint \vec{n} \, d\vec{a} . \tag{18}$$

For a normal system, of course, it follows that $\oint \vec{n} \, d\vec{a} = 0$.

The two remaining functions in (6) are dual to the functions studied before. They make use of the fact that $\vec{a}$ and $\vec{n}$ appear symmetric in the fundamental formulas (3). It is apparent from the duality inherent in the symmetry of equation (3) that the $\vec{n}$ and $\vec{a}$ vectors are essentially interchangeable. It is easy to imagine a thought experiment in which the former are the positional vectors and the latter the normal vectors. This would lead to a different variation principle, but to the same system of partial differential equations

$$dK = \frac{d}{dt} \vec{a} \, d\vec{n} \, ,$$

$$V = \int K \, dt \, , \qquad\qquad (19)$$

$$dV = \vec{a}' \, d\vec{n}' - \vec{a} \, d\vec{n} \, .$$

The investigation of the last equation in (6) and its geometrical significance will be a problem for future consideration.

If we assume (3) to be rigorously fulfilled, but, as we must, allow the transversal curves to have surfaces of discontinuity, our equation (10) gives

$$(\vec{n}' - \vec{n}) \, d\vec{a} = 0 \qquad\qquad (20)$$

or

$$(\vec{n}' \times \vec{0}) = (\vec{n} \times \vec{0}) \, . \qquad\qquad (21)$$

Thus we have derived again the refraction law, using only the fundamental formula (3).

If a point of discontinuity exists, equation (2) must be fulfilled for all values of $d\vec{a}$. That leads to $\vec{n} = \vec{n}'$, i.e., the wave front is continuous in an isolated point of discontinuity of the rays.

This paper does not claim to present new material but to derive certain well known mathematical facts by simple and straightforward reasoning. It attempts to show how different fields of mathematical endeavor tell the same story in different languages and it aims to make available the knowledge gained in one field to the students of another.

Complete understanding of these problems can only be given in the frame of a generalized vector analysis of n-dimensions.

The advantages of starting from Lagrange's bracket instead of the Hamiltonian function will be obvious if we generalize our methods to include as normal vectors bivectors or K-vectors corresponding to higher dimensional manifolds, or, regarded in another way, if we go from the investigation of systems of curves to the investigation of systems of surfaces, etc. This leads to variation principles with more than one independent variable and to partial differential equations for more than one unknown function. Here most of our results can still be obtained, whereas it does not seem easy to generalize Hamilton's method.

The aim of this paper is fulfilled if some of you feel that it might be worthwhile to reëxamine the basic ideas in different fields of mathematics and physics with the express aim of finding the simplest model common to diversified investigation.

SUPPLEMENTARY NOTE NO. IV

SYMMETRY AND ASYMMETRY IN OPTICAL IMAGES

M. Herzberger, Kodak Research Laboratories

Lecture given at Brown University, August 25, 1944

Every physicist knows Gaussian optics and most of them know the five Seidel aberrations, i.e., spherical aberration, coma, astigmatism, distortion, and field curvature. The general assumption is that these typical errors are sufficient to describe the quality of a lens with finite aperture and field.

This is decidedly incorrect. The analysis of the complex image of an off-axis point in a system with axial symmetry requires a deeper understanding of image formation. It is the aim of this paper to amplify this statement.

Another weakness of the ordinary method of lens designing seems to me to be the fact that only rays in the meridian plane of an optical system are traced, whereas obviously the skew ray errors are of the utmost importance for image formation. The difficulty has not been the tracing of skew rays; methods have been known for a long time for tracing such rays, and these methods have been perfected in recent years, but interpretation of the results of the trace has offered serious difficulty. Here, too, this paper will try to make helpful suggestions.

The powerful new tool which we shall apply is the consistent use of the fact that an ordinary optical system has an axis of symmetry. Especially helpful will be the concept of "diapoint" introduced by the author in 1935, and repeated herein.

An object point off axis and the axis of a rotation-symmetric optical system form a plane called the meridian plane of the object point. A ray from the object point in the meridian plane is called a meridian ray; a ray not in the meridian plane is called a skew ray. A meridian ray obviously remains a meridian ray after refraction according to the refraction law, since the meridian plane is the plane of incidence, and a skew ray remains a skew ray.

The optical system which we investigate co-ordinates to each object ray an image ray. The point where the image ray intersects the meridian plane is called the diapoint of the object.

Since we have a two-dimensional manifold of rays from a given object point, we shall in general have a two-dimensional manifold of diapoints. If we

have a one-dimensional manifold of diapoints, i.e., if the diapoints form a
segment of a curve, we shall say that the object point has a half-symmetric
image; if the diapoints form the segment of a straight line, we speak of a
symmetric image; if all the diapoints fall together into a single point, we call
it a sharp image. The importance of these conceptions will be better understood if we state that optical laws demand that if all rays coming from an
object point intersect a curve, then the rays meeting at each point of the curve
form the same angle with the curve. Neglecting vignetting, we can therefore
say that the image rays are split up into cones which have the curve tangent
at their vertices as a structural axis (Fig. 1).

It is easy to see that in this case the image in any plane perpendicular
to the axis can be considered as originated by superimposing a one-dimensional
number of eccentric ellipses, formed by the intersection of these cones with
the image plane. An example of this kind of image formation is given by the
well known coma figure for a point near the axis of an optical system for
which the sine condition is not fulfilled. There the image consists of a series
of eccentric circles with two common tangents forming an angle of 60 degrees
(Fig. 2). There is no doubt that such an image formation is undesirable, but
one who has seen images of off-axis points will agree that there are still more
undesirable image errors in a lens system. We shall speak of these as errors
of deformation and we shall say that a point which is half symmetrically
imaged is free of errors of deformation.

An object point is called symmetrically imaged if its diapoints form a
straight line-segment. In this case all the rays can be separated into cones
with the same axis, and the image consists of the superposition of a number
of concentric ellipses. The image is still not a sharp image, but it has no
more errors than an axial point. We say that such a point has no asymmetry
errors.

If all the rays that come from an object point intersect in a sharp image,
then we can say that this image point is the diapoint for all object rays. The
image is free from spherical aberration.

Let us now assume that each point of an object has a sharp image. We
still have two kinds of errors. These sharp points might or might not lie in
a plane. The deviation from a plane is then a measure of field curvature.
Even for a flat field we still need not have the same magnification as we have
for rays near the axis. This leads to distortion errors.

Before giving the analytical foundation of these facts let us take as an
example the case of an infinite object, a case usually encountered in handling
a photographic objective. An "infinite object point" stands for a bundle of
parallel rays of a given direction. For such a point the plane through the
axis parallel to the direction of the incident rays is defined as the meridian
plane. If the image followed the laws of Gaussian optics, there would be only

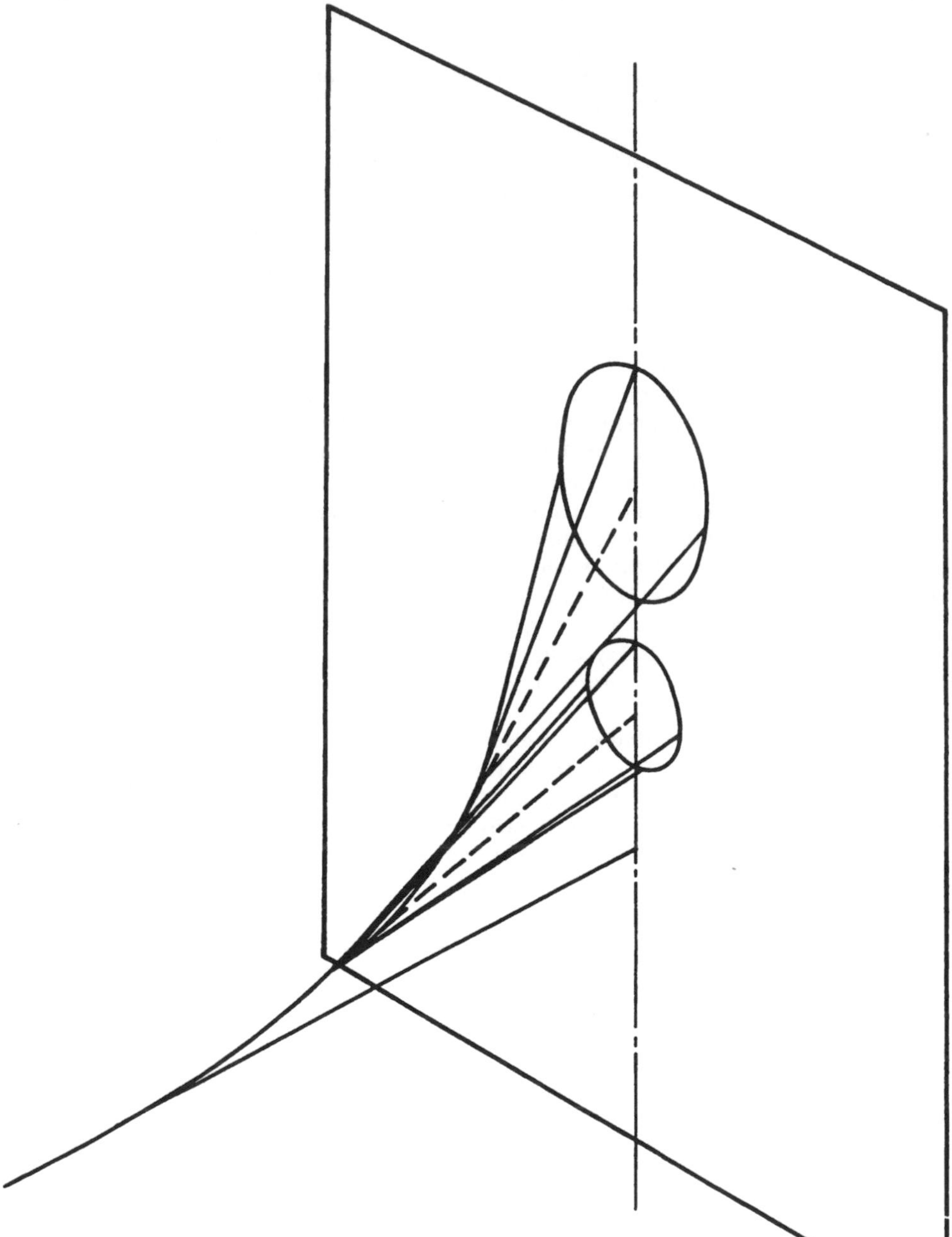

THE HALF-SYMMETRIC IMAGE

Figure 1

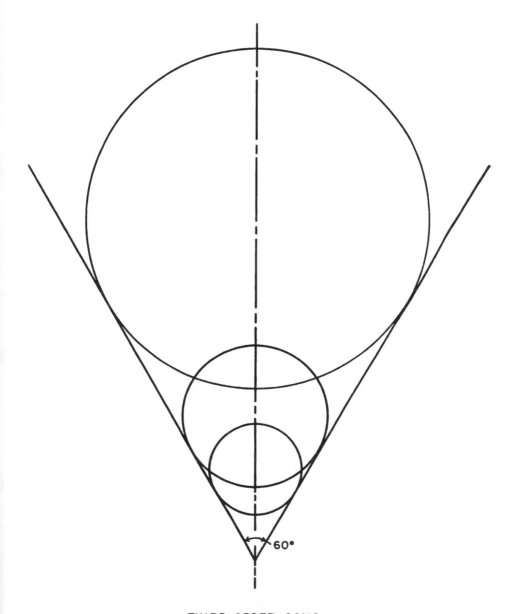

THIRD ORDER COMA

Figure 2

one diapoint. This point would lie in the focal plane and would have the
distance h from the axis, where

$$h = f \tan \sigma \, , \tag{1}$$

σ being the angle of the incident ray with the axis, f being the Gaussian focal
length of the system.

We suggest putting the image co-ordinate origin at the focal point and
choosing the zy plane as the meridian plane. The diapoints then have the coordinates z and y, and if we plot z against $\Delta f = \frac{h}{\tan \sigma} - f$, we obtain the aberration vector for a single ray. For $\frac{h}{\tan \sigma}$ we might introduce the name <u>diafocal
length</u>.

Tracing a finite manifold of rays, we can determine: (a) the best curve
through the diapoints, (b) the best straight line through the diapoints, (c) the
center of gravity of diapoints.

The deviation from the best curve gives us a measure for deformation
errors. The deviation of this curve from a straight line gives a measure for
the asymmetry, the length of the straight line is a measure for the spherical
aberration, and the two co-ordinates of the center of gravity tell us about
field curvature and distortion.

Let us now find the mathematical formulation of these ideas and the
differential equations that have to be fulfilled for special kinds of image formation. Moreover, let us try to find the connecting link between the new kind
of image theory and the one to which you are accustomed.

The concept of the angle characteristic is well known. Let 0, 0' characterize two axis points, one in object and one in image space. Let us choose
a cartesian co-ordinate system such that the x,x' and y,y' axes, respectively, are
parallel and that the z,z' axis coincides with the optical axis. Let P(P') be the
foot of the perpendicular drawn from origin to object (image) ray. Let T be
the optical path length (sum of lengths in each medium multiplied by its refractive index) from P to P'. W. R. Hamilton has proved that under these
circumstances T is a function alone of the direction cosines $\xi, \eta, \zeta = \sqrt{1 - \xi^2 - \eta^2}$ of
object and $\xi', \eta', \zeta' = \sqrt{1 - \xi'^2 - \eta'^2}$ of image ray.

The co-ordinates (x,y,x',y') of the intersection points with the plane
z = 0, respectively, z' - 0, are, according to Bruns, given by

$$nx = - \frac{\partial T}{\partial \xi} , \qquad ny = - \frac{\partial T}{\partial \eta} ,$$

$$n'x' = \frac{\partial T}{\partial \xi'} , \qquad n'y' = \frac{\partial T}{\partial \eta'} ; \tag{2}$$

where n and n' are the refractive indices of object and image space.

It can be shown in the case of an optical system with rotation symmetry that T can be considered a function of only three variables, namely the symmetric functions of ξ, η, ξ', η':

$$a = \frac{1}{2}(\xi^2 + \eta^2) \ ,$$

$$b = \xi\xi' + \eta\eta' \ , \tag{3}$$

$$c = \frac{1}{2}(\xi'^2 + \eta'^2) \ .$$

We find then, instead of (2), the equations

$$- nx = T_a\xi + T_b\xi' \ , \qquad - ny = T_a\eta + T_b\eta' \ ,$$

$$n'x' = T_b\xi + T_c\xi' \ , \qquad n'y' = T_b\eta + T_c\eta' \ ; \tag{4}$$

where T_a is an abbreviation for $\frac{\partial T}{\partial a}$. Let us now consider an infinite object point (ξ, η constant, ξ', η' variable, or $\underline{a}$ constant, $\underline{b}$, $\underline{c}$ variable).

An arbitrary point on the image ray (distance z* from the origin, coordinates x'*, y'*, z'*) is obviously given by

$$n'x'^* = n'x' + \lambda\xi \ ,$$

$$n'y'^* = n'y' + \lambda\eta \ , \tag{5}$$

$$n'z'^* = \qquad \lambda\zeta = \lambda\sqrt{1 - 2c} \ .$$

Elimination of λ gives

$$n'x'^* = n'x' + \frac{n'z'^*}{\sqrt{1-2c}}\xi' = T_b\xi + \left(T_c + \frac{n'z'^*}{\sqrt{1-2c}}\right)\xi' \ ,$$

$$n'y'^* = n'y' + \frac{n'z'^*}{\sqrt{1-2c}}\eta' = T_b\eta + \left(T_c + \frac{n'z'^*}{\sqrt{1-2c}}\right)\eta' \ . \tag{6}$$

The point in question is the diapoint, if $x'^*/y'^* = \xi/\eta$. The necessary and sufficient condition is, of course,

$$n'z'^* = - T_c\sqrt{1 - 2c} \ ,$$

$$n'x'^* = T_b\xi \ , \tag{7}$$

$$n'y'^* = T_b\eta \ .$$

The tangent of the entering ray is $\tan \sigma = \sqrt{\dfrac{2a}{1 - 2a}}$, the intersection height is $h' = \sqrt{x'^2 + y'^2}$. We therefore find for the diafocal length

$$f_s = \frac{h'}{\tan \sigma} = T_b \sqrt{1 - 2a} \; , \; z' = - T_c \sqrt{1 - 2a} \tag{8}$$

to recapitulate formula (7).

For a given infinite object point (a = const.) f and z are functions of two variables b and c.

The condition that our system is free of deformation errors means that

$$\frac{df'}{db} \frac{dz'}{dc} - \frac{df'}{dc} \frac{dz'}{db} = 0 \tag{9}$$

which leads to

$$T_{bb} \left(T_{cc} - \frac{T_c}{1 - 2c} \right) = T_{bc}^2 \; . \tag{10}$$

If equation (10) is fulfilled for all values of a, then every point of the infinite plane is imaged half-symmetrically. Integration of (10) leads of course to

$$T_b = f \left(T_c \sqrt{1 - c}, a \right) \; . \tag{11}$$

The object point has a symmetric image if in (11) f is a linear function. This leads to

$$T_b = A(a) \, T_c \sqrt{1 - 2c} + B(a) \; . \tag{12}$$

Elimination of A(a) and B(a) gives the differential equations

$$\frac{T_{bb}}{T_{bc}} = \frac{T_{bc}}{T_{cc} - T_c \dfrac{1}{1 - 2c}} = \frac{T_{bbc}}{T_{bcc} - \dfrac{T_{bc}}{1 - 2c}} = \frac{T_{bbb}}{T_{bbc}} \; . \tag{13}$$

Every infinite point has a sharp image, if for all values of a

$$T_b = \delta(a) \; ,$$
$$T_c \sqrt{1 - 2c} = z(a) \; . \tag{14}$$

Equation (14) is equivalent to the differential equations

$$T_{bb} = T_{bc} = T_{cc} - \frac{T_c}{1 - 2c} = 0 \; . \tag{15}$$

$\delta(a)$ is a measure for the distortion, $z(a)$ for the curvature of the image. The image is undistorted if

$$\delta(a) = \text{const.}$$

or   (16)

$$T_{ba} = 0 \ .$$

The image is plane if

$$T_c \sqrt{1 - 2c} = \text{const.}$$

or   (17)

$$T_{ca} = \frac{c}{\sqrt{1 - 2c}} \ .$$

Let us now investigate how this new image error theory is connected with the old image error theory.

In the ordinary theory we calculate the intersection point of the rays with the plane through the Gaussian focal point. Let us designate by the suffix zero to a function the value of the function for $a = b = 0$.

The intersection with the Gaussian focal plane is then given by inserting into (6)

$$n'z' = - T_c^0 \ .$$

That leads to

$$n'x'* = T_b \zeta + \left[T_c - \frac{T_c^0}{1 - 2c} \right] \zeta' \ ,$$

                                                                                    (18)

$$n'y'* = T_b \eta + \left[T_c - \frac{T_c^0}{1 - 2c} \right] \eta' \ .$$

If Gaussian optics held we would have

$$n'x_0'* = T_b^0 \frac{\xi}{\zeta} \ ,$$

                                                                                    (19)

$$n'y_0'* = T_b^0 \frac{\eta}{\zeta}$$

with $\zeta = \sqrt{1 - 2a}.$

The differences from Gaussian optics, the so-called image error, are therefore given by

$$n'x'^* - n'x_0^! = \left[T_b \sqrt{1 - 2a} - T_b^0\right]\frac{\xi}{\zeta} + \left[T_c \sqrt{1 - 2c} - T_c^0\right]\frac{\xi'}{\zeta'} \ ,$$

$$n'y'^* - n'y_0^! = \left[T_b \sqrt{1 - 2a} - T_b^0\right]\frac{\eta}{\zeta} + \left[T_c \sqrt{1 - 2c} - T_c^0\right]\frac{\eta'}{\zeta'} \ . \tag{20}$$

The brackets, however, are exactly the functions which we investigated previously, so that we have

$$n'x' - n'x_0^! = \Delta f_d' \ \frac{\xi}{\zeta} - \Delta z_b' \ \frac{\xi'}{\zeta'} \ ,$$

$$n'y' - n'y_0^! = \Delta f_d' \ \frac{\eta}{\zeta} - \Delta z_b' \frac{\eta'}{\zeta'} \ . \tag{21}$$

Equation (21) shows that the investigation of $\Delta f'$ and $\Delta z'$ is sufficient and necessary for investigating the image errors.